SMART ELECTRONIC MATERIALS

Smart materials respond rapidly to external stimuli to alter their physical properties. They are used in devices that are driving advances in modern information technology and have applications in electronics, optoelectronics, sensors, memories and other areas.

This book fully explains the physical properties of these materials, including semiconductors, dielectrics, ferroelectrics, and ferromagnetics. Fundamental concepts are consistently connected to their real-world applications. It covers structural issues, electronic properties, transport properties, polarization-related properties, and magnetic properties of a wide range of smart materials.

The book contains carefully chosen worked examples to convey important concepts and has many end-of-chapter problems.

It is written for first year graduate students in electrical engineering, material sciences, or applied physics programs. It is also an invaluable book for engineers working in industry or research laboratories. A solution manual and a set of useful viewgraphs are also available for instructors by visiting http://www.cambridge.org/0521850274.

JASPRIT SINGH obtained his Ph.D. in Solid State Physics from the University of Chicago. He is currently a professor in the Applied Physics Program and in the Department of Electronic and Computer Science at the University of Michigan, Ann Arbor. He has held visiting positions at the University of California in Santa Barbara. He has authored over 250 technical articles. He has also authored eight textbooks in the area of applied physics and technology. His area of expertise is novel materials for applications in intelligent devices.

SMART ELECTRONIC MATERIALS

Fundamentals and Applications

JASPRIT SINGH
University of Michigan

CAMBRIDGE
UNIVERSITY PRESS

University Printing House, Cambridge CB2 8BS, United Kingdom

Cambridge University Press is part of the University of Cambridge.

It furthers the University's mission by disseminating knowledge in the pursuit of education, learning and research at the highest international levels of excellence.

www.cambridge.org
Information on this title: www.cambridge.org/9780521850278

© Cambridge University Press 2005

This publication is in copyright. Subject to statutory exception
and to the provisions of relevant collective licensing agreements,
no reproduction of any part may take place without the written
permission of Cambridge University Press.

First published 2005

A catalogue record for this publication is available from the British Library

ISBN 978-0-521-85027-8 Hardback

Cambridge University Press has no responsibility for the persistence or accuracy
of URLs for external or third-party internet websites referred to in this publication,
and does not guarantee that any content on such websites is, or will remain,
accurate or appropriate.

CONTENTS

PREFACE	page	xi
INTRODUCTION		xiii
1	SMART MATERIALS: AN INTRODUCTION	xiii
2	INPUT--OUTPUT DECISION ABILITY	xiv
	2.1 Device based on conductivity changes	xiv
	2.2 Device based on changes in optical response	xv
3	BIOLOGICAL SYSTEMS: NATURE'S SMART MATERIALS	xix
4	ROLE OF THIS BOOK	xxii

1 STRUCTURAL PROPERTIES — 1

1.1	INTRODUCTION	1
1.2	CRYSTALINE MATERIALS	1
	1.2.1 Basic lattice types	2
	1.2.2 Some important crystal structures	5
	1.2.3 Notation to denote planes and points in a lattice: Miller indices	12
	1.2.4 Artificial structures: superlattices and quantum wells	16
	1.2.5 Surfaces: ideal versus real	17
	1.2.6 Interfaces	19
1.3	DEFECTS IN CRYSTALS	20
1.4	HETEROSTRUCTURES	23
1.5	NON-CRYSTALLINE MATERIALS	24
	1.5.1 Polycrystalline materials	25
	1.5.2 Amorphous and glassy materials	26
	1.5.3 Liquid crystals	27
	1.5.4 Organic materials	31
1.6	SUMMARY	31

| 1.7 | Problems | 33 |
| 1.8 | Further reading | 37 |

2 QUANTUM MECHANICS AND ELECTRONIC LEVELS — 39

2.1	Introduction	39
2.2	Need for quantum description	40
	2.2.1 Some experiments that ushered in the quantum age	40
2.3	Schrodinger equation and physical observables	48
	2.3.1 Wave amplitude	52
	2.3.2 Waves, wavepackets, and uncertainty	54
2.4	Particles in an attractive potential: bound states	57
	2.4.1 Electronic levels in a hydrogen atom	58
	2.4.2 Particle in a quantum well	62
	2.4.3 Harmonic oscillator problem	67
2.5	From atoms to molecules: coupled wells	69
2.6	Electrons in crystalline solids	77
	2.6.1 Electrons in a uniform potential	80
	2.6.2 Particle in a periodic potential: Bloch theorem	85
	2.6.3 Kronig–Penney model for bandstructure	87
2.7	Summary	93
2.8	Problems	93
2.9	Further reading	99

3 ELECTRONIC LEVELS IN SOLIDS — 100

3.1	Introduction	100
3.2	Occupation of states: distribution function	100
3.3	Metals, insulators, and superconductors	104
	3.3.1 Holes in semiconductors	104
	3.3.2 Bands in organic and molecular semiconductors	107
	3.3.3 Normal and superconducting states	108
3.4	Bandstructure of some important semiconductors	110
	3.4.1 Direct and indirect semiconductors: effective mass	111

3.5	MOBILE CARRIERS		116
	3.5.1 Electrons in metals		117
	3.5.2 Mobile carriers in pure semiconductors		120
3.6	DOPING OF SEMICONDUCTORS		126
3.7	TAILORING ELECTRONIC PROPERTIES		131
	3.7.1 Electronic properties of alloys		131
	3.7.2 Electronic properties of quantum wells		132
3.8	LOCALIZED STATES IN SOLIDS		136
	3.8.1 Disordered materials: extended and localized states		138
3.9	SUMMARY		141
3.10	PROBLEMS		141
3.11	FURTHER READING		146

4 CHARGE TRANSPORT IN MATERIALS   148

4.1	INTRODUCTION	148
4.2	AN OVERVIEW OF ELECTRONIC STATES	149
4.3	TRANSPORT AND SCATTERING	151
	4.3.1 Scattering of electrons	154
4.4	MACROSCOPIC TRANSPORT PROPERTIES	162
	4.4.1 Velocity--electric field relations in semiconductors	162
4.5	CARRIER TRANSPORT BY DIFFUSION	173
	4.5.1 Transport by drift and diffusion: Einstein's relation	175
4.6	IMPORTANT DEVICES BASED ON CONDUCTIVITY CHANGES	178
	4.6.1 Field effect transistor	179
	4.6.2 Bipolar junction devices	184
4.7	TRANSPORT IN NON-CRYSTALLINE MATERIALS	186
	4.7.1 Electron and hole transport in disordered systems	187
	4.7.2 Ionic conduction	191
4.8	IMPORTANT NON-CRYSTALLINE ELECTRONIC DEVICES	193
	4.8.1 Thin film transistor	193
	4.8.2 Gas sensors	195
4.9	SUMMARY	195
4.10	PROBLEMS	199
4.11	FURTHER READING	200

5 LIGHT ABSORPTION AND EMISSION — 202

5.1	INTRODUCTION	202
5.2	IMPORTANT MATERIAL SYSTEMS	204
5.3	OPTICAL PROCESSES IN SEMICONDUCTORS	207
	5.3.1 Optical absorption and emission	210
	5.3.2 Chargei injection, quasi-Fermi levels, and recombination	219
	5.3.3 Optical absorption, loss, and gain	225
5.4	OPTICAL PROCESSES IN QUANTUM WELLS	226
5.5	IMPORTANT SEMICONDUCTOR OPTOELECTRONIC DEVICES	231
	5.5.1 Light detectors and solar cells	231
	5.5.2 Light emitting diode	238
	5.5.3 Laser diode	243
5.6	ORGANIC SEMICONDUCTORS: OPTICAL PROCESSES & DEVICES	251
	5.6.1 Excitonic state	252
5.7	SUMMARY	255
5.8	PROBLEMS	255
5.9	FURTHER READING	262

6 DIELECTRIC RESPONSE: POLARIZATION EFFECTS — 264

6.1	INTRODUCTION	264
6.2	POLARIZATION IN MATERIALS: DIELECTRIC RESPONSE	265
	6.2.1 Dielectric response: some definitions	265
6.3	FERROELECTRIC DIELECTRIC RESPONSE	273
6.4	TAILORING POLARIZATION: PIEZOELECTRIC EFFECT	275
6.5	TAILORING POLARIZATION: PYROELECTRIC EFFECT	285
6.6	DEVICE APPLICATIONS OF POLAR MATERIALS	287
	6.6.1 Ferroelectric memory	287
	6.6.2 Strain sensor and accelerometer	288
	6.6.3 Ultrasound generation	289
	6.6.4 Infrared detection using pyroelectric devices	289

6.7	Summary	291
6.8	Problems	291
6.9	Further reading	295

7 OPTICAL MODULATION AND SWITCHING — 296

7.1	Introduction	296
7.2	Light propagation in materials	297
7.3	Modulation of optical properties	302
	7.3.1 Electro-optic effect	303
	7.3.2 Electro-absorption modulation	309
7.4	Optical modulation devices	312
	7.4.1 Electro-optic modulators	316
	7.4.2 Interferroelectric modulators	318
7.5	Summary	323
7.6	Problems	325
7.7	Further reading	325

8 MAGNETIC EFFECTS IN SOLIDS — 326

8.1	Introduction	326
8.2	Magnetic materials	326
8.3	Electromagnetic field magnetic materials	327
8.4	Physical basis for magnetic properties	331
8.5	Coherent transport: quantum interference	335
	8.5.1 Aharonov Bohm effect	335
	8.5.2 Quantum interference in superconducting materials	338
8.6	Diamagnetic and paramagnetic effects	340
	8.6.1 Diamagnetic effect	340
	8.6.2 Paramagnetic effect	341
	8.6.3 Paramagnetism in the conduction electrons in metals	345

	8.7	FERROMAGNETIC EFFECTS	346
		8.7.1 Exchange interaction and ferromagnetism	346
		8.7.2 Antiferromagnetic ordering	348
	8.8	APPLICATIONS IN MAGNETIC DEVICES	352
		8.8.1 Quantum interference devices	352
		8.8.2 Application example: cooling by demagnetization	354
		8.8.3 Magneto-optic modulators	355
		8.8.4 Application example: magnetic recording	357
		8.8.5 Giant magnetic resistance (GMR) devices	359
	8.9	SUMMARY	359
	8.10	PROBLEMS	359
	8.11	FURTHER READING	362
A	**IMPORTANT PROPERTIES OF SEMICONDUCTORS**		**363**
B	**P–N DIODE: A SUMMARY**		**368**
	B.1	INTRODUCTION	368
	B.2	P–N JUNCTION	368
		B.2.1 P–N Junction under bias	372
C	**FERMI GOLDEN RULE**		**380**
D	**LATTICE VIBRATIONS AND PHONONS**		**386**
E	**DEFECT SCATTERING AND MOBILITY**		**393**
	E.1	ALLOY SCATTERING	393
	E.2	SCREENED COULOMBIC SCATTERING	396
	E.3	IONIZED IMPURITY LIMITED MOBILITY	400
	E.4	ALLOY SCATTERING LIMITED MOBILITY	402
INDEX			**404**

PREFACE

Semiconductor-based devices such, as transistors and diodes enabled technologies that have ushered in the information age. Computation, communication, storage, and display have all been impacted by semiconductors. The importance of semiconductors is recognized if we examine the number of undergraduate and graduate courses that cater to the physics and devices based on these materials. In nearly all electrical engineering departments there are one to two undergraduate courses on the general topic of "physics of semiconductor devices." There are similarly two to three courses in graduate programs on semiconductor physics and devices. In many materials science departments and in physics (or applied physics) departments there are one or two courses where the focus is on semiconductors.

Semiconductors have achieved dominance in information technology because it is possible to rapidly alter their conductivity and optical properties. However, there are other materials that can also rightfully claim to be "smart." New applications and needs are now making these other materials increasingly important. Devices that are usually called sensors or actuators are based on ceramics or insulators which have some properties that traditional semiconductors cannot match. Similarly, organic polymers can provide low-cost alternatives to traditional semiconductors in areas like image display, solar energy conversion, etc.

Increasingly we have to view intelligent devices as being made from a wide variety of materials – semiconductors, piezoelectric materials, pyroelectric materials, ferroelectrics, ferromagnetics, organic semiconductors, etc. Currently some electrical engineering departments and some materials science departments offer courses on "sensors and actuators" or "ceramics." Some physics departments also offer courses on general "solid state physics," which cover some aspects of ceramics. In this book I have attempted to offer material where "traditional" semiconductors, "traditional" smart ceramics, and newly emerging organic semiconductors are discussed in a coherent manner. The book covers structural issues, electronic properties, transport properties, polarization-related properties, and magnetic properties of a wide range of smart materials. We also discuss how these properties are exploited for device applications.

This book is written for first year graduate students in an electrical engineering, material science, or applied physics program.

I am grateful to my editor, Phil Meyler, for his support and encouragement. The design, figures, and layout of the book was done by Teresa Singh, my wife. She also provided the support without which this book would not be possible.

JASPRIT SINGH
Ann Arbor, MI

INTRODUCTION

I.1 SMART MATERIALS: AN INTRODUCTION

Humans have used smart materials – materials that respond to input with a well-defined output – for thousands of years. The footprint on a soft trail in a jungle can tell a well-trained human (and almost all wild animals) what kind of animal recently passed and even how much it weighed. In this case the soft mud acts as a smart material – responding to and storing information about a passing animal. A reader of Sherlock Holmes is undoubtedly familiar with all kinds of information stored in intelligent materials that the clever detective was able to exploit. Over the last couple of decades the role of smart materials in our lives has become so widespread that (at least, in the industrial countries) most of us would be lost without these materials guiding us.

Let us follow Mr. XYZ (of course, it could also be a Ms. XYZ), a super salesman for a medical supplies company, as he gets up one morning and goes about his business. He checks his schedule on his laptop (semiconductor-based devices process the information, liquid crystals help display the information, ferromagnetic- and polymer-based materials store the information, a laser using semiconductors reads the information...). Mr. XYZ sees that he has to catch a flight in an hour to make a presentation. As he drives to the airport he sees on his car map that there is an accident on his normal route. The car computer hooked up to a satellite system gives him an alternate route, which gets him to the airport on time.

On the way to the terminal he has used a smart parking ticket on his cell phone. As he goes through airport security he is scanned by a battery of machines, which have used electromagnetic radiation of several frequencies, chemical sensors, ultrasound images...

The airplane he takes is, of course, a marvel packed with smart materials – sensors and computers fly most of the flight. Mr. XYZ deplanes and gets a rental car with his credit card (another smart device). He makes a very successful presentation with his smart audiovisual card, which he carries in his wallet. A dozen managers in plants located all over the world also participate in the presentation.

As Mr. XYZ is heading back he falls and suffers a gash on his hand. It does not look serious, but he stops by a clinic to have it checked. His health card is scanned, giving the nurse a full history of his allergies, drugs he cannot take, current medication, etc. His gash is patched up and he is given a pill, which will speed up the healing.

Mr. XYZ makes it safely to his home to enjoy a nice movie and some playtime with his family.

Semiconductors, ferroelectrics, ferromagnetics, piezoelectrics, tailor-made polymers – a plethora of smart materials have allowed Mr. XYZ to sail through the day. As he sleeps soundly his two-year-old has a nightmare and screams out. He spends the rest of the night consoling the toddler. Although he does not have a smart technology that will substitute for his hugs, perhaps after another 20 years...who knows!

In this book we will focus on the several classes of materials which have led to modern information age devices. The list of materials being exploited for intelligent devices is continuously increasing. However, there are certain common physical effects that will form the underlying foundations for the materials we will examine.

I.2 INPUT–OUTPUT DECISION ABILITY

A key reason why some materials can be used in intelligent devices is the nature of response that can be generated in some physical property of the device to input. For example a voltage pulse applied across a copper wire does not produce a response (in current) that can be used for digital or analog applications. However, a voltage pulse across transistor made from silicon creates a response that can be exploited for intelligent devices. Later in the book we will discuss what makes an input – output response usable for decision making.

In Fig. 1 we show a typical input – output response in an intelligent device. There are many other forms of the input – output relations that can be exploited for decision making and we will discuss them later. In the response shown in Fig. 1 we see that output has a "thresholding" behavior; i.e., it is low for a range of input and then over a small range of input change it becomes high. This is a response that can be exploited for "switching" applications or memory applications.

The input that a device may respond to may be an optical or a microwave signal, a poisonous gas, a pressure pulse (a sound pulse for example), an electrical voltage pulse, etc. The output response also depends upon a wide range of physical phenomena that alter the state of the device. The most commonly used physical phenomena for smart devices are the following: (i) Conductivity changes or current flow in the device. (ii) Optical properties that may involve light emission, light absorption, light amplification, etc. The effects may involve changes in the refractive index, including absorption coefficient or gain, of the material. (iii) Polarization changes. Many sensor technologies exploit changes that occur in the polarization of a material when subjected to pressure or strain or other inputs. The change in polarization produces a voltage change that can be used to make decisions. (iv) Magnetization changes are exploited in technologies such as a recording medium. In addition to these basic physical phenomena (charges) there may be other changes such as temperature changes, volume changes, etc., which can also be exploited for devices.

The materials that are used for modern information devices are varied and complex and come from many different categories of solids. In Figs. 2 – 5 we show an overview of the devices and materials that are driving the modern information age.

I.2.1 Devices based on conductivity changes

Devices that are based on materials where conductivity can be changed rapidly form the bulk of modern information-processing devices. In Fig. 2 we show an overview of the various devices, materials, and technologies that exploit changes in conductivity. As

I.2. Input–output decision ability

shown in the figure electronic transport in material can be incoherent or coherent. Most present devices are based on incoherent transport, where the wave nature of electrons (i.e., the quantum nature of electrons behaving as propagating waves with well-defined phase coherence) is not exploited. Conductivity changes arise primarily due to an increase or decrease in the number of current-carrying particles. Devices such as diodes and field effect transistors that form the basis of modern semiconductor technology rely on being able to alter conductivity rapidly by an input signal. Materials that have properties that allow large changes (up to orders of magnitude) in conductivity are usually semiconductors, such as Si, Ge, GaAs, InP, etc. Recently organic materials have also shown great promise.

Coherent transport devices
In classical physics, electrons, which are responsible for carrying current in solids, are particles described by their mass, momentum, and position. In the more accurate quantum description, the electrons are described by waves with a certain wavelength and phase. In most cases, as electrons move in a solid, they suffer scattering, causing loss of phase coherence. However, in very small devices as well as in superconductors, the scattering is essentially absent and phase information is retained. In such cases, coherent transport occurs and effects such as interference and diffraction can be exploited to design devices.

As fabrication technologies improve, coherent transport-based devices will become easier to fabricate for room temperature operation. At present such devices can only operate at low temperatures. As shown in Fig. 2, such devices can be made from semiconductors, metals, superconductors, etc.

I.2.2 Devices based on changes in optical response

The electromagnetic spectrum, in general, and visible light, in particular, are an important part of the human experience. We use sight and sense (heat/cold) to survive and thrive in nature. It is not surprising that technologies that involve generation or detection of light are very important. Optically active (i.e., optical properties can be altered) materials form the basis of light emitters (for displays, optical communication, optical readout, publishing, etc.), light detectors (for imaging and coding/decoding), light switches (for communication, image projection), and many other optical technologies, such as medical diagnostics, crime scene analysis, etc.

A vast range of materials is used to design optical devices. These include traditional semiconductor polymers (such as GaAs, InGaAs, InN, and GaN).

Devices based on polar materials
There are a number of materials in which there is net polarization. The polarization that causes a detectable electric field (or a voltage signal) can be exploited for a range of applications. As shown in Fig. 4, several interesting physical phenomena involve polarization effects. In ferroelectric materials the polarization can be altered by an external electric field. The electric field–polarization relation shows a hysteresis curve, so that the direction of polarization at a zero applied field can be switched. Such an effect can be used for memory devices and is used widely for "smart cards." A number

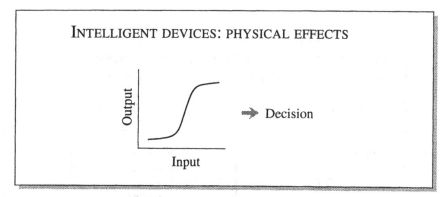

Input: Optical/Electromagnetic; Pressure/Acoustic; Voltage/Current; Chemical species

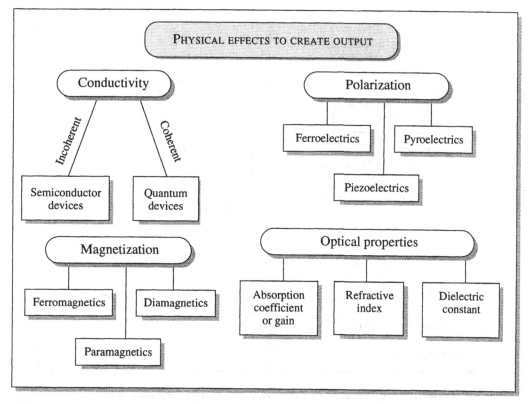

Figure I.1: An overview of the input–output response of intelligent devices and various physical effects that can be exploited for device design.

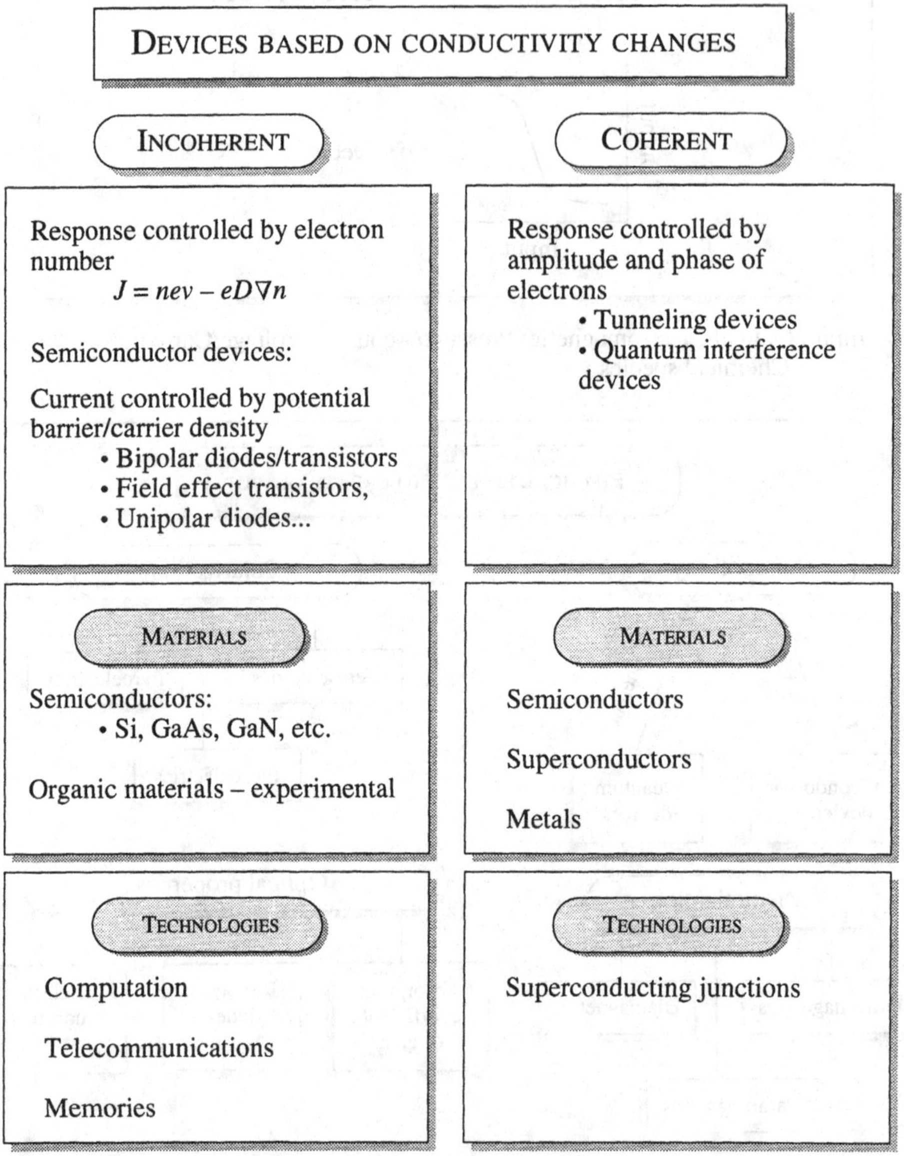

I.2. Input–output decision ability

DEVICES BASED ON CONDUCTIVITY CHANGES

INCOHERENT

Response controlled by electron number
$$J = nev - eD\nabla n$$

Semiconductor devices:

Current controlled by potential barrier/carrier density
- Bipolar diodes/transistors
- Field effect transistors,
- Unipolar diodes...

MATERIALS

Semiconductors:
- Si, GaAs, GaN, etc.

Organic materials – experimental

TECHNOLOGIES

Computation

Telecommunications

Memories

COHERENT

Response controlled by amplitude and phase of electrons
- Tunneling devices
- Quantum interference devices

MATERIALS

Semiconductors

Superconductors

Metals

TECHNOLOGIES

Superconducting junctions

Figure I.2: Devices and technologies based on the control of conductivity of materials.

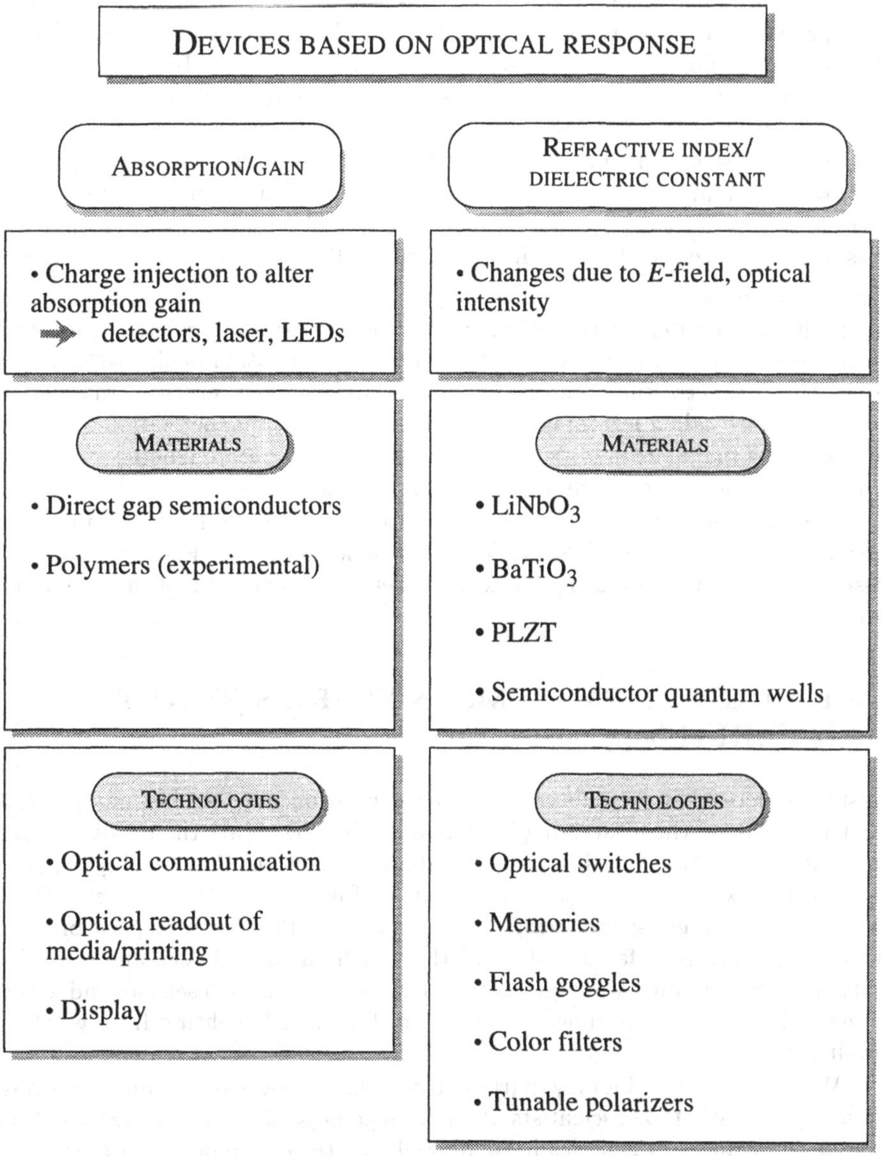

DEVICES BASED ON OPTICAL RESPONSE

ABSORPTION/GAIN

- Charge injection to alter absorption gain
  ⇒ detectors, laser, LEDs

MATERIALS

- Direct gap semiconductors
- Polymers (experimental)

TECHNOLOGIES

- Optical communication
- Optical readout of media/printing
- Display

REFRACTIVE INDEX/ DIELECTRIC CONSTANT

- Changes due to E-field, optical intensity

MATERIALS

- $LiNbO_3$
- $BaTiO_3$
- PLZT
- Semiconductor quantum wells

TECHNOLOGIES

- Optical switches
- Memories
- Flash goggles
- Color filters
- Tunable polarizers

Figure I.3: Materials and devices that are based on control of optical response.

of ferroelectric materials are used in modern technology and rapid advances in synthesis techniques promise more applications based on the ferroelectric effect.

Another physical effect based on polarization is the piezoelectric effect, where the polarization depends upon the strain applied to the sample. A potential signal can also produce strain in a piezoelectric material. Materials like quartz and PZT are widely used for technologies based on the piezoelectric effect. Technologies that use the piezoelectric effect include sensors/actuators (including developments in the micro-electro-mechanical systems or MEMs technology) and ultrasonics.

An interesting and important effect based on polarization is the pyroelectric effect in which a temperature change causes a polarization change in a material. This allows us to convert a thermal signal into a voltage signal (or vice versa). The pyroelectric effect is primarily used for thermal imaging, especially for night vision applications.

Mangetic materials

Magnetic effects arise in materials in which there is a net spin (intrinsic angular momentum associated with electrons) so that there is a magnetization in the system. In some materials the magnetization can exist in the absence of any external magnetic field. Such materials are called ferromagnets. In other materials magnetization only arises in the presence of a magnetic field. Such materials are called paramagnetic or diamagnetic (depending upon whether the magnetization is parallel or opposite of the field).

Magnetic materials have been an important part of the recording media industry. For memories the hysteresis curve for ferromagnets shown in Fig. 5 is used to create a two-state system, whereby using an external field the orientation of the magnetization is altered.

I.3 BIOLOGICAL SYSTEMS: NATURE'S SMART MATERIALS

Scientists have looked at nature's creations for inspiration since the beginning of civilization. Scientific laws – the underlying basis of all technology – are the result of observing nature in action and then developing a consistent description. Biological systems – from a cell to a complex nervous system – are a source of inspiration for scientists. The flight of birds has inspired aerospace technology, neural networks derive inspiration from the brain (although nature is far ahead), and the way living objects see and sense has inspired technologies in microwave and optics. The list goes on with sensors and actuators, micro machines, robots, pharmaceuticals, chemistry ... all benefiting from what nature has produced.

With advances in biology, particularly with advances in chemical and physical probes and diagnostic tools, scientists are able to go beyond mere observation of biological systems. Advances in genetics have allowed scientists to understand how biological systems function and how they can be manipulated. Human intervention in the manipulation of biological systems (genetically modified foods, cloning, selective breeding of species, etc.) is a highly charged area, with ethics, religious beliefs, legal systems, and local customs all being important factors in making decisions about whether technology should be allowed to advance.

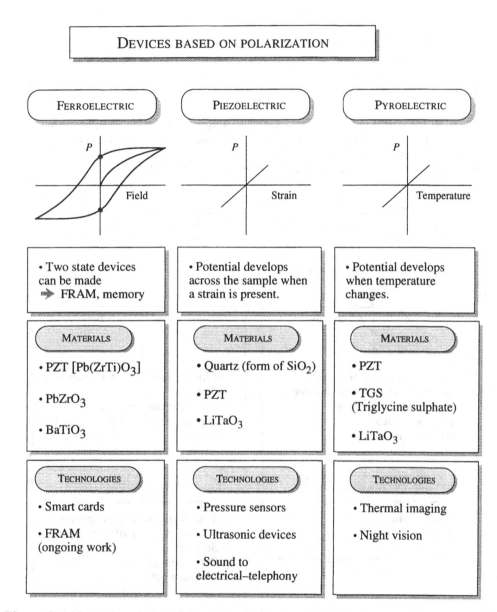

Figure I.4: Materials and devices based on changes in polarization in materials.

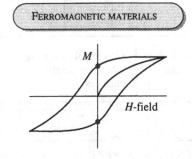

Figure I.5: Magnetic changes in materials and device technologies based on them.

Modern information-processing devices can certainly better biological systems in many areas of information processing. Even gifted mathematicians cannot keep up with a simple calculator when it comes to number crunching. Similarly a low-end computer can "memorize" and accurately recall millions of names and phone numbers. However, pertaining to real-life, "hard" problems, biological systems are well beyond what technology can accomplish. In the area of problems, such as recognition, conversation, associative memories, etc., technology and software are not even at the level of an insect. In fact many imaginative ideas about future technology are based on observing what living organisms can do.

A critical component of technology is the ability to synthesize a structure repeatedly. A dream of processing engineers is to simply assemble the materials together and let things just "self-assemble" on their own. This does happen in chemical reactions – provided the right thermodynamic conditions are maintained. Information-processing devices, however, are still far from this point of self-assembly. Electronic devices, for example, require a lot of processes, such as masking, etching, regrowth, undercutting, etc., to form the final device.

A process engineer, even in the most advanced fabrication facility, can only wonder at nature's ability to produce enormously complex organs. The ultimate incredibly complex self-assembly is the multiplication of cells. In this process nature makes exact copies of DNA. As we will see in this text, even in state-of-the-art facilities, devices are made using essentially "hammer and chisel" approaches. However, advances are being made in the synthesis of at least some parts of devices through self-assembly.

I.4 ROLE OF THIS BOOK

This book has been prepared for a one-semester course on the physics of smart materials. The book would be ideal for courses taught at a senior level or beginning graduate level in departments of applied physics, material science, or electrical engineering. The approach used in this book takes the reader from basic physics towards applications. Many important devices, such as field effect transistors, bipolar transistors, organic transistors, light emitters, memory devices, sensors, and actuators are discussed in the context of the physical phenomena examined.

The user of this book will see that in every chapter and in most sections there is a liberal use of pedagogical tools, such as flow charts, tables, figures, and solved examples. The solved examples would be useful for the student, since they involve examining realistic numbers. Some topics are such that a simple explanation can be given for the underlying physics. For example, issues related to semiconductor devices can be explained on the basis of band theory. However, the physics behind effects such as ferroelectricity, ferromagnetism, etc. is quite a bit more complex. In such cases we provide a motivation for the phenomena, but avoid rigorous derivations. There are also several effects that require knowledge of advanced quantum mechanics for their understanding (e.g., the phenomena of spontaneous and stimulated emissions). In such cases we use the results from quantum mechanics and apply them. Simple arguments are presented to explain the results, but a rigorous derivation is avoided, given the overall level of this book.

Chapter 1

STRUCTURAL PROPERTIES

1.1 INTRODUCTION

In this text we will discuss a variety of physical properties, which form the basis for intelligent devices. These properties are closely linked to the physical structure of the materials. The arrangements of the atoms/molecules determine important symmetries in the system that, in turn, influence the electronic and optical properties. For example, the presence or absence of inversion symmetry determines properties such as the piezoelectric effect used for sensors and ultrasonic applications. Ferroelectric materials depend upon special crystalline properties of ionic crystals. Valence band properties in semiconductors are determined by the cubic symmetry in the crystals.

In addition to the arrangement of atoms in crystals, it is also important to understand the nature of surfaces and interfaces. Many devices are based on phenomena that are unique to surfaces or interfaces. Finally, we have to realize that most materials are far from perfect crystals. Polycrystalline materials, amorphous materials, and materials with defects are also used in making smart devices.

In this chapter we will examine the structural properties of a variety of materials used for smart device applications. We will start with perfect crystals.

1.2 CRYSTALLINE MATERIALS

Almost all high-performance devices are based on crystalline materials. Although, as we will see later in the chapter, there are some devices that use low-cost amorphous or polycrystalline semiconductors, their performance is quite poor. Crystals are made up of identical building blocks, the block being an atom or a group of atoms. While in "natural" crystals the crystalline symmetry is fixed by nature, new advances in crystal growth techniques are allowing scientists to produce artificial crystals with modified crystalline structures. These advances depend upon atomic layers being placed with exact precision and control during growth, leading to "superlattices." To define the crystal structure, two important concepts are introduced. The *lattice* represents a set of points in space, which form a periodic structure. Each point sees an exact similar environment. The lattice is by itself a mathematical abstraction. A building block of atoms called the *basis* is then attached to each lattice point, yielding the crystal structure.

The properties of a lattice are defined by three vectors $\mathbf{a}_1$, $\mathbf{a}_2$, $\mathbf{a}_3$, chosen so that any lattice point $\mathbf{R}'$ can be obtained from any other lattice point $\mathbf{R}$ by a translation

$$\mathbf{R}' = \mathbf{R} + m_1\mathbf{a}_1 + m_2\mathbf{a}_2 + m_3\mathbf{a}_3 \tag{1.1}$$

where m_1, m_2, m_3 are integers. Such a lattice is called a Bravais lattice. The entire lattice can be generated by choosing all possible combinations of the integers m_1, m_2, m_3. The translation vectors $\mathbf{a}_1$, $\mathbf{a}_2$, and $\mathbf{a}_3$ are called primitive vectors if the volume of the cell formed by them is the smallest possible. There is no unique way to choose the primitive vectors. One choice is to pick

$\mathbf{a}_1$ to be the shortest period of the lattice
$\mathbf{a}_2$ to be the shortest period not parallel to $\mathbf{a}_1$
$\mathbf{a}_3$ to be the shortest period not coplanar with $\mathbf{a}_1$ and $\mathbf{a}_2$

It is possible to define more than one set of primitive vectors for a given lattice, and often the choice depends upon convenience. The volume cell enclosed by the primitive vectors is called the *primitive unit cell*. The crystalline structure is now produced by attaching the basis to each of these lattice points.

$$\text{lattice} + \text{basis} = \text{crystal structure} \tag{1.2}$$

Because of the periodicity of a lattice, it is useful to define the symmetry of the structure. The symmetry is defined via a set of point group operations, which involve a set of operations applied around a point. The operations involve rotation, reflection, and inversion. The symmetry plays a very important role in the electronic properties of the crystals. For example, the inversion symmetry is extremely important and many physical properties of semiconductors are tied to the absence of this symmetry. As will be clear later, in the diamond structure (Si, Ge, C, etc.), inversion symmetry is present, while, in the zinc blende structure (GaAs, AlAs, InAs, etc.), it is absent. Because of this lack of inversion symmetry, these semiconductors are piezoelectric; i.e., when they are strained an electric potential is developed across the opposite faces of the crystal. In crystals with inversion symmetry, where the two faces are identical, this is not possible.

1.2.1 Basic lattice types

The various kinds of lattice structures possible in nature are described by the symmetry group that describes their properties. Rotation is one of the important symmetry groups. Lattices can be found which have a rotation symmetry of 2π, $\frac{2\pi}{2}$, $\frac{2\pi}{3}$, $\frac{2\pi}{4}$, $\frac{2\pi}{6}$. The rotation

1.2. Crystalline materials

h

System	Number of lattices	Restrictions on conventional cell axes and singles
Triclinic	1	$a_1 \neq a_2 \neq a_3$ $\alpha \neq \beta \neq \gamma$
Monoclinic	2	$a_1 \neq a_2 \neq a_3$ $\alpha = \gamma = 90° \neq \beta$
Orthorhombic	4	$a_1 \neq a_2 \neq a_3$ $\alpha = \beta = \gamma = 90°$
Tetragonal	2	$a_1 = a_2 \neq a_3$ $\alpha = \beta = \gamma = 90°$
Cubic	3	$a_1 = a_2 = a_3$ $\alpha = \beta = \gamma = 90°$
Trigonal	1	$a_1 = a_2 = a_3$ $\alpha = \beta = \gamma < 120°, \neq 90°$
Hexagonal	1	$a_1 = a_2 \neq a_3$ $\alpha = \beta = 90°$ $\gamma = 120°$

Table 1.1: The 14 Bravais lattices in three-dimensional systems and their properties.

symmetries are denoted by 1, 2, 3, 4, and 6. No other rotation axes exist; e.g., $\frac{2\pi}{5}$ or $\frac{2\pi}{7}$ are not allowed because such a structure could not fill up an infinite space.

There are 14 types of lattices in 3D. These lattice classes are defined by the relationships between the primitive vectors a_1, a_2, and a_3, and the angles α, β, and γ between them. The general lattice is triclinic ($\alpha \neq \beta \neq \gamma, a_1 \neq a_2 \neq a_3$) and there are 13 special lattices. Table 1.1 provides the basic properties of these three-dimensional lattices and Fig. 1.1 shows a schematic.

Most materials forming the basis of modern information technologies have an underlying cubic or hexagonal lattice. There are three kinds of cubic lattices: simple cubic, body-centered cubic, and face-centered cubic.

Simple cubic: The simple cubic lattice shown in Fig. 1.2 is generated by the primitive vectors

$$a\mathbf{x}, a\mathbf{y}, a\mathbf{z} \quad (1.3)$$

where the **x**, **y**, **z** are unit vectors.

Body-centered cubic: The bcc lattice shown in Fig. 1.3 can be generated from the simple cubic structure by placing a lattice point at the center of the cube. If $\hat{\mathbf{x}}, \hat{\mathbf{y}}$, and $\hat{\mathbf{z}}$ are three orthogonal unit vectors, then a set of primitive vectors for the body-centered

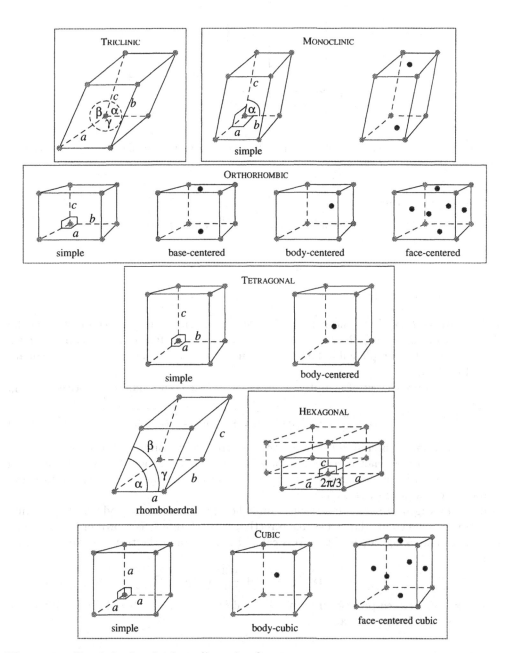

Figure 1.1: Bravis lattices in three-dimensional systems.

1.2. Crystalline materials

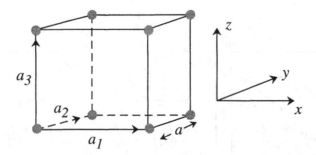

Figure 1.2: A simple cubic lattice showing the primitive vectors. The crystal is produced by repeating the cubic cell through space.

cubic lattice could be

$$a_1 = a\hat{x}, a_2 = a\hat{y}, a_3 = \frac{a}{2}(\hat{x} + \hat{y} + \hat{z}) \tag{1.4}$$

A more symmetric set for the bcc lattice is

$$a_1 = \frac{a}{2}(\hat{y} + \hat{z} - \hat{x}), a_2 = \frac{a}{2}(\hat{z} + \hat{x} - \hat{y}), a_3 = \frac{a}{2}(\hat{x} + \hat{y} - \hat{z}) \tag{1.5}$$

Face-centered cubic: Another equally important lattice for semiconductors is the *face-centered cubic* (fcc) Bravais lattice. To construct the face-centered cubic Bravais lattice add to the simple cubic lattice an additional point in the center of each square face (Fig. 1.4).

A symmetric set of primitive vectors for the face-centered cubic lattice (see Fig. 1.4) is

$$a_1 = \frac{a}{2}(\hat{y} + \hat{z}), a_2 = \frac{a}{2}(\hat{z} + \hat{x}), a_3 = \frac{a}{2}(\hat{x} + \hat{y}) \tag{1.6}$$

The face-centered cubic and body-centered cubic Bravais lattices are of great importance since an enormous variety of solids crystallize in these forms, with an atom (or ion) at each lattice site. Essentially all semiconductors of interest for electronics and optoelectronics have the fcc structure.

Simple hexagonal structure: The simple hexagonal lattice is produced by stacking two-dimensional triangular structures directly over each other, as shown in Fig. 1.5. The direction of stacking (a_3 in Fig. 1.5) is called the c-axis and the three primitive vectors are

$$a_1 = a\hat{x}, a_2 = \frac{a}{2}\hat{x} + \frac{\sqrt{3}a}{2}\hat{y}; a_3 = c\hat{z} \tag{1.7}$$

The hexagonal closed-packed structure, to be discussed later, is based on two interpenetrating simple hexagonal lattices.

1.2.2 Some important crystal structures

Many of the materials employed to create devices used for electronics, optoelectronics, and sensing are given category names, such as metals, insulators, and semiconductors.

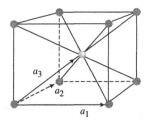

Material	Lattice constant (a)
Ba	5.02
Cr	2.88
Cs	6.05
Fe	2.87
Nb	3.30
Rb	5.59
Ta	3.31
W	3.16

Some materials which crystallize in monoatomic bcc structures

Figure 1.3: The body-centered cubic lattice along with a choice of primitive vectors. Also shown are lattice constants of some materials that crystallize in the monoatomic bcc structure.

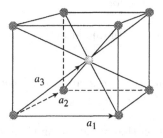

Material	Lattice constant (a)
Ag	4.09
Al	4.05
Au	4.08
Ca	5.58
Ce	5.16
Cu	3.61
La	5.30
Ni	3.52
Pb	4.95
Pd	3.89
Pt	3.92
Th	5.08

Materials with monoatomic fcc structures

Figure 1.4: Primitive basis vectors for the face-centered cubic lattice. Also shown are some materials that crystallize in the monoatomic fcc structure.

1.2. Crystalline materials

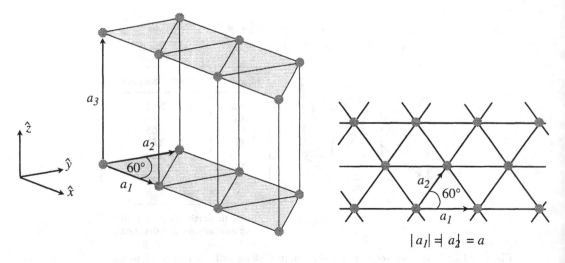

Figure 1.5: The simple hexagonal Bravais lattice. Two-dimensional triangular nets (shown in inset) are stacked directly above one another, a distance c apart. Also shown are the three unit vectors.

Depending upon applications, they are also categorized as ceramics, polar materials, ferroelectrics, ferromagnetics, etc. These materials have a crystal structure, ranging from the very simple with one atom per basis to complex ones with several atoms on a basis. Also in many materials the positions of atoms in the structure are not ideal, due to "spontaneous" effects arising from charges on the ions.

Monoatomic body-centered cubic
There are many metals which have the bcc lattice with one atom per basis. In Fig. 1.3 we show some of these materials.

Monoatomic face-centered cubic
Many metals crystallize in the fcc lattice and have just one atom per basis. In Fig. 1.4 we show some of the important metals that fall into this category.

Sodium chloride structure
The sodium chloride (NaCl) structure is based on the fcc lattice and a basis of one Na atom and one Cl atom separated by half of the body diagonal of the cube. The basis atoms are at 0 and $a/2(\hat{x} + \hat{y} + \hat{z})$. The structure is shown in Fig. 1.6, along with some materials which crystallize in this structure.

Cesium chloride structure
The cesium chloride structure is shown in Fig. 1.7. The cesium and chloride atoms are placed on the points of a bcc lattice so that each atom has eight neighbors. The underlying lattice is simple cubic with two atoms per basis. The atoms are at 0 and $a/2(\hat{x} + \hat{y} + \hat{z})$. In Fig. 1.7 we show some important materials which have the CsCl structure.

Diamond and zinc blende structures
Most semiconductors of interest for electronics and optoelectronics have an underlying fcc lattice. However, they have two atoms per basis. The coordinates of the two basis

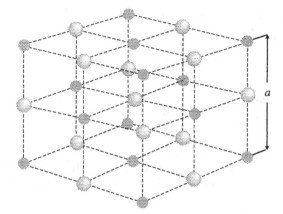

Material	Lattice Constant (a) Å
AgBr	5.77
KCl	6.29
LiH	4.08
MgO	4.20
MnO	4.43
NaCl	5.63
PbS	5.92

Figure 1.6: The sodium chloride crystal structure. The space lattice is fcc, and the basis has one Na$^+$ ion at 0 0 0 and one Cl$^-$ ion at $\frac{1}{2}\frac{1}{2}\frac{1}{2}$. The table shows some materials with NaCl structure.

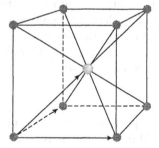

Material	Lattice constant (a) Å
AlNi	2.88
BeCu	2.7
CsCl	4.11
LiHg	3.29

Some materials that have the cesium chloride structure.

Figure 1.7: The cesium chloride crystal structure. The space lattice is simple cubic, and the basis has one Cs$^+$ ion and one Cl$^-$ ion at $\frac{1}{2}\frac{1}{2}\frac{1}{2}$. The table shows some materials with the cesium chloride structure.

1.2. Crystalline materials

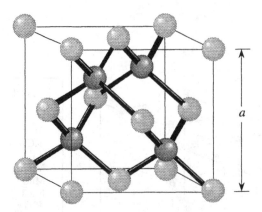

Figure 1.8: The zinc blende crystal structure. The structure consists of the interpenetrating fcc lattices, one displaced from the other by a distance $(\frac{a}{4}\frac{a}{4}\frac{a}{4})$ along the body diagonal. The underlying Bravais lattice is fcc with a two atom basis. The positions of the two atoms is (000) and $(\frac{a}{4}\frac{a}{4}\frac{a}{4})$.

atoms are
$$(000) \text{ and } \left(\frac{a}{4}, \frac{a}{4}, \frac{a}{4}\right) \tag{1.8}$$

Since each atom lies on its own fcc lattice, such a two atom basis structure may be thought of as two interpenetrating fcc lattices, one displaced from the other by a translation along a body diagonal direction $(\frac{a}{4}\frac{a}{4}\frac{a}{4})$.

Figure 1.8 gives details of this important structure. If the two atoms of the basis are identical, the structure is called diamond. Semiconductors such as Si, Ge, C, etc. fall into this category. If the two atoms are different, the structure is called the zinc blende structure. Semiconductors such as GaAs, AlAs, CdS, etc. fall into this category. Semiconductors with the diamond structure are often called elemental semiconductors, while the zinc blende semiconductors are called compound semiconductors. The compound semiconductors are also denoted by the position of the atoms in the periodic chart, e.g., GaAs, AlAs, InP are called III–V (three–five) semiconductors, while CdS, HgTe, CdTe, etc., are called II–VI (two–six) semiconductors.

Hexagonal close-pack structure

The hexagonal close-pack (hcp) structure is an important lattice structure and many metals have this underlying lattice. Some semiconductors, such as BN, AlN, GaN, SiC, etc., also have this underlying lattice (with a two-atom basis). The hcp structure is formed as shown in Fig. 1.9a. Imagine that a close-packed layer of spheres is formed. Each sphere touches six other spheres, leaving cavities, as shown. A second close-packed layer of spheres is placed on top of the first one so that the second-layer sphere centers are in the cavities formed by the first layer. The third layer of close-packed spheres can now be placed so that centers of the spheres do not fall on the centers of the starting spheres (left side of Fig. 1.9a) or coincide with the centers of the starting spheres (right side of Fig. 1.9b). These two sequences, when repeated, produce the fcc and hcp lattices.

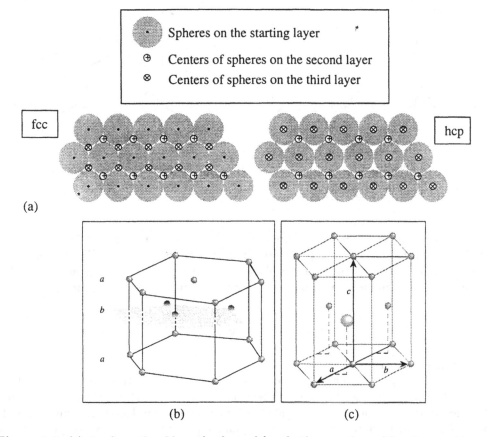

Figure 1.9: (a) A schematic of how the fcc and hcp lattices are formed by close packing of spheres. (b) The hcp structure is produced by two interpenetrating simple hexagonal lattices with a displacement, as discussed in the text. Arrangement of lattice points on an hcp lattice.

1.2. Crystalline materials

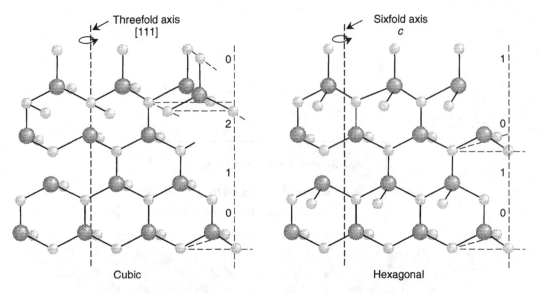

Figure 1.10: The stacking of tetrahedral layers in cubic and hexagonal ZnS. The large atoms are S; the small atoms are Zn. The vertical axis of hexagonal ZnS is a six-fold screw axis involving a translation of one-half c for each 60 degrees of rotation.

Underlying the hcp structure is a simple hexagonal lattice (discussed earlier). The hcp structure consists of two interpenetrating simple hexagonal lattices as shown in Fig. 1.9b. The two lattices are displaced from each other by $\mathbf{a}_1/3 + \mathbf{a}_2/3 + \mathbf{a}_3/2$ as shown. The magnitude of $\mathbf{a}_3$ is denoted by c and in an ideal hcp structure

$$\frac{c}{a} = \sqrt{\frac{8}{3}} \tag{1.9}$$

Wurtzite structures
A number of important semiconductors crystallize in the hcp structure with two atoms per lattice site. The coordination of the atoms is the same as in the diamond or zinc blende structures. The nearest neighbor bonds are tetrahedral and are similar in both zinc blende and wurtzite structures. The symmetry of rotation is, however, different as shown in Fig. 1.10.

In Tables 1.2 and 1.3 we show the structural properties of some important materials that crystallize in the diamond, zinc blende, and wurtzite structures.

Perovskite structure
Materials like $CaTiO_3$, $BaTiO_3$, $SrTiO_3$, etc., have the perovskite structure using $BaTiO_3$ as an example. The structure is cubic with Ba^{2+} ions at the cube corners and O^{2-} ions at the face centers. The Ti^{4+} ion is at the body center.

Perovskites show a ferroelectric effect below a temperature called Curie temperature and have spontaneous polarization due to relative movements of the cations and anions. As shown in Fig. 1.11 the Ba^{2+} ions and Ti^{4+} ions are displaced relative

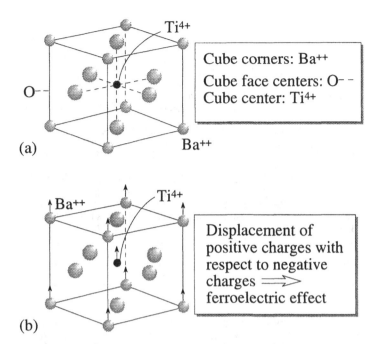

Figure 1.11: (a) The structure of a typical perovskite crystal illustrated by examining barium titanante. (b) The ferroelectric effect is produced by a net displacement of the positive ions with respect to the negative ions.

to the O^{2-} ions creating a dipole moment. As will be discussed later in this book the polarization can be controlled by an external electric field.

For many applications, one uses alloys made from two or more different materials. The lattice constant of the alloy is given by Vegard's law, according to which the alloy lattice constant is the weighted mean of the lattice constants of the individual components.

$$a_{\text{alloy}} = x a_A + (1-x) a_B \qquad (1.10)$$

where a_{alloy} is the lattice constant of the alloy $A_x B_{1-x}$, and a_A and a_B are the lattice constants of materials A and B, respectively.

1.2.3 Notation to denote planes and points in a lattice: Miller indices

A simple scheme is used to describe lattice planes, directions and points. For a plane, we use the following procedure:

(1) Define the x, y, z axes (primitive vectors).
(2) Take the intercepts of the plane along the axes in units of lattice constants.

1.2. Crystalline materials

Zinc blende and wurtzite

MATERIAL	CRYSTAL STRUCTURE	BANDGAP (EV)	STATIC DIELECTRIC CONSTANT	LATTICE CONSTANT (Å)	DENSITY (gm-cm^{-3})
C	DI	5.50, I	5.570	3.56683	3.51525
Si	DI	1.1242, I	11.9	5.431073	2.329002
SiC	ZB	2.416, I	9.72	4.3596	3.166
Ge	DI	0.664, I	16.2	5.6579060	5.3234
BN	HEX	5.2, I	$\varepsilon_{\parallel} = 5.06$ $\varepsilon_{\perp} = 6.85$	$a = 6.6612$ $c = 2.5040$	2.18
BN	ZB	6.4, I	7.1	3.6157	3.4870
BP	ZB	2.4, I	11.	4.5383	2.97
BAs	ZB	—	—	4.777	5.22
AlN	W	6.2,D	$\bar\varepsilon = 9.14$	$a = 3.111$ $c = 4.981$	3.255
AlP	ZB	2.45,I	9.8	5.4635	2.401
AlAs	ZB	2.153,I	10.06	5.660	3.760
AlSb	ZB	1.615,I	12.04	6.1355	4.26
GaN	W	3.44,D	$\varepsilon_{\parallel}=10.4$ $\varepsilon_{\perp}=9.5$	$a = 3.175$ $c = 5.158$	6.095
GaP	ZB	2.272,I	11.11	5.4505	4.138
GaAs	ZB	1.4241,D	13.18	5.65325	5.3176
GaSb	ZB	0.75,D	15.69	6.09593	5.6137
InN	W	1.89,D		$a = 3.5446$ $c = 8.7034$	6.81
InP	ZB	1.344,D	12.56	5.8687	4.81
InAs	ZB	0.354,D	15.15	6.0583	5.667
InSb	ZB	0.230,D	16.8	6.47937	5.7747
ZnO	W	3.44,D	$\varepsilon_{\parallel}= 8.75$ $\varepsilon_{\perp}=7.8$	$a = 3.253$ $c = 5.213$	5.67526
ZnS	ZB	3.68,D	8.9	5.4102	4.079
ZnS	W	3.9107,D	$\bar\varepsilon = 9.6$	$a = 3.8226$ $c = 6.2605$	4.084
ZnSe	ZB	2.8215,D	9.1	5.6676	5.266
ZnTe	ZB	2.3941,D	8.7	6.1037	5.636
CdO	R	0.84,I	21.9	4.689	8.15
CdS	W	2.501,D	$\bar\varepsilon = 9.83$	$a = 4.1362$ $c = 6.714$	4.82
CdS	ZB	2.50,D	—	5.818	—
CdSe	W	1.751,D	$\varepsilon_{\parallel}=10.16$ $\varepsilon_{\perp}= 9.29$	$a = 4.2999$ $c = 7.0109$	5.81
CdSe	ZB	—	—	6.052	
CdTe	ZB	1.475,D	10.2	6.482	5.87
PbS	R	0.41,D*	169.	5.936	7.597
PbSe	R	0.278,D*	210.	6.117	8.26
PbTe	R	0.310,D*	414.	6.462	8.219

Data are given at room temperature values (300 K).
Key: DI: diamond; HEX: hexagonal; R: rocksalt; W: wurtzite; ZB: zinc blende;
*: gap at L point; D: direct; I: indirect $\varepsilon_{\parallel}$: parallel to c-axis; $\varepsilon_{\perp}$: perpendicular to c-axis.

Table 1.2: Structural properties of some important semiconductors.

Material	$a(\text{Å})$	$c(\text{Å})$	c/a
Be	2.29	3.58	1.56
Cd	2.98	5.62	1.89
Mg	3.21	5.21	1.62
Ti	2.95	4.69	1.59
Zn	2.66	4.95	1.86
Zr	3.23	5.15	1.59

Table 1.3: Materials with hcp closed-packed structure. The "ideal" c/a ratio is 1.6.

(3) Take the reciprocal of the intercepts and reduce them to the smallest integers.

The notation (hkl) denotes a family of parallel planes.

The notation (hkl) denotes a family of equivalent planes.

To denote directions, we use the smallest set of integers having the same ratio as the direction cosines of the direction.

In a cubic system, the Miller indices of a plane are the same as the direction perpendicular to the plane. The notation [ ] is for a set of parallel directions; < > is for a set of equivalent direction. Fig. 1.12 shows some examples of the use of the Miller indices to define planes.

EXAMPLE 1.1 The lattice constant of silicon is 5.43 Å. Calculate the number of silicon atoms in a cubic centimeter. Also calculate the number density of Ga atoms in GaAs which has a lattice constant of 5.65 Å.

Silicon has a diamond structure, which is made up of the fcc lattice with two atoms on each lattice point. The fcc unit cube has a volume a^3. The cube has eight lattice sites at the cube edges. However, each of these points is shared with eight other cubes. In addition, there are six lattice points on the cube face centers. Each of these points is shared by two adjacent cubes. Thus the number of lattice points per cube of volume a^3 are

$$N(a^3) = \frac{8}{8} + \frac{6}{2} = 4$$

In silicon, there are two silicon atoms per lattice point. The number density is, therefore

$$N_{Si} = \frac{4 \times 2}{a^3} = \frac{4 \times 2}{(5.43 \times 10^{-8})^3} = 4.997 \times 10^{22} \text{ atoms/cm}^3$$

In GaAs, there is one Ga atom and one As atom per lattice point. The Ga atom density is, therefore

$$N_{Ga} = \frac{4}{a^3} = \frac{4}{(5.65 \times 10^{-8})^3} = 2.22 \times 10^{22} \text{ atoms/cm}^3$$

There are an equal number of As atoms.

EXAMPLE 1.2 In semiconductor technology, a Si device on a VLSI chip represents one of the smallest devices, while a GaAs laser represents one of the larger devices. Consider a

1.2. Crystalline materials

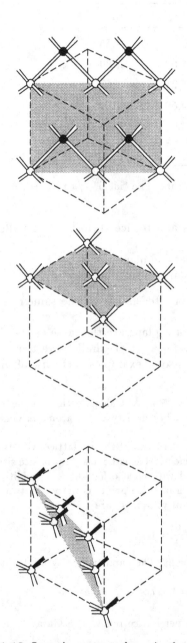

ATOMS ON THE (110) PLANE

Each atom has 4 bonds:
• 2 bonds in the (110) plane
• 1 bond connects each atom to adjacent (110) planes

⟹ Cleaving adjacent planes requires breaking 1 bond per atom

ATOMS ON THE (001) PLANE

2 bonds connect each atom to adjacent (001) plane

Atoms are either Ga or As in a GaAs crystal

⟹ Cleaving adjacent planes requires breaking 2 bonds per atom

ATOMS ON THE (111) PLANE

Could be either Ga or As

1 bond connecting an adjacent plane on one side

3 bonds connecting an adjacent plane on the other side

Figure 1.12: Some important planes in the zinc blende or diamond structure along with their Miller indices. This figure also shows how many bonds connect adjacent planes. This number determines how easy or difficult it is to cleave the crystal along these planes.

Si device with dimensions $(5 \times 2 \times 1)$ μm^3 and a GaAs semiconductor laser with dimensions $(200 \times 10 \times 5)$ μm^3. Calculate the number of atoms in each device.

From Example 1.1 the number of Si atoms in the Si transistor are

$$N_{Si} = (5 \times 10^{22} \text{ atoms/cm}^3)(10 \times 10^{-12} \text{ cm}^3) = 5 \times 10^{11} \text{ atoms}$$

The number of Ga atoms in the GaAs laser are

$$N_{Ga} = (2.22 \times 10^{22})(10^4 \times 10^{-12}) = 2.22 \times 10^{14} \text{ atoms}$$

An equal number of As atoms are also present in the laser.

EXAMPLE 1.3 Calculate the surface density of Ga atoms on a Ga terminated (001) GaAs surface.

In the (001) surfaces, the top atoms are either Ga or As leading to the terminology Ga terminated (or Ga stabilized) and As terminated (or As stabilized), respectively. A square of area a^2 has four atoms on the edges of the square and one atom at the center of the square. The atoms on the square edges are shared by a total of four squares. The total number of atoms per square is

$$N(a^2) = \frac{4}{4} + 1 = 2$$

The surface density is then

$$N_{Ga} = \frac{2}{a^2} = \frac{2}{(5.65 \times 10^{-8})^2} = 6.26 \times 10^{14} \text{ cm}^{-2}$$

EXAMPLE 1.4 Calculate the height of a GaAs monolayer in the (001) direction.

In the case of GaAs, a monolayer is defined as the combination of a Ga and As atomic layer. The monolayer distance in the (001) direction is simply

$$A_{ml} = \frac{a}{2} = \frac{5.65}{2} = 2.825 \text{ Å}$$

1.2.4 Artificial structures: superlattices and quantum wells

It is known that electronic and optical properties can be altered by using heterostructures; i.e., combinations of more that one semiconductor. Epitaxial techniques allow monolayer (~3 Å) control in the chemical composition of the growing crystal. Nearly every semiconductor extending from zero bandgap (α-Sn,HgCdTe) to large bandgap materials, such as ZnSe,CdS, etc., has been grown by epitaxial techniques.

Heteroepitaxial techniques allow one to grow heterostructures with atomic control, we can change the periodicity of the crystal in the growth direction. This leads to the concept of superlattices where two (or more) semiconductors A and B are grown alternately with thicknesses d_A and d_B respectively. The periodicity of the lattice in the growth direction is then $d_A + d_B$. A $(GaAs)_2$ $(AlAs)_2$ superlattice is illustrated in Fig. 1.13. It is a great testimony to the precision of the new growth techniques that values of d_A and d_B as low as monolayer have been grown.

It is important to point out that the most widely used heterostructures are not superlattices but quantum wells, in which a single layer of one semiconductor is

1.2. Crystalline materials

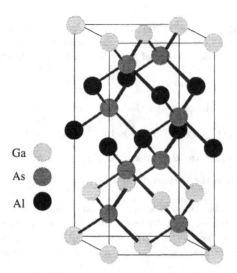

Figure 1.13: Arrangement of atoms in a (GaAs)$_2$(AlAs)$_2$ superlattice grown along (001) direction.

sandwiched between two layers of a larger bandgap material. Such structures allow one to exploit special quantum effects that have become very useful in electronic and optoelectronic devices.

1.2.5 Surfaces: ideal versus real

The crystalline and electronic properties are quite different from the properties of the bulk material. The bulk crystal structure is decided by the internal chemical energy of the atoms forming the crystal with a certain number of nearest neighbors, second nearest neighbors, etc. At the surface, the number of neighbors is suddenly altered. Thus the spatial geometries which were providing the lowest energy configuration in the bulk may not provide the lowest energy configuration at the surface. Thus, there is a readjustment or "reconstruction" of the surface bonds towards an energy-minimizing configuration.

An example of such a reconstruction is shown for the GaAs surface in Fig. 1.14. The figure (a) shows an ideal (001) surface, where the topmost atoms form a square lattice. The surface atoms have two nearest neighbor bonds (Ga–As) with the layer below, four second neighbor bonds (e.g., Ga–Ga or As–As) with the next lower layer, and four second neighbor bonds within the same layer. In a "real" surface, the arrangement of atoms is far more complex. We could denote the ideal surface by the symbol C(1×1), representing the fact that the surface periodicity is one unit by one unit along the square lattice along the [110] and [$\bar{1}$10]. The reconstructed surfaces that occur in nature are generally classified as C(2×8) or C(2×4) etc., representing the increased periodicity along the [$\bar{1}$10] and [110] respectively. The C(2×4) case is shown schematically in Fig. 1.14b, for an arsenic stabilized surface (i.e., the top monolayer is

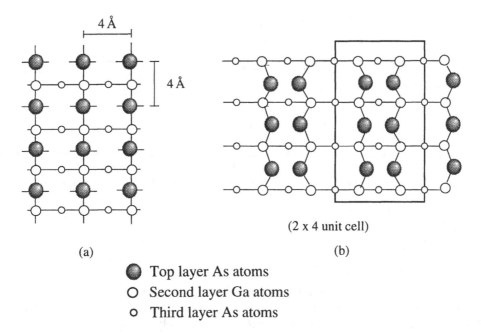

Figure 1.14: The structure (a) of the unreconstructed GaAs (001) arsenic-rich surface. The missing dimer model (b) for the GaAs (001) (2×4) surface. The As dimers are missing to create a 4 unit periodicity along one direction and a two unit periodicity along the perpendicular direction.

As). The As atoms on the surface form dimers (along [$\bar{1}$10] on the surface to strengthen their bonds. In addition, rows of missing dimers cause a longer range ordering as shown to increase the periodicity along the [110] direction to cause a C(2×4) unit cell. The surface periodicity is directly reflected in the x-ray diffraction pattern.

A similar effect occurs for the (110) surface of GaAs. This surface has both Ga and As atoms (the cations and anions) on the surface. A strong driving force exists to move the surface atoms and minimize the surface energy. Reconstruction effects also occur in silicon surfaces, where depending upon surface conditions a variety of reconstructions are observed. Surface reconstructions are very important since often the quality of the epitaxial crystal growth depends critically on the surface reconstruction.

EXAMPLE 1.5 Calculate the planar density of atoms on the (111) surface of Ge.

As can be seen from Fig. 1.12, we can form a triangle on the (111) surface. There are three atoms on the tips of the triangle. These atoms are shared by six other similar triangles. There are also three atoms along the edges of the triangle, which are shared by two adjacent triangles. Thus the number of atoms in the triangle are

$$\frac{3}{6} + \frac{3}{2} = 2$$

The area of the triangle is $\sqrt{3}a^2/2$. The density of Ge atoms on the surface is then 7.29×10^{14} cm^{-2}.

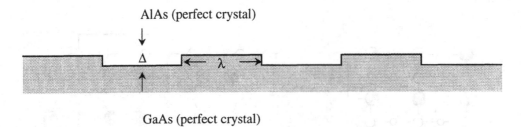

Figure 1.15: A schematic picture of the interfaces between materials with similar lattice constants such as GaAs/AlAs. No loss of crystalline lattice and long-range order is suffered in such interfaces. The interface is characterized by islands of height Δ and lateral extent λ.

1.2.6 Interfaces

Like surfaces, interfaces are an integral part of semiconductor devices. We have already discussed the concept of heterostructures and superlattices, which involve interfaces between two semiconductors. These interfaces are usually of high quality with essentially no broken bonds, except for dislocations in strained structures (to be discussed later). There is, nevertheless, an *interface roughness* of one or two monolayers which is produced because of either non-ideal growth conditions or imprecise shutter control in the switching of the semiconductor species. The general picture of such a rough interface is as shown in Fig. 1.15 for epitaxially grown interfaces. The crystallinity and periodicity in the underlying lattice is maintained, but the chemical species have some disorder on interfacial planes. Such a disorder is quite important in many electronic and opto-electronic devices.

One of the most important interfaces in electronics is the Si/SiO_2 interface. This interface and its quality is responsible for essentially all of the modern consumer electronic revolution. This interface represents a situation where two materials with very different lattice constants and crystal structures are brought together. However, in spite of these large differences, the interface quality is quite good. In Fig. 1.16 we show a TEM cross-section of a Si/SiO_2 interface. It appears that the interface has a region of a few monolayers of amorphous or disordered Si/SiO_2 region, creating fluctuations in the chemical species (and consequently in potential energy) across the interface. This interface roughness is responsible for reducing the mobility of electrons and holes in MOS devices. It can also lead to "trap" states, which can seriously deteriorate device performance if the interface quality is poor.

Finally, we have the interfaces formed between metals and semiconductors. Structurally, these important interfaces are hardest to characterize. These interfaces are usually produced in the presence of high temperatures and involve diffusion of metal elements along with complex chemical reactions. The "interfacial region" usually extends over several hundred Angstroms and is a complex non-crystalline region.

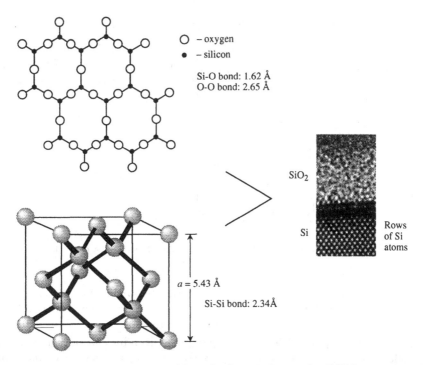

Figure 1.16: The tremendous success of Si technology is due to the Si/SiO$_2$ interface. In spite of the very different crystal structure of Si and SiO$_2$, the interface is extremely sharp, as shown in the TEM picture in this figure.

1.3 DEFECTS IN CRYSTALS

In the previous section we have discussed the properties of the perfect crystalline structure. In real semiconductors, there are invariably some defects that are introduced due to either thermodynamic considerations or the presence of impurities during the crystal growth process. In general, defects in crystalline semiconductors can be characterized as: (i) point defects; (ii) line defects; (iii) planar defects, and (iv) volume defects. These defects are detrimental to the performance of electronic and optoelectronic devices and are to be avoided as much as possible. We will give a brief overview of the important defects.

Point defects
A point defect is a highly localized defect that affects the periodicity of the crystal only in one or a few unit cells. There are a variety of point defects, as shown in Fig. 1.17. Defects are present in any crystal and their concentration is given roughly by the thermodynamics relation

$$\frac{N_\text{d}}{N_\text{Tot}} = k_d \exp\left(-\frac{E_\text{d}}{k_B T}\right) \tag{1.11}$$

where N_d is the vacancy density, N_Tot the total site density in the crystal, E_d the

1.3. Defects in crystals

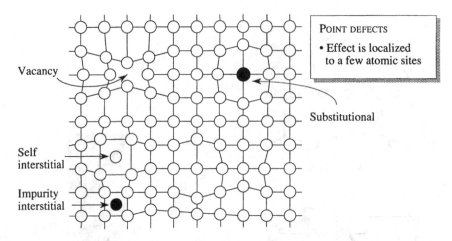

Figure 1.17: A schematic showing some important point defects in a crystal.

defect formation energy, k_d is a dimensionless parameter with values ranging from 1 to 10 in semiconductors, and T, is the crystal growth temperature. The vacancy formation energy is in the range of an eV for most semiconductors.

An important point defect in compound semiconductors such as GaAs is the anti-site defect in which one of the atoms, say Ga, sits on the arsenic sublattice instead of the Ga sublattice. Such defects (denoted by Ga_{As}) can be a source of reduced device performance.

Other point defects are interstitials in which an atom is sitting in a site that is in between the lattice points as shown in Fig. 1.17, and impurity atoms which involve a wrong chemical species in the lattice. In some cases the defect may involve several sites forming a defect complex.

Line defects or dislocations
In contrast to point defects, line defects (called dislocations) involve a large number of atomic sites that can be connected by a line. Dislocations are produced if, for example, an extra half plane of atoms are inserted (or taken out) of the crystal as shown in Fig. 1.18. Such dislocations are called edge dislocations. Dislocations can also be created if there is a slip in the crystal so that part of the crystal bonds are broken and reconnected with atoms after the slip.

Dislocations can be a serious problem, especially in the growth of strained heterostructures (to be discussed later). In optoelectronic devices, dislocations can ruin the device performance and render the device useless. Thus the control of dislocations is of great importance.

Planar defects and volume defects
Planar defects and volume defects are not important in single crystalline materials, but can be of importance in polycrystalline materials. If, for example, silicon is grown on a glass substrate, it is likely that polycrystalline silicon will be produced. In the polycrystalline material, small regions of Si ($\sim$ a few microns in diameter) are perfectly

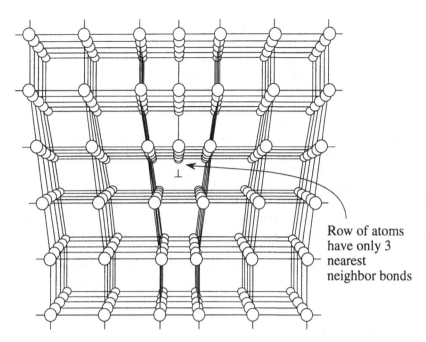

Figure 1.18: A schematic showing the presence of a dislocation. This line defect is produced by adding an extra half plane of atoms. At the edge of the extra plane, the atoms have a missing bond.

crystalline, but are next to microcrystallites with different orientations. The interface between these microcrystallites are called grain boundaries. Grain boundaries may be viewed as an array of dislocations.

Volume defects can be produced if the crystal growth process is poor. The crystal may contain regions that are amorphous or may contain voids. In most epitaxial techniques used in modern optoelectronics, these defects are not a problem. However, the developments of new material systems such as diamond (C) or SiC are hampered by such defects.

EXAMPLE 1.6 Consider an equilibrium growth of a semiconductor at a temperature of 1000 K. The vacancy formation energy is 2.0 eV. Calculate the vacancy density produced if the site density for the semiconductor is 2.5×10^{22} cm^{-3}. Assume that $k_d = 1$.

The vacancy density is

$$\begin{aligned} N_{\text{vac}} &= N_{\text{Tot}} \exp\left(-\frac{E_{\text{vac}}}{k_B T}\right) \\ &= (2.5 \times 10^{22} \text{ cm}^{-3}) \exp\left(-\frac{2.0 \text{ eV}}{0.0867 \text{ eV}}\right) \\ &= 2.37 \times 10^{12} \text{ cm}^{-3} \end{aligned}$$

This is an extremely low density and will have little effect on the properties of the semiconductor. The defect density would be in mid 10^{15} cm^{-3} range if the growth temperature was 1500 K. At such values, the defects can significantly affect device performance.

1.4 HETEROSTRUCTURES

Nearly all modern devices involve combinations of two or more materials. The "active" regions of a device are usually grown on a substrate, which is a thick material and provides not only a rigid foundation on which the device material is grown but is often also the basis of the crystalline lattice structure. The most important substrates in semiconductor devices are silicon, GaAs, and InP. These crystals can be grown with high quality by bulk crystal growth techniques and sliced into 300–400 μm thick wafers. Other substrates used in various device technologies include AlN, SiC, Al_2O_3, etc.

There are two important categories of growth of active device layers on a substrate. In one case (the ideal case) the active device material has the same (or very similar) lattice structure as the substrate. In this case the crystal can be grown by epitaxial techniques such as molecular beam epitaxy (MBE) and metal organic chemical vapor deposition (MOCVD). These techniques allow growth of layers to one monolayer precision. In this case it is possible to grow high-quality, defect-free layers. The active device region can be made up of multiple layers (quantum wells and superlattices) with very abrupt interfaces with essentially no defects. Most high-performance devices are based on this approach. In silicon-based devices, such as MOSFETs and bipolar devices, epitaxial Si is grown on Si substrates. Similarly GaAs and AlAs/GaAs based optoelectronic devices are grown on GaAs substrates.

Another manner in which substrates are used is one where the overlayer has a lattice mismatch with the substrate. In this case the following scenarios occur: (i) If the lattice mismatch is small (typically less than 3–4%), the initial epilayer grows "coherently" with a substrate, i.e., it adjusts its inplane lattice constant to fit the substrate. This produces a strain energy in the overlayer, which is proportional to the film thickness. Once the film thickness reaches a thickness called the critical thickness, it is energetically favorable to create dislocation in the overlayer. (ii) If the lattice mismatch is large, ($> 5\%$) dislocations are generated as soon as growth progresses on the substrate.

In lattice mismatched growth a key to the success of the growth technology is to ensure that if dislocations are generated, they do not propagate through the active layers. As shown schematically in Fig. 1.19, ideally we would like to have the dislocations "bend" so as to be confined near the substrate.

Crystal growers develop a lot of personal recipes to improve the quality of the overlayer, although there are several proven approaches to minimizing the dislocation density. Nevertheless, it is a challenge to grow epitaxial layers on mismatched substrates.

Heterostructures in active devices come in several categories. We will briefly list several specific examples below.

• In the information age's most important device–the metal oxide semiconductor field effect transistor, or MOSFET–the heterostructure between SiO_2 and Si is critical. Silicon dioxide serves as an insulator between the gate and the channel. The quality of the interface is crucial to device performance, since electrons (holes) move near this interface.

• Heterostructure field effect transistors (HFETs) are used for very high frequency applications, where Si based devices cannot operate. These devices involve a

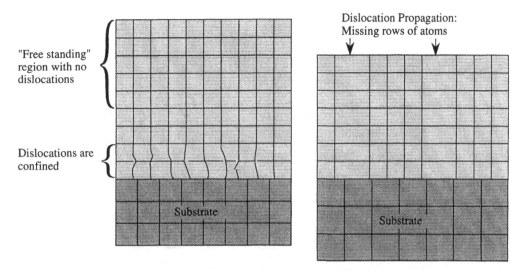

Figure 1.19: Strained epitaxy above critical thickness. On the left hand side is shown a structure in which the dislocations are confined near the overlayer-substrate interface. This is a desirable mode of eptiaxy. On the right hand side, the dislocations are penetrating the overlayer, rendering it useless for most optoelectronic applications.

combination of a large bandgap semiconductor, such as AlAs (or the alloy AlGaAs), and a narrower gap semiconductor such as GaAs. Unlike SiO_2 on Si, AlGaAs and GaAs have very similar lattice structures and the heterostructure has essentially no interface defects.

• In quantum well lasers, differences in the bandgap of various semiconductors is exploited to create two-dimensional electron-hole systems with very low lasing threshold currents. In strained quantum well structures, thin regions (~ 100 Å) with a lattice constant different from that of the substrate are sandwiched between other semiconductors. The built-in strain can be exploited for optoelectronic devices with special properties.

• Multi-layer structures based on crystalline and non-crystalline materials are used for optical and waveguides. These structures are used to "bend" and "switch" optical signals. Since optical wavelengths are about 1 μm and light does not interact strongly with defects, the constraints on the precision and quality of such structures are less severe than those on electronic devices.

1.5 NON-CRYSTALLINE MATERIALS

In the sections discussed above we have focused on crystalline structures. Most high performance devices are based on crystalline materials with as few defects as possible. However, for a variety of reasons (some discussed later) non-crystalline material based devices are also very important. In non-crystalline structures the long ranged structural order present in a crystal is absent. Non-crystalline materials can be grown on the

1.5. Non-crystalline materials

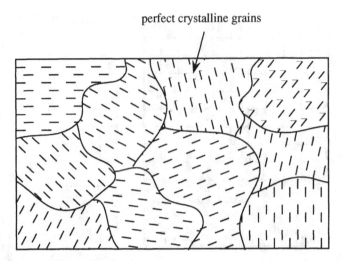

Figure 1.20: A schematic description of a polycrystalline material. Atoms are arranged periodically in a grain, but there is no order between the various grains. The grain boundaries represent regions where defects produced by broken or unfilled bonds are present.

usual substrates used for crystalline materials, but more often they are grown on non-crystalline substrates, such as glass (SiO_2) or even flexible substrates (plastics).

1.5.1 Polycrystalline materials

Polycrystalline materials are widely used in electronic and optoelectronic technologies. Polycrystalline structures are produced when a material is deposited on a substrate which does not have a similar crystal structure. For example, if a metal film is deposited on a semiconductor, the film grows in a polycrystalline form. Also, if silicon is deposited on a glass substrate, it grows in a polycrystalline form.

Polycrystalline films are described by their average grain size as shown in Fig. 1.20. Within a grain the atoms are arranged as in crystal; i.e., with perfect order. However, each grain is surrounded by a grain boundary, which is a region with a high density of defects. The defects arise due to broken or unfulfilled bonds between atoms. In some cases, chemical impurities may also gather at these grain boundaries. Different grains in the polycrystal have essentially no order between their constituent atoms.

Depending upon the growth process and the differences between the substrate and the deposited film, the grain size of a polycrystal can range from 0.1 μm to 10 μm or more. If the grain size exceeds 10 μm, for some device applications the material can be considered to be crystalline. The presence of grain boundaries has serious consequences on the electrical and optical properties of the material. Indeed, certain devices, such as light emitting diodes (LEDs) or laser diodes (LDs), cannot be made from polycrystalline materials. However, some electronic devices used as control transistors for displays can be made from polycrystalline materials. The key advantages of polycrystalline materials is the low cost of the film deposition and the large area of the film possible. Thus

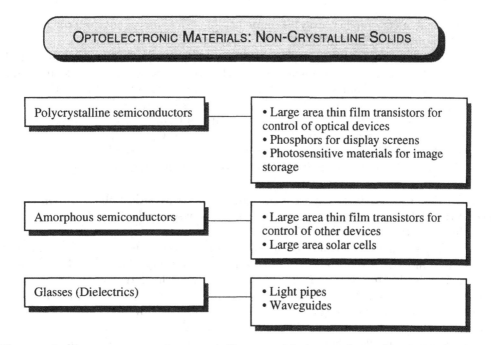

Figure 1.21: Important uses of non-crystalline materials in optoelectronic technology.

polycrystalline technology is an important technology for displays (high-density TV, portable and personal computers, etc.). In Fig. 1.21 the important uses of non-crystalline materials are outlined.

It is essential to mention $PbZrO_3$–$PbTiO_3$–La_2O_3 (PLZT), an important polycrystalline material that is finding extensive use in optoelectronics. This material is a ceramic oxide with ferroelectric properties. While single crystal electro-optic materials, such as potassium dihydrogen phosphate (KDP), $BaTiO_3$, and $Gd(M_0O_4)_3$, are important materials, their applications are limited by cost, size, and susceptibility to moisture (especially for KDP). In contrast, polycrystalline ceramics are not subject to these limitations. The fabrication technology of PLZT is now highly developed and this ceramic is used for a variety of electro-optic devices.

1.5.2 Amorphous and glassy materials

In amorphous materials (sometimes also called glasses) the order among atoms is even lower than that in polycrystalline materials. The most important amorphous materials in optoelectronic technology are glasses based on SiO_x (with different dopants) and amorphous semiconductors, such as amorphous silicon (a-Si). These materials find important uses, as shown in Fig. 1.21.

The amorphous materials are characterized by good short-range order, but poor long-range order. Thus the nearest neighbor and even second neighbor coordination is quite good in amorphous materials. However, the arrangement of atoms which are third

1.5. Non-crystalline materials

nearest neighbors (or further away) is unpredictable. The amorphous material may also have a high density of broken bonds.

The most important amorphous material in optics is, of course, "glass," which is used in all kinds of optical elements, such as lenses, prisms, etc., as well as in optical fibers. Glass is made from some of the most abundant elements on the earth's crust, vis. oxygen (which forms 62% of earth's crust) and silicon (which forms 21% of the crust). In silica based glass, silicon and oxygen atoms form a lattice network, which is not crystalline but has a good nearest neighbor ordering. Glass is used in a wide range of applications ranging from optical fibers to the oxide in MOSFETs. In Fig. 1.16 we have shown the structure of Si/SiO_2, one of the most important heterostructures in technology. The SiO_2 is amorphous in nature.

The most widely used glass is silica based, as far as optical fibers are concerned, due to the high purity level that is possible. Glasses based on B_2O_3, N_2O_3, etc. are used for other industrial applications.

Of particular importance is amorphous silicon, a-Si, which is perhaps the most important amorphous semiconductor material due to its importance in solar cell technology and display technology. In Fig. 1.22, we show a schematic comparison between crystalline Si and a-Si. We note that as in crystalline silicon, in a-Si, the Si atoms are four-fold coordinated; i.e., they have four nearest neighbors. However, some of the atoms have broken or dangling bonds. A high density of dangling bonds can render the material useless electronically. Thus in the growth of a-Si, we ensure that a large fraction of H (or F) is incorporated into the film. The H atoms "tie up" the dangling bonds and thus improve the properties of a-Si. Hydrogenated amorphous silicon is usually denoted by a-Si:H.

An important difference between the crystalline and amorphous materials is in the macroscopic symmetry of the material. The crystals are anisotropic due the precise arrangement of atoms. The amorphous materials, however, are isotropic.

1.5.3 Liquid crystals

In the liquid form, usually there is no short or long-range order among the atoms or molecules. Thus, even though there is a weak interaction between neighboring atoms, (molecules) we do not describe a liquid by any type of order. However, liquid crystals are one of the most fascinating material systems in nature, having properties of liquids (such as low viscosity and ability to conform to the shape of a container) as well as of a solid crystal. Their ability to modulate light when an applied electrical signal is used has made them invaluable in flat panel display technology.

Due to the ordered arrangement of atoms, crystalline materials have anisotropic properties (they look different from different directions), while non-crystalline materials and liquids are isotropic. Liquid crystals have anisotropic properties similar to solid crystals because of the ordered way in which some of the constituent molecules are arranged. However, the liquid crystals have low viscosity and can flow. The liquid crystals are a stable phase of matter called the mesophase existing between the solid and the liquid.

There are an essentially unlimited number of liquid crystals that can be formed.

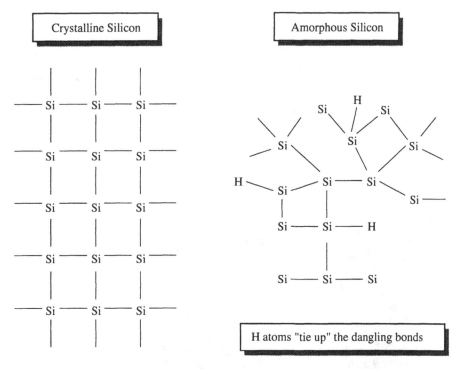

Figure 1.22: A schematic arrangement of the atoms in crystalline and amorphous silicon. The a-Si is hydrogenated to tie up dangling bonds, which would otherwise make the material electrically inactive.

The crystal is made up of organic molecules which are rod-like in shape with a length of ~ 20Å - 100Å. In Fig. 1.23a, we show a typical organic molecule, p-azoxyanisole, that can lead to a liquid crystal. This rod-like molecule is about 20 Å long and about 5 Å wide. A perfectly ordered arrangement of such a molecule can lead to a solid crystal, as shown in Fig. 1.23b. However, at high temperatures, a (disordered) liquid state is produced. The orientation of the rod-like molecule defines the "director" of the liquid crystal. The different arrangements of these rod-like molecules leads to three main categories of liquid crystals.

The three categories of liquid crystals can be understood by referring to Fig. 1.24. In Fig. 1.24a we show a structure which is referred to as smectic. In this structure the rod-like molecules are arranged in layers, and within each layer there is orientational order over a long range. Thus, in a given layer, the rods are all oriented in the same direction. Also, in the smectic liquid crystals, the molecules of different layers are ordered as shown in Fig. 1.24a. *Thus both orientation order and positional order is present in the smectic crystals.*

The second class of liquid crystal structure is called the nematic structure and is shown in Fig. 1.24b. In the nematic structure the positional order between layers of molecules is lost, but the orientation order is maintained.

1.5. Non-crystalline materials

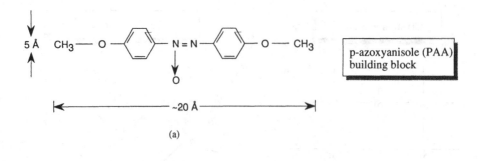

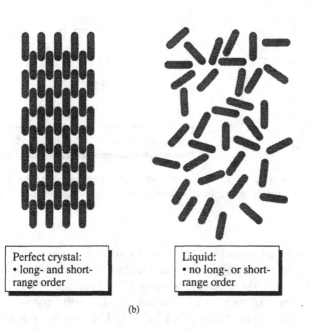

Figure 1.23: (a) A typical building block for liquid crystals. (b) A schematic description of a perfect crystal and a liquid. Liquid crystals form a phase of nature in between these extremes.

A third class of liquid crystals has the structure shown in Fig. 1.24c and is called cholesteric. In these crystals the rod-like molecules in each layer are oriented at a different angle within each layer. Orientation order is maintained within each layer. The cholesteric liquid crystal is related to the nematic crystal, with the difference being the twist of the molecules as we go from one layer to another.

In addition to the orientational order present in each layer an additional parameter defining subclasses of a smectic crystal is the chirality (i.e., relative twist) between molecules. The optical activity of the crystal depends upon the orientation and the twist present in the molecular layers.

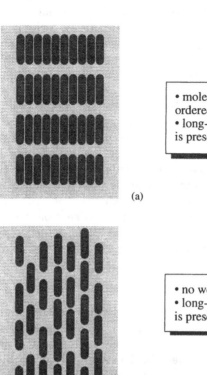

- molecules within a layer are ordered
- long-range orientation order is present

(a)

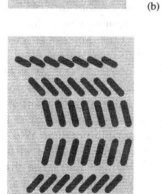

- no well defined layer order
- long-range orientation order is present

(b)

- well defined order within layers
- long-range "twist" between molecules on each layer

(c)

Figure 1.24: A schematic description of the arrangement of molecules in: (a) smectic; (b) nematic; and (c) cholesteric liquid crystals.

1.6. Summary

To fully exploit the potential of liquid crystals, an important feature regarding the interaction of the liquid with surfaces is exploited. It is found that, if the surface of a glass plate is rubbed along a certain direction (with, say, a cloth), then when the liquid crystal comes in contact with the surface, the surface molecules align themselves along the rubbed section. Now consider that a second rubbed glass plate is placed so that the spacing between the two plates is $\sim$ 5-20 μm. The orientation of the liquid crystal surface molecules can be prechosen by simply orienting the rubbed direction of the two plates. This produces a twist in the liquid crystal molecules as we go from one plate to another as shown in Fig. 1.25. Such liquid crystal systems are called *twisted nematic* and a total rotation of 90° can be produced. If the twist angle is increased to enhance the effect, the film becomes unstable if normal nematic films are used. For example, if a twist of 270° is desired, the stable state is one with a $-90°$ twist. However, if cholesteric liquid crystals are used in which there is already a built-in twist, the 270° twist is possible. Such structures are called *supertwisted*.

The unusual orientation dependence of the rod-like molecules of liquid crystals can be modified by an electric field. This in turn modifies their optical properties resulting in their efficient use as light valves.

1.5.4 Organic materials

Organic materials have formed the basis of many important technologies including plastics, drugs, chemicals, etc. Their uses in information processing applications have, in the past, largely relied on chemical reactions with the environment. However, over the past few years rapid progress has been made in the structural quality of many organic materials so that they can be used with tailorable electronic and optical properties much like traditional semiconductors. We will see in Chapter 3 that organic semiconductors can have properties similar to the inorganic semiconductors in some important ways and can thus be used for light emission and absorption as well as for switching applications. An added advantage of organic materials is that they can be grown on flexible substrates (various types of plastics) as well as on glass and silicon.

Organic semiconductors (polymers) are formed from long chains of molecules. If there is a good fit between the molecules, the materials can crystallize upon drawing or cooling. Advances in crystallization techniques have led to high-quality polymer crystals, which display electronic bands similar to those shown by "traditional" semiconductors. In Fig. 1.26 we show chains of polyacetylene–an important polymer. While it is possible to get crystalline polymers, usually the material is non-crystalline with a long-range order of a few microns. The long-range order has a strong influence on charge transport which is orders of magnitude poorer (in mobility) compared to crystalline inorganic semiconductors.

1.6 SUMMARY

In this chapter we have discussed the important structural properties of semiconductors and their heterostructures. The semiconductors we will be dealing with in most optoelectronic devices have a zinc blende or diamond structure. We have discussed the important

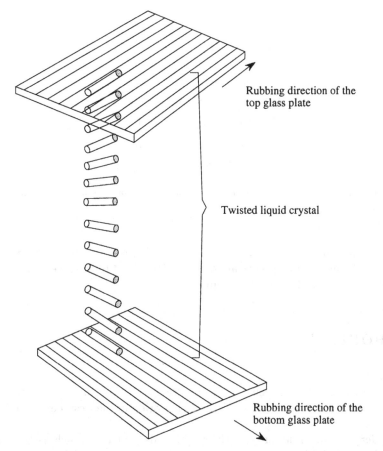

Figure 1.25: A schematic of the twisted nematic liquid crystal produced by using two rubbed glass plates as a container.

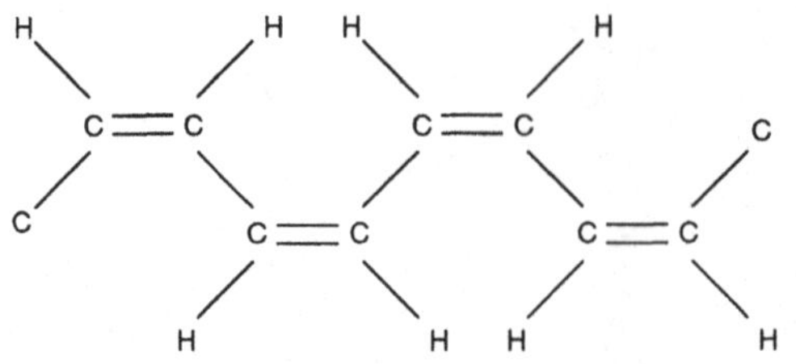

CIS-POLYACETYLENE

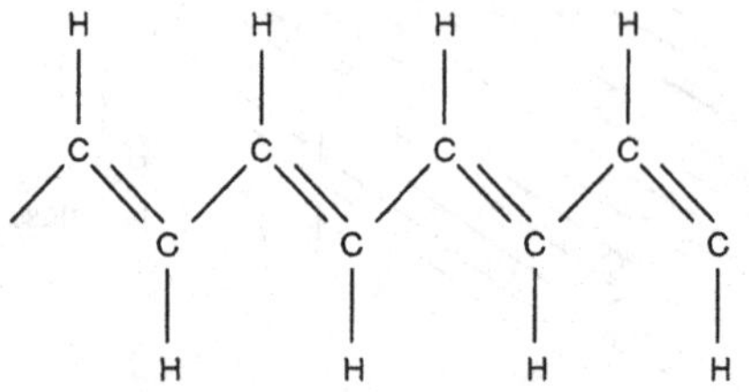

TRANS-POLYACETYLENE

Figure 1.26: Schematic of chains forming the polymer polyacetylene. Two forms of this polymer are shown.

growth techniques used in producing the semiconductors. We have also identified the important techniques used to fabricate devices. Tables 1.4 to 1.6 give an overview of the issues that have emerged from this chapter.

1.7 PROBLEMS

Section 1.2

1.1 (a) Find the angles between the tetrahedral bonds of a diamond lattice.
(b) What are the direction cosines of the (111) oriented nearest neighbor bond along the **x**,**y**,**z** axes?
1.2 Consider a semiconductor with the zinc blende structure (such as GaAs).
(a) Show that the (100) plane is made up of either cation or anion type atoms.
(b) Draw the positions of the atoms on a (110) plane assuming no surface reconstruction.
(c) Show that there are two types of (111) surfaces: one where the surface atoms are bonded to three other atoms in the crystal and another where the surface atoms are bonded to only one. These two types of surface are called the A and B surfaces, respectively.
1.3 Suppose that identical solid spheres are placed in space so that their centers lie on the atomic points of a crystal, and the spheres on the neighboring sites touch each other. Assuming that the spheres have unit density, show that the density of such spheres is

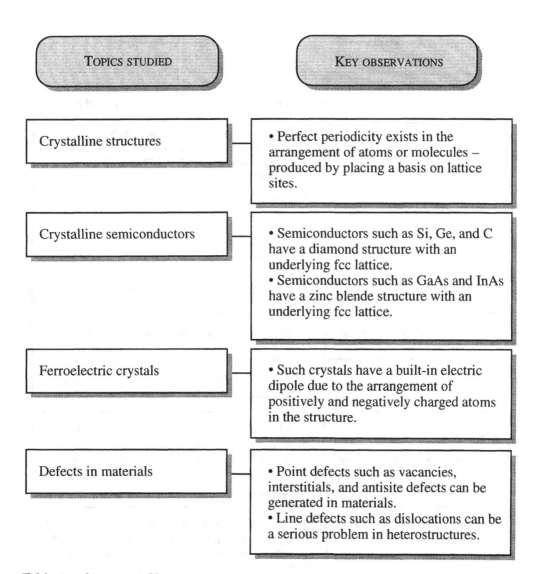

Table 1.4: Summary table

1.7. Problems

Topics Studied	Key Findings
Non-crystalline materials	Polycrystalline materials and amorphous materials are relatively inexpensive and are becoming very important in technolgy. Polycrystalline materials have good long-range order within a grain size. Amorphous materials have good short-range, but poor long-range order.
Polycrystalline materials	• Perfect short- and long-range order exists over the grain size, which can be several microns. • The structure loses order as we go from one grain to another.
Amorphous or glassy materials	• Good short-range order exists, but there is no long-range order. • The materials also have broken or dangling bonds and impurities.
Liquid crystals	• Liquid crystals are a phase of matter which can flow like a liquid, but unlike other liquids, the molecules have long-range order in their orientation.
Organic materials	Advances in synthesis techniques have allowed organic materials to become "smart"; i.e., their optical and electronic properties can be made tailorable.

Table 1.5: Summary table

the following for the various crystal structures:

$$\begin{aligned} \text{fcc} &: \sqrt{2}\pi/6 = 0.74 \\ \text{bcc} &: \sqrt{3}\pi/8 = 0.68 \\ \text{sc} &: \pi/6 = 0.52 \\ \text{diamond} &: \sqrt{3}\pi/16 = 0.34 \end{aligned}$$

1.4 Calculate the number of cells per unit volume in GaAs (a = 5.65 Å). Si has a 4% larger lattice constant. What is the unit cell density for Si? What is the number of atoms per unit volume in each case?

1.5 A Si wafer is nominally oriented along the (001) direction, but is found to be cut 2° off, towards the (110) axis. This off-axis cut produces "steps" on the surface which are 2 monolayers high. What is the lateral spacing between the steps of the 2° off-axis wafer?

1.6 Conduct a literature search to find out what the lattice mismatch is between GaAs and AlAs at 300 K and 800 K. Calculate the mismatch between GaAs and Si at the same temperatures.

1.7 In high purity Si crystals, defect densities can be reduced to levels of 10^{13} cm^{-3}. On an average what is the spacing between defects in such crystals? In heavily doped Si, the dopant density can approach 10^{19} cm^{-3}. What is the spacing between defects for such heavily doped semiconductors?

1.8 A GaAs crystal, which is nominally along the (001) direction, is cut θ off towards the (110) axis. This produces one monolayer high step. If the step size is to be no more than 100 Å, calculate θ.

1.9 Assume that a Ga–As bond in GaAs has a bond energy of 1.0 eV. Calculate the energy needed to cleave GaAs in the (001) and (110) planes.

1.10 Show that in the hcp structure the ratio $c/a, (a = a_1, a_2)$ is given by $\sqrt{8/3} = 1.633$. The values of these lattice constants for several semiconductors are given in the text.

1.11 The lattice constant of BaTiO$_3$ is 3.99 Å at room temperature with a relative displacement (discussed in the text) of O ions of 0.03 Å. Draw the position of atoms in a BaTiO$_3$ cell.

1.12 Do a literature search to find the lattice parameters for LiNbO$_3$, LiTaO$_3$, SrTiO$_3$.

Section 1.3

1.13 A serious problem in the growth of a heterostructure made from two semiconductors is due to the difficulty in finding a temperature at which both semiconductors can grow with high quality. Consider the growth of HgTe and CdTe, which is usually grown at $\sim$ 600 K. Assume that the defect formation energy in HgTe is 1.0 eV and in CdTe is 2.0 eV. Calculate the density of defects in the heterostructure with equal HgTe and CdTe.

1.14 Calculate the defect density in GaAs grown by LPE at 1000 K. The defect formation energy is 2.0 eV.

1.15 Why are entropy considerations unimportant in dislocation generation?

Section 1.5

1.16 A silicon polycrystalline film has a grain size of 2.0 μm. How many atoms are in the grain, assuming that the grain is a cube?

1.17 A GaAs film is grown on a glass substrate and heat treated to produce a polycrystalline film of grain size 10.0 μm. If a 2 μm ×2 μm diode is fabricated on the film, what is the probability that the device has high performance? Assume that the grain size is square on the surface.

1.18 When a polycrystalline film is heat treated (annealed), the grain size usually increases. Comment on why this occurs.

1.19 Using symmetry arguments and energy minimization arguments, discuss why in a nematic crystal, it is not possible to achieve a 270° twist.

1.20 Some liquid crystals are used as temperature detectors. Using simple thermodynamic arguments, discuss why the long-range order of the liquid crystal is destroyed as temperature increases.

1.21 A typical thickness of a liquid crystal cell used in laptop computers is 5 μm. If the thickness of the nematic crystal molecule is 5 Å and the average spacing between the molecules is 10 Å, how many molecules are stacked in a typical display cell?

1.8 FURTHER READING

- **Crystal structures**
    - J. M. Buerger, *Introduction to Crystal Geometry*, McGraw-Hill (1971).
    - M. Lax, *Symmetry Principles in Solid State and Molecular Physics*, J. Wiley (1974). Has a good description of the Brillouin zones of several structures in Appendix E.
    - J. F. Nye, *Physical Properties of Crystals*, Oxford (1985).
    - F. C. Phillips, *An Introduction of Crystallography*, J. Wiley (1971).

- **Defects in semiconductors**
    - P.K. Bhattacharya and S. Dhar, *Deep Levels in III-V Compound Semiconductors Grown by Molecular Beam Epitaxy*, Semiconductors and Semimetals, eds. A.C. Willardson and C. Beer, Academic Press, New York, vol. 26 (1988).
    - E.N. Economou, *Green's Functions in Quantum Physics*, Springer Verlag, Berlin (1979).
    - G.F. Foster and J.C. Slater, *Phys. Rev.*, **96**, 1208 (1954).
    - H.F. Matare, *Defect Electronics in Semiconductors*, Wiley-Interscience, New York (1971).
    - S. Pantelides, *Rev. Mod. Phys.*, **50**, 797 (1978).

- **Dislocations and lattice mismatched epitaxy**
  - S. Amelinckx, *Dislocations in Solids*, ed. F.R.N. Nabarro, North-Holland, New York (1988).
  - C.A.B Ball and J.H. van der Merwe, *Dislocations in Solids*, ed. F.R.N. Nabarro, North-Holland, New York, vol. 5 (1983).
  - H.F. Matare, *Defect Electronics in Semiconductors*, Wiley-Interscience, New York (1971).

Chapter 2

QUANTUM MECHANICS AND ELECTRONIC LEVELS

2.1 INTRODUCTION

Essentially all smart devices depend upon the electronic properties of materials and how these properties are influenced by external perturbations which may be electromagnetic, or mechanical, or magnetic, etc. The simplest approach to understanding such properties would be to use classical physics. Based on classical physics the general problem could be solved by using Newton's equation

$$\frac{d\mathbf{p}}{dt} = e\left(\dot{\mathbf{E}} + \mathbf{v}\times\mathbf{B}\right)$$

where $\mathbf{p}$ is the electron momentum, $\mathbf{v}$ the velocity, and $\mathbf{E}$ and $\mathbf{B}$ are the electrical and magnetic fields, respectively. Additional forces, if present, can be added on the right-hand side of the equation. Similarly, in classical physics, we could use the Maxwell equation to represent properties of electromagnetic waves in solids.

Although classical physics has been successful in describing many of nature's phenomena, it fails completely when it is used to describe electrons in solids. To understand the underlying physical properties that form the basis of modern intelligent information devices, we need to use quantum mechanics, which is a more accurate description of nature than classical physics. According to quantum mechanics, entities that are particles in the classical description behave as waves under certain conditions. To the level needed in device physics, the wave equation that is capable of describing particles is the Schrödinger equation. A second aspect of quantum mechanics says that classical waves sometimes behave as particles. Thus wave energy becomes "quantized" or appears in discrete steps. Both of these aspects of quantum mechanics are critical to an understanding of solids and their physical properties.

In this chapter we will start with a brief historical description on the origin of quantum mechanics. We will then establish the Schrödinger equation and discuss its outcome for several important problems.

2.2 NEED FOR QUANTUM DESCRIPTION

2.2.1 Some experiments that ushered in the quantum age

Classical physics has proven to be adequate to describe most physical phenomena. It is successful in describing planetary motion, trajectories of particles, wave propagation, etc. At the beginning of the twentieth century, some observations were made which started shaking the foundations of classical thinking. In this section we will examine some of the critical experiments that eventually led to quantum mechanics. In this text we will focus only on non-relativistic quantum mechanics. This means that we will consider situations where the speed of particles (other than photons) is much slower than the speed of light.

Waves behaving as particles: blackbody radiation

An extremely important experimental discovery which played a central role in the development of quantum mechanics was the problem of the spectral density of a blackbody radiation. If we take a body with a surface that absorbs any radiation (a blackbody) we find that it emits radiation at different wavelengths. The intensity (power per unit area) of the radiation emitted between wavelengths λ and $\lambda + d\lambda$ is defined as

$$dI = R(\lambda)d\lambda \qquad (2.1)$$

where $R(\lambda)$ is called the radiancy. The spectral dependence of $R(\lambda)$ is found to have a certain dependence on the wavelength and temperature of the blackbody. In Fig. 2.1 we show how the experimentally observed $R(\lambda)$ behaves at different temperatures. Several interesting experimental observations are made in regard to the emitted radiation:

- The total intensity has the behavior

$$I = \int_0^\infty R(\lambda)d\lambda \propto T^4$$

or

$$I = \sigma T^4 \qquad (2.2)$$

This is known as Stefan's law. The constant σ is called the Stefan–Boltzmann constant. It is found to have a value

$$\sigma = 5.67 \times 10^{-8} \text{ Wm}^{-2}\text{K}^{-4}$$

- The radiancy versus wavelength plot shows that there is a maximum at a certain

2.2. Need for quantum description

wavelength λ_{max} as can be seen from Fig. 2.1. The temperature dependence of this wavelength is given by

$$\lambda_{max} \propto \frac{1}{T} \qquad (2.3)$$

The proportionality constant is given by the relation

$$\lambda_{max} T = 2.898 \times 10^{-3} \text{ meter-Kelvin}$$

This relation is known as Wien's displacement law.

Using classical physics a formalism has been developed to understand blackbody radiation. According to classical physics, radiancy is given by the Rayleigh–Jeans law,

$$R(\lambda) = \frac{8\pi}{\lambda^4} k_B T \frac{c}{4} \qquad (2.4)$$

However, when careful experiments were carried out and the spectral density tabulated, it was found that the Rayleigh–Jeans law was applicable only in a small frequency range. In fact, as can be seen from the equation, the classical law predicts an infinite energy density at very short wavelengths – an obviously unphysical result. It can be seen from Fig. 2.1 that while the classical law gives a reasonable fit to experiments for long wavelengths, it completely fails at short wavelengths. The entire spectrum was only understood when Planck suggested that an electromagnetic wave with frequency ω exchanges energy with matter in a "quantum" given by

$$E = h\nu = \hbar\omega \qquad (2.5)$$

Here h is a universal constant called Planck's constant. The symbol $\hbar$ stands for $h/2\pi$. This assumption seemed to suggest that light waves have a well-defined energy just as particles do.

The quantity h or $\hbar$ that has been introduced is called Planck's constant and has a value

$$\begin{aligned} h &= 6.6261 \times 10^{-34} \text{ J} \cdot \text{s} \\ \hbar &= \frac{h}{2\pi} = 1.05 \times 10^{-34} \text{ J} \cdot \text{s} \end{aligned} \qquad (2.6)$$

Waves behaving as particles: photoelectric effect

It is well known that electromagnetic waves are described by Maxwell's equations. Phenomena such as interference and diffraction are well explained by Maxwell's equations. An important outcome of the wave theory is that the energy of a light beam can change continuously. As the intensity of the wave increases, the energy carried by a light beam increases. This seems quite intuitive and in most experiments this expectation is indeed verified. However, in 1887 Heinrich Hertz carried out an experiment which the wave theory of light was unable to explain. The experiment is known as the photoelectric effect and was the basis for Einstein's model for how light behaves.

In the photoelectric experiment light falls upon a material system and electrons are knocked out due to the interaction of the light with electrons. A typical experimental

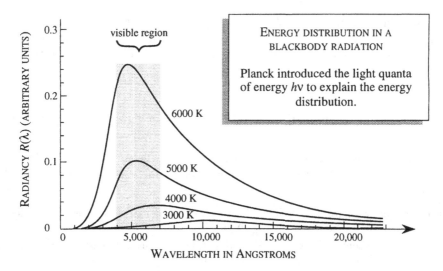

Figure 2.1: Measured spectral energy distribution of a blackbody radiation. The explanation of such experimental observations forced Planck to introduce a constant h (now called Planck's constant).

arrangement used is shown in Fig. 2.2. A potential is applied between the emitter and the collector. If the impinging light cannot knock an electron out of the metal emitter, there will be no photocurrent. If electrons are knocked out these electrons can make it to the collector if their energy is larger than the potential energy eV_{ext} between the emitter and the collector.

Let us assume that the electrons in the metal need to overcome an energy $e\phi$ in order to escape from the metal. The quantity $e\phi$ is called the work function of the metal and arises from the binding of the electrons to the metal ion. If the impinging light beam gives the emitted electrons an energy E_{em}, the electrons will emerge from the metal with an energy $E_{\text{em}} - e\phi$. An opposing bias is applied to the emitter–collector and the value of this bias is adjusted so that the electrons emitted are just unable to make it to the collector, i.e., the photocurrent becomes zero. This value of the applied voltage V_s is called the stopping voltage. The current will go to zero when

$$E_{\text{em}} - e\phi = eV_s \tag{2.7}$$

Experimentally we can measure V_s as a function of the intensity and frequency of light.

Classical wave theory suggests that the following observations should be made in the photoelectric effect:

- The energy with which the electrons should emerge from the metal should be proportional to the intensity of the light beam. Thus, as the intensity increases, the stopping voltage should also increase.
- The electron emission should occur at any frequency provided the intensity of the light beam is sufficiently high.

2.2. Need for quantum description

- There should be a time interval Δt between the switching on of the light beam and the emission of electrons. If A is the area over which the electron is confined (roughly equal to the area of an atom or 10^{-19} m^2), the time it should take the electron to gain an energy ΔE is

$$\Delta t = \frac{\Delta E}{IA}$$

where I is the light intensity. If we use $\Delta E \sim 1.0$ eV and $I \sim 1$ W/cm^2, we find that $\Delta T \sim 10^{-3}$ s.

The three expectations from classical physics all seem consistent with our physical intuition. However, actual experiments show them to be incorrect. Instead, the following occurs:

- If the frequency of light is below a cutoff value, there is no emission of electrons, regardless of intensity as shown in Fig. 2.2b.
- The stopping potential is completely independent of the intensity of light. A typical result is shown in Fig. 2.2c. As can be seen from this figure, the stopping voltage is unaffected by intensity, although the photocurrent scales with intensity.
- The initial electrons are emitted within a nanosecond or so of the light being turned on. There is essentially no delay between the impingement of light and electron emission.

The experimental observations were thus completely opposed to what was expected on the basis of the wave theory for electromagnetic radiation. It was clear that a radical new interpretation of light was needed. As noted in the previous subsection, Max Planck had developed his formalism to explain the spectral density of blackbody radiation. Based on Planck's ideas, Einstein saw that the photoelectric effect could be explained if light was regarded as made up of particles with energy

$$E = \hbar\omega = h\nu \tag{2.8}$$

Thus light was to be regarded not as waves but discrete bundles or *quanta* of energy. These quanta were later called photons.

In Einstein's theory electrons are emitted by a single photon knocking the electron out. Thus the kinetic energy of the emitted electron is

$$E_e(K.E) = h\nu - e\phi = eV_s \tag{2.9}$$

There is no dependence of the electron energy on the intensity of light. A beam with higher intensity has more photons, but each photon has the same energy. The cutoff frequency for electron emission is given by the relation

$$h\nu = e\phi \tag{2.10}$$

In Fig. 2.3 we show the work function values for several metals. Also shown are the dependence of stopping voltage on the frequency of light. The slope of this curve is h/e.

Particles behaving as waves: atomic spectra
An area where experiments baffled classical physics was atomic spectra and properties

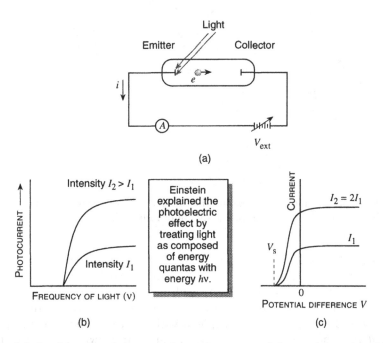

Figure 2.2: (a) A schematic of the experimental setup used for studying the photoelectric effect; (b) photocurrent as a function of the frequency of the impinging light for a fixed applied bias; and (c) photocurrent versus applied bias for when the optical signal frequency is above the threshold frequency.

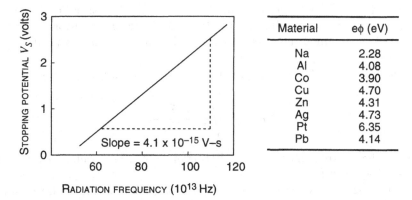

Material	$e\phi$ (eV)
Na	2.28
Al	4.08
Co	3.90
Cu	4.70
Zn	4.31
Ag	4.73
Pt	6.35
Pb	4.14

Figure 2.3: Stopping voltage versus frequency results for sodium. The slope of the curve is h/e. Also shown are the work functions of several metals.

2.2. Need for quantum description

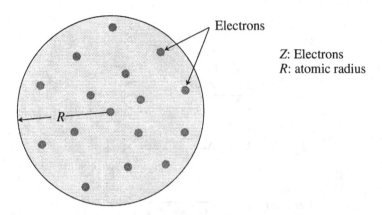

Figure 2.4: Thomson model for an atom. The Z electron with charge $-e$ are point particles embedded in a uniformly charged positive sphere of radius R.

of atoms. The question of relevance is: What are the electron energies and trajectories in an atom? One of the earliest models for the atom was proposed by J. J. Thomson, who was the first to identify the electron and measure the ratio of its charge to mass. Thomson built the atomic model on the basis of classical physics. He assumed that the atom was made up of a uniform sphere with positive charge Ze and radius R in which negatively charged electrons were embedded, as shown in Fig. 2.4. The size of the positively charged sphere is assumed to be of the order of an Angstrom or so and the electrons are assumed to be embedded in the uniform charge at various distances. If r is the distance of an electron from the center, the force on it is (from classical electrostatics)

$$\mathbf{F} = \frac{Ze^2 \mathbf{r}}{4\pi\epsilon_0 R^3} \qquad (2.11)$$

At equilibrium there would be a balancing force from the other negatively charged electrons. Away from the equilibrium position there is a linear restoring force on the electron and it is assumed the electron will oscillate about its mean position just as a pendulum does.

The frequency of the oscillation is (according to classical physics)

$$\nu = \frac{1}{2\pi}\sqrt{\frac{Ze^2}{4\pi\epsilon_0 R^3 m_0}} \qquad (2.12)$$

where m_0 is the electron mass. The oscillating electron would radiate electromagnetic radiation of frequency ν.

Experiments showed that the frequencies of radiation emitted from atoms were not in agreement with what the model predicted. Also, scattering experiments, in which scattering of alpha particles from atoms was studied, showed that most of the atom was *empty*. It was found that the positive charge was not distributed over an Angstrom but over a much smaller region.

Advances in optical spectroscopic techniques made possible direct measurements of the frequencies of the emitted and absorbed radiation by atoms. When atoms are excited (say, by electromagnetic radiation), they can absorb the radiation. Once they absorb radiation, they can emit radiation as well. It was found experimentally that emitted and absorbed spectra from a species of atoms consisted of several series of sharp lines; i.e., discrete frequencies. It was possible to fit simple relations to the positions of these lines. For example, Johannes Balmer found that the emission wavelengths of hydrogen in the visible regime could be fitted to the relation (the Balmer formula)

$$\lambda_n = 364.5 \frac{n^2}{n^2 - 4} \text{ nm}; \quad n = 3, 4, \ldots \tag{2.13}$$

In fact, other groups of lines in H-spectra were fitted to other expressions. For example, we have the following sequences of optical wavelengths:
Paschen Series:

$$\lambda_n = 820.1 \frac{n^2}{n^2 - 3^2} \text{ nm}; \quad n = 4, 5, \ldots \tag{2.14}$$

Lyman Series:

$$\lambda_n = 91.35 \frac{n^2}{n^2 - 1} \text{ nm}; \quad n = 2, 3, \ldots \tag{2.15}$$

Other atoms were found to have spectra which satisfied similar relations. In Fig. 2.5 we show series of lines observed in atomic spectra of hydrogen.

The observation of atomic spectra showed that for some reason electrons inside an atom can only have certain well-defined energies–not a continuum of energies. Without fully explaining why this should occur, Bohr came up with a model that explained the results shown in Fig. 2.5. Bohr assumed that the nucleus was essentially a point particle and the electrons spun around the nucleus, just as planets orbit the sun. *However, unlike the planets, the electrons can only go around in orbits in which the angular momentum was an integral multiple of $\hbar$.* If r is the radius of an orbit, we must have

$$m_0 v r = n\hbar; n = 1, 2 \ldots \tag{2.16}$$

By proposing this postulate, Bohr made a daring leap. He was able to fit the emission and absorption spectra of the hydrogen atom and he was able to explain why the electron does not radiate continuously, even though it is orbiting the nucleus. The electron cannot radiate electromagnetic energy unless it *jumps from one allowed orbit to another.*

Based on the postulate that the electron orbits were *quantized* Bohr was able to calculate the *allowed energies* of electrons in an atom. Equating the centripetal force and the Coulombic force we get

$$F = \frac{1}{4\pi\epsilon_0} \frac{e^2}{r^2} = \frac{m_0 v^2}{r} \tag{2.17}$$

The kinetic energy is now

$$K = \frac{1}{2} m_0 v^2 = \frac{1}{8\pi\epsilon_0} \frac{e^2}{r} \tag{2.18}$$

2.2. Need for quantum description

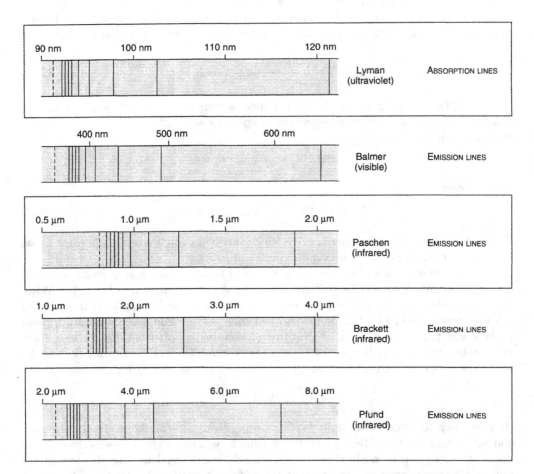

Figure 2.5: Emission and absorption lines in the hydrogen atom spectra. There is a regularity in the spacings of the spectral lines and the lines get closer as they reach the upper limit of each series (denoted by the dashed lines).

The potential energy is the Coulombic energy

$$U = -\frac{1}{4\pi\epsilon_0}\frac{e^2}{r} \tag{2.19}$$

and the total energy is

$$E = -\frac{1}{8\pi\epsilon_0}\frac{e^2}{r} \tag{2.20}$$

From Eqs. 2.16 and 2.17, we have

$$(m_0 v)v = \frac{e^2}{4\pi\epsilon_0 r} \quad \text{or} \quad v = \frac{e^2}{4\pi\epsilon_0} \cdot \frac{1}{n\hbar} \tag{2.21}$$

From this equation and Eq. 2.16, we have, for the allowed orbit radii

$$r_n = \frac{4\pi\epsilon_0}{m_0 e^2}(n\hbar)^2 \tag{2.22}$$

Substituting this equation into Eq. 2.20 for the electron energy, we have

$$\begin{aligned} E_n &= -\frac{m_0 e^4}{32\pi^2\epsilon_0^2\hbar^2}\frac{1}{n^2}; \quad n = 1, 2, \ldots \\ &= -\frac{13.6}{n^2} \text{ eV} \end{aligned} \tag{2.23}$$

The allowed energy levels are shown in Fig. 2.6. The allowed radii of the orbits are, from Eq. 2.22

$$r_n = \frac{4\pi\epsilon_0 \hbar^2}{m_0 e^2} n^2 = a_0 n^2 \tag{2.24}$$

where

$$a_0 = \frac{4\pi\epsilon_0 \hbar^2}{m_0 e^2} = 0.529 \text{ Å} \tag{2.25}$$

Based on his model Bohr was able to provide a model which was consistent with observations made on the hydrogen atom. For example, the emission lines resulted from an electron jumping from a higher energy level to a lower energy level, as shown in Fig. 2.8b. Absorption lines resulted from reverse transitions. The Bohr model, although pioneering, was not found to be adequate to describe the spectra of other atoms. It also failed to explain many other experiments.

2.3 SCHRÖDINGER EQUATION AND PHYSICAL OBSERVABLES

As noted in the previous section Bohr used the idea of angular momentum "quantization" to understand the H-atom spectra. It is well known that when we deal with waves, the idea of quantization is not so unusual. For example, in musical instruments only

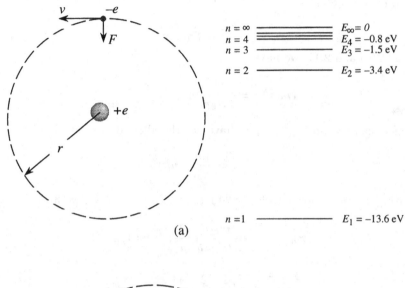
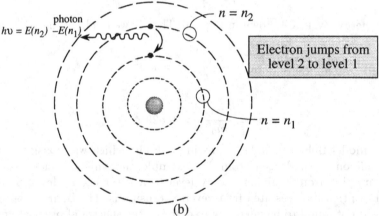

Figure 2.6: (a) A conceptual picture of the Bohr model along with the energy levels for a hydrogen atom; and (b) discrete spectral lines are explained by transitions of the electron from one level to another.

certain wavelengths are allowed, while others are forbidden. This emerges mathematically from the solution of wave equations subject to appropriate boundary conditions. Is it possible that classical particles, such as electrons, can be represented by a wave equation? The question that we now ask is the following: if particles have a wave-like behavior, what is the nature of the equation they satisfy? This question was first answered by Schrödinger.

The Schrödinger equation, given below, is no more intuitive than Newton's equations or Maxwell's equations. It is important to note that the Schrödinger equation describing the non-relativistic behavior of particles cannot be derived from any fundamental principles – just as Newton's equation cannot be derived. It is an equation that gives us solutions that can explain experimentally observed physical phenomena.

A clue to the form of the Schrödinger equation comes from the relation between a particle momentum and its wavelength as given by de Broglie. According to this relation, for a particle in free space the kinetic energy is (replacing the momentum by wavelength or wavevector)

$$K = \frac{p^2}{2m} = \frac{\hbar^2 k^2}{2m}; \quad k = \frac{2\pi}{\lambda} \text{ or } \lambda = \frac{h}{p} \tag{2.26}$$

If we examine the identity

$$K + U = E$$

where U is the potential energy and E is the total energy, we can write a *wave equation* for a wave with amplitude ψ

$$(K + U)\psi = E\psi$$

In case the particle is free (i.e., the potential energy is zero) we have

$$K\psi = E\psi$$

Using a hint from Eq. 2.26 for the form of the kinetic energy, we write the kinetic energy as an operator (say, for a one-dimensional case)

$$K = -\frac{\hbar^2}{2m} \frac{d^2}{dx^2} \tag{2.27}$$

We now see that the wave equation takes the form

$$-\frac{\hbar^2}{2m} \frac{d^2}{dx^2} \psi = E\psi$$

and the general solution to the equation is

$$\psi = Ae^{ikx} + Be^{-ikx}$$

The kinetic energy is

$$K = E = \frac{\hbar^2 k^2}{2m}$$

which is consistent with the relation we get from the de Broglie relation.

2.3. Schrödinger equation and physical observables

The Schrödinger equation is written by expressing the kinetic energy as a second-order operator, so that the identity

$$(K + U)\psi = E\psi$$

becomes

$$\left[-\frac{\hbar^2}{2m}\nabla^2 + V(r)\right]\psi = E\psi \qquad (2.28)$$

To write the time dependence of the particle wave we use the analogy for the phase of the particle wave, which goes as

$$\psi(t) \sim e^{iwt}$$

We can write the time dependence of the wave as

$$\psi(t) \sim \exp\left(-\frac{iEt}{\hbar}\right) \qquad (2.29)$$

so that the knowledge of ψ at any time allows us to predict the value of ψ at all times via the equation

$$\frac{\partial \psi}{\partial t} = \frac{-i}{\hbar} E\psi$$

or

$$i\hbar \frac{\partial \psi}{\partial t} = E\psi \qquad (2.30)$$

Note that, in this development, the quantity h or $\hbar$ has been introduced to define the proportionality between the energy and the particle wave frequency. In classical physics, this quantity is assumed to be zero. As noted earlier in this chapter, experiments carried out in the early twentieth century showed that $\hbar$ was not zero but had a value of 1.055×10^{-34} J.s.

It is important to emphasize that the derivation given above *does not constitute a proof for the Schrödinger equation*. Only experiments could determine the validity of such an extension.

In our "derivation" of the Schrödinger equation we see that the observable properties of the particle such as momentum and energy appear as *operators* operating on the particle wavefunction. From Eqs. 2.27 and 2.30 we see the operator form of these observables

$$p_x \rightarrow -i\hbar \frac{\partial}{\partial x}$$
$$p_y \rightarrow -i\hbar \frac{\partial}{\partial y}$$
$$p_z \rightarrow -i\hbar \frac{\partial}{\partial z} \qquad (2.31)$$
$$E \rightarrow i\hbar \frac{\partial}{\partial t} \qquad (2.32)$$

This observation that physical observables are to be treated as operators is quite generic in the quantum description. The energy operator is called the *Hamiltonian* in quantum mechanics.

EXAMPLE 2.1 Calculate the wavelength associated with a 1 eV (a) photon, (b) electron, and (c) neutron.

(a) The relation between the wavelength and the energy of a photon is

$$\lambda_{ph} = \frac{hc}{E} = \frac{(6.6 \times 10^{-34} \text{ J.s})(3 \times 10^8 \text{ m/s})}{(1.6 \times 10^{-19} \text{ J})} = 1.24 \times 10^{-6} \text{ m}$$
$$= 1.24 \text{ } \mu\text{m}$$

(b) The relation between the wavelength and energy for an electron is ($k = 2\pi/\lambda$)

$$\lambda_e = \frac{h}{\sqrt{2m_0 E}} = \frac{6.6 \times 10^{-34} \text{ J.s}}{[2(0.91 \times 10^{-30} \text{ kg})(1.6 \times 10^{-19} \text{ J})]^{1/2}}$$
$$= 12.3 \text{ Å}$$

(c) For the neutron using the same relation, we have

$$\lambda_n = \lambda_e \left(\frac{m_0}{m_n}\right)^{1/2} = \lambda_e \left(\frac{1}{1824}\right)^{1/2} = 0.28 \text{ Å}$$

The wavelengths of different "particles" play an important role when these particles are used to "see" atomic phenomena in a variety of microscopic techniques.

2.3.1 Wave amplitude

The Schrödinger equation is a second-order differential equation which can be solved analytically or numerically. When we solve a wave equation describing electromagnetic waves, we get a solution that gives the amplitude of the fields and the frequency of vibration. Similarly, when we solve the Schrödinger equation for a given potential energy term, we get a set of solutions $\{E_n, \psi_n\}$, which give us the allowed energy E_n of the particle along with the wavefunction ψ_n. The wavefunction provides a complete quantum mechanical description of the behavior of a particle of mass m in a potential energy V. We now develop a mathematical and physical interpretation of the wave amplitude. This interpretation must be consistent with the experimental observations.

Let us briefly recall the meaning of the wave amplitude $\phi(r,t)$ in the wave equations describing sound waves or light waves; i.e., in classical physics. The energy of the wave at a particular point in space and time is related to $|\phi(r,t)|^2$. It is "more likely" that the wave is found in regions where ϕ is large. Thus the quantity $|\phi(r,t)|^2$ represents some sort of probability function for the wave. A similar interpretation has to be developed for a wave ψ describing a particle of mass m. The interpretation has to be statistical in nature. It is thus natural to regard ψ as the measure of finding the particle at a particular point and space. We assume, therefore, that the product of ψ and its complex conjugate ψ^* is the probability density

$$P(r,t) = \psi^*(r,t)\psi(r,t) = |\psi(r,t)|^2 \tag{2.33}$$

2.3. Schrödinger equation and physical observables

Thus the probability of finding the particle in a region of volume $dx\,dy\,dz$ around the position r is simply

$$P(r,t)\,dx\,dy\,dz = |\psi(r,t)|^2\,dx\,dy\,dz \tag{2.34}$$

Normalization of the wavefunction

If we assume that the particle is confined to a certain volume V, we know that the probability of finding it somewhere in the region must be unity, so that we must have the following condition satisfied

$$\int_V |\psi(r,t)|^2\,d^3r = 1 \tag{2.35}$$

The volume over which the integral is carried out is sometimes arbitrary and the coefficient of the wavefunction must be chosen to satisfy the normalization integral. The normalization factor does not change the fact that ψ is a solution of the Schrödinger equation due to the homogeneous (linear) form of the equation in ψ. Thus, if ψ is a solution, and if A is a constant, then $A\psi$ is also a solution.

Expectation values

In classical physics, when a measurement is made on a particle, the outcome is information on the particle energy position, momentum, etc. We need to know how to relate a physical observation to what we calculate by solving the Schrödinger equation. In view of the probabilistic nature of quantum mechanics, we can only define the "expectation value" of an observable in a physical measurement. Using the probability function $P(r,t) = |\psi(r,t)|^2$, we define the *expectation value* for the measurement of the position of the particle as

$$\langle r \rangle = \int rP(r,t)\,d^3r = \int \psi^*(r,t)r\psi(r,t)\,d^3r \tag{2.36}$$

For the expectation value along different axes we have

$$\langle x \rangle = \int \psi^* x \psi\,d^3r$$

$$\langle y \rangle = \int \psi^* y \psi\,d^3r$$

$$\langle z \rangle = \int \psi^* z \psi\,d^3r$$

In all these equations, ψ is normalized to the volume of the integral.

In a similar manner, the expectation value for the potential energy of the particle is

$$\langle V \rangle = \int \psi^*(r,t)V(r,t)\psi(r,t)\,d^3r \tag{2.37}$$

To find the expectation values of momentum and energy, we need to express them in their operator form. We note that the expectation value of energy is

$$\langle E \rangle = \left\langle \frac{p^2}{2m} \right\rangle + \langle V \rangle \tag{2.38}$$

or, in the form of the differential operators

$$\left\langle i\hbar \frac{\partial}{\partial t} \right\rangle = \left\langle -\frac{\hbar^2}{2m} \nabla^2 \right\rangle + \langle V \rangle \tag{2.39}$$

This equation is consistent with Schrödinger equations with the following expectation values

$$\langle E \rangle = \int \psi^* i\hbar \frac{\partial \psi}{\partial t} \, d^3r \tag{2.40}$$

$$\langle p \rangle = \int \psi^* (-i\hbar) \nabla \psi \, d^3r \tag{2.41}$$

The different momentum components are

$$\langle p_x \rangle = \int \psi^* (-i\hbar) \frac{\partial \psi}{\partial x} \, d^3r$$

$$\langle p_y \rangle = \int \psi^* (-i\hbar) \frac{\partial \psi}{\partial y} \, d^3r$$

$$\langle p_z \rangle = \int \psi^* (-i\hbar) \frac{\partial \psi}{\partial z} \, d^3r$$

Once we solve a Schrödinger equation, we will get one or more allowed wavefunctions. The expressions given above can then be used to determine the expectation values of various physical observables in each of these allowed states. Since all quantities calculated from the wavefunction, which can be related to direct physical meaning, have a form involving $\psi^* \psi$ product, it is clear that the wavefunction is only determined within a phase factor of the form $e^{i\alpha}$, where α is a real number. This indeterminacy in the wavefunction is unimportant, since it has no effect on physical results.

An important and useful aspect of quantum mechanics is the principle of superposition of states, which is directly related to the quantum mechanics wave equation being linear in the wavefunction ψ. If $\psi_1(q)$ is a state leading to measurement R_1 and $\psi_2(q)$ is a state leading to R_1, then every function of the form $a_1\psi_1 + a_2\psi_2$ (where c_1 and c_2 are constants) gives a state in which the results of measurements are known from the knowledge ψ_1 and ψ_2.

2.3.2 Waves, wavepackets, and uncertainty

When we consider a wave, one of the notions we have is that the wave is spread over some regions in space. Its position is not well defined. In fact waves satisfy an uncertainty relation

$$\Delta x \Delta k \sim 1$$

where Δx represents the region where the wave energy is localized and Δk represents the error in specifying the wavevector $\mathbf{k}$ of the wave, where

$$\mathbf{k} = \frac{2\pi}{\lambda} \tag{2.42}$$

2.3. Schrödinger equation and physical observables

defines the wavelength λ. An example of this is seen when you look at a wave moving across an ocean beach. The position of the wave extends over a wide space.

A particle in quantum mechanics also displays an uncertainty relation that is related to its wave nature. The relations are known as Heisenberg uncertainty relation and are

$$\Delta p_x \Delta x \geq \hbar/2$$
$$\Delta E \Delta t \geq \hbar/2 \qquad (2.43)$$

The first of these relations imply that it is not possible to specify the values of both momentum and position of a particle with complete certainty. The second uncertainty relation involves a subtler concept. In quantum mechanics a particle has an energy, E, and also a "lifetime" or how long it stays in the "state with energy E." The energy-time uncertainty relation says that, if the energy is defined with a precision of ΔE, the lifetime has to be greater than $\hbar/(2\Delta E)$. Thus the greater the lifetime, the more precisely we can define the particle energy.

Note that, for the momentum position, the uncertainty only exists between p_i, r_i i.e., between (p_x, x), (p_y, y) and (p_z, z). There is no uncertainty in the simultaneous measurement of, say, p_x and y.

Equations of motion: Ehrenfest theorem

If particles behave as waves, how do forces influence that particle wave? It can be shown that if the potential energy is smoothly varying (on the scale of the particle wavelength) the equation of motion for the particle wave is given by an equation that looks very much like Newton's equations or

$$\frac{d}{dt} \langle p_x \rangle = \left\langle -\frac{\partial V}{\partial x} \right\rangle \qquad (2.44)$$

The spatial derivative of the potential energy is the force applied, and thus we obtain a classical-like equation for the rate of change of the expectation value of the momentum.

The result above showing that the properties of a wavepacket can be determined from classical equations is called the Ehrenfest theorem. This is an extremely useful theorem in applied quantum mechanics, especially as used in describing various devices. In such cases an appropriate Schrödinger equation is solved to describe the allowed energy and momentum states of a particle. *Once the quantum problem is solved, the response of the particle to slowly varying external forces can be treated as if the particle is obeying classical equations.* This approach is widely used when we use quantum mechanics in describing devices such as semiconductor transistors and lasers.

EXAMPLE 2.2 In an experiment known as the Frank and Hertz experiment, electrons are raised to an excited state by colliding with a beam of electrons. It is found that, even if the beam of electrons is monoenergetic, after the *collision with the atomic electrons* the energy of the electrons in the beam has a certain spread. Explain this spread. In a particular experiment the energy spread is found to be 10^{-6} eV. What is the lifetime of the excited state?

The electronic levels in an atom have a certain lifetime, which produces an uncertainty in the energy needed to excite the electrons

$$\Delta E \, \Delta t = \frac{\hbar}{2}$$

For the given case

$$\Delta t = \frac{\hbar}{2\Delta E} = \frac{1.05 \times 10^{-34} \text{ J.s}}{2\left(10^{-6} \times 1.6 \times 10^{-19} \text{ J}\right)} = 3.28 \times 10^{-10} \text{ s}$$

EXAMPLE 2.3 Use the uncertainty relation to evaluate the ground state energy of the hydrogen atom.

Let us assume that the electron is confined to a radius r_o in the hydrogen atom. We will find the value of r_o that gives the lowest energy. The uncertainty principle tells us that the momentum associated with the electron is

$$p \sim \frac{\hbar}{r_o}$$

The electron–proton system energy is

$$\begin{aligned}
E(r_o) &= \frac{p^2}{2m} + V(r) \\
&= \frac{\hbar^2}{2mr_o^2} - \frac{e^2}{4\pi\epsilon_o r_o}
\end{aligned}$$

To find the lowest energy we need to minimize E. This is minimized when

$$\begin{aligned}
r_o &= \frac{4\pi\epsilon_o \hbar^2}{me^2} \\
&= \frac{4\pi\left(8.84 \times 10^{-12} \text{ F/m}\right)\left(1.05 \times 10^{-34} \text{ J.s}\right)^2}{\left(9.1 \times 10^{-31} \text{ kg}\right)\left(1.6 \times 10^{-19} \text{ C}\right)^2} \\
&= 5.24 \times 10^{-11} \text{ m} \\
&= 0.524 \text{ Å}
\end{aligned}$$

With this value of r_o, the energy becomes

$$\begin{aligned}
E_1 &= -\frac{1}{2} \frac{me^4}{(4\pi\epsilon_o)^2 \hbar^2} \\
&= -\frac{\left(9.1 \times 10^{-31} \text{ kg}\right)\left(1.6 \times 10^{-19} \text{ C}\right)^4}{2\left(4\pi \times 8.84 \times 10^{-12} \text{ F/m}\right)^2 \left(1.05 \times 10^{-34} \text{ J.s}\right)^2} \\
&= -2.16 \times 10^{-18} \text{ J} \\
&= -13.6 \text{ eV}
\end{aligned}$$

These results are quite accurate, considering the crude approximations used.

2.4. Particles in an attractive potential: bound states

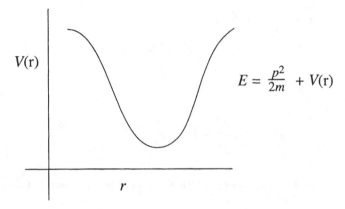

Figure 2.7: Particle in an attractive potential well. In classical physics a particle can have any energy larger than the potential energy. The energy can change continuously.

2.4 PARTICLES IN AN ATTRACTIVE POTENTIAL: BOUND STATES

We will now examine several important quantum problems that have impact on materials and physical phenomena useful for device applications. We note that in classical physics the energy of a particle is simply given by the sum of the kinetic and potential energies

$$E = \frac{p^2}{2m} + V(r) \tag{2.45}$$

One outcome of this equation is that a particle can have any energy greater than the potential energy starting from the zero momentum value (see Fig. 2.7). The particle energy can have continuously changing values. In quantum mechanics the "obvious" observation made above does not hold. In attractive potentials (such a schematic is shown in Fig. 2.7) a particle can only have certain select energies consistent with the Schrödinger equation.

In this section we will discuss three important problems with implications for electronic and optical properties in devices (i) electron in a hydrogen atom; (ii) particle in a "square" quantum well; and (iii) particle in a harmonic oscillator potential.

While the details of these problems are different there are several common outcomes in the solutions to these problems:
• Allowed energy levels are not continuous, but there are a series of discrete allowed energies separated by "energy gaps."
• Associated with the allowed energies are wavefunctions (or eigenfunctions) which describe the spread of the particle wave in space. In some cases there may be more than one wavefunction associated with the same energy. Such wavefunctions are called degenerate (doubly degenerate, triply degenerate, etc.).

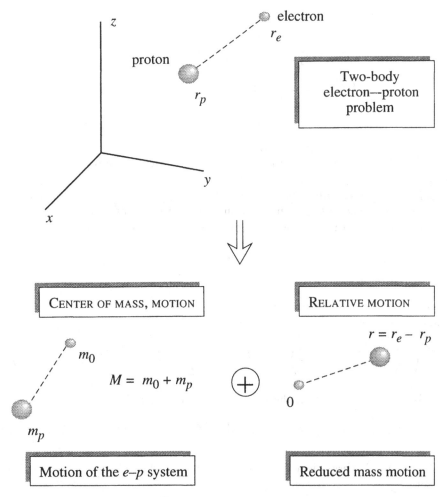

Figure 2.8: A schematic of the hydrogen atom problem. The two-body problem can be represented by the center-of-mass problem and a one-body problem.

2.4.1 Electronic levels in a hydrogen atom

In this book we are not directly interested in atoms and atomic spectra. However, the insight provided by the solution of the H-atom problem is very useful in understanding a number of technology-related phenomenon. The hydrogen atom consists of an electron and a proton interacting with the Coulombic interaction. The problem can be solved exactly and provides insight into how electrons behave inside atoms. Mathematically the H-atom problem is similar to the problem of dopant atoms and of excitons and the solutions, appropriately modified, are extremely useful.

The hydrogen atom problem involves an electron and a proton interacting with each other via the Coulombic interaction, as shown in Fig. 2.8. The Schrödinger equation

2.4. Particles in an attractive potential: bound states

is

$$\left[\frac{-\hbar^2}{2m_0}\left(\frac{\partial^2}{\partial x_e^2}+\frac{\partial^2}{\partial y_e^2}+\frac{\partial^2}{\partial z_e^2}\right)-\frac{\hbar^2}{2m_p}\right.$$
$$\left.\left(\frac{\partial^2}{\partial x_p^2}+\frac{\partial^2}{\partial y_p^2}+\frac{\partial^2}{\partial z_p^2}\right)+V\left(x_e,y_e,z_e,x_p,y_p,z_p\right)\right]$$
$$\psi\left(x_e,y_e,z_e,x_p,y_p,z_p\right)=E_{Tot}\psi\left(x_e,y_e,z_e,x_p,y_p,z_p\right) \quad (2.46)$$

where the subscripts e and p refer to the electron and the proton. Note that the mass of the electron is denoted by m_0. It is well known in classical physics that, if the potential energy depends only upon the relative coordinates; i.e., $V = V(x_e - x_p, y_e - y_p, z_e - z_p)$, the problem can be separated into two one-body problems, as shown schematically in Fig. 2.8. A similar separation is possible in quantum mechanics. This can be done by using the relative coordinates and the center-of-mass coordinates defined by

$$x = x_e - x_p$$
$$MX = m_e x_e + m_p x_p \quad (2.47)$$

etc. Here $M = m_0 + m_p$ is the mass of the electron and the proton. With the new coordinates, Eq. 2.46 can be rewritten as

$$\left[\frac{-\hbar^2}{2M}\left(\frac{\partial^2}{\partial X^2}+\frac{\partial^2}{\partial Y^2}+\frac{\partial^2}{\partial Z^2}\right)-\frac{\hbar^2}{2\mu}\left(\frac{\partial^2}{\partial x^2}+\frac{\partial^2}{\partial y^2}+\frac{\partial^2}{\partial z^2}\right)+V(x,y,z)\right]\psi=E_{Tot}\psi \quad (2.48)$$

where μ is the reduced mass of the electron–proton system, i.e.

$$\mu = \frac{m_0 m_p}{m_0 + m_p} \quad (2.49)$$

Since the potential energy depends only upon the relative coordinate, we can make the separation

$$\Psi = \psi(x,y,z)U(X,Y,Z) \quad (2.50)$$

and get the two equations

$$\left[\frac{-\hbar^2}{2\mu}\nabla^2+V(r)\right]\psi(r)=E\psi(r) \quad (2.51)$$

$$\frac{-\hbar^2}{2M}\nabla^2 U = E'U \quad (2.52)$$

and the total energy is

$$E_{Tot} = E + E' \quad (2.53)$$

The solution to Eq. 2.52 is straightforward and simply represents the "free" motion of the atom

$$E' = \frac{\hbar^2 K^2}{2M} \quad (2.54)$$

We are not interested in the motion of the atom and will assume that the atom is at rest.

We will now discuss the one-body problem described by Eq. 2.51 with

$$V(r) = -\frac{e^2}{4\pi\epsilon_0 r} \tag{2.55}$$

The time-independent wave equation can be written in the spherical coordinate system as

$$\frac{-\hbar^2}{2\mu}\left[\frac{1}{r^2}\frac{\partial}{\partial r}\left(r^2\frac{\partial}{\partial r}\right) + \frac{1}{r^2 \sin\theta}\frac{\partial}{\partial \theta}\left(\sin\theta\frac{\partial}{\partial \theta}\right) + \frac{1}{r^2 \sin^2\theta}\frac{\partial^2}{\partial \phi^2}\right]\psi + V(r)\psi = E\psi \tag{2.56}$$

where ψ is the wavefunction and E is the energy of the system. Let us now separate the radial and angular parts of the solution by the substitution

$$\psi(r, \theta, \phi) = R(r)F(\theta)G(\phi)$$

When this problem is solved, the eigenfunctions and eigenvalues that result are described by three quantum numbers (the quantum numbers are 3 because this is a three-dimensional problem). The eigenfunction is given by

$$\psi_{n\ell m}(r, \theta, \phi) = R_{n\ell}(r)F_{\ell m}(\theta)G_m(\phi) \tag{2.57}$$

The symbols n, ℓ, m are the three quantum numbers describing the solution. The three quantum numbers have the following allowed values:

$$\begin{aligned} \text{principle number}, n &: \text{Takes values } 1, 2, 3, \ldots \\ \text{angular momentum number}, \ell &: \text{Takes values } 0, 1, 2, \ldots n-1 \\ \text{magnetic number}, m &: \text{Takes values } -\ell, -\ell+1, \ldots \ell \end{aligned} \tag{2.58}$$

The principle quantum number specifies the energy of the allowed electronic levels. The energy eigenvalues are given by

$$E_n = -\frac{\mu e^4}{2\left(4\pi\epsilon_0\right)^2 \hbar^2 n^2} \tag{2.59}$$

Note that the energy levels obtained here have the same values as those obtained by applying Bohr's quantization rules discussed earlier. The spectrum is shown schematically in Fig. 2.9. Due to the much larger mass of the nucleus as compared with the mass of the electron, the reduced mass μ is essentially the same as the electron mass m_0. The ground state of the hydrogen atom is given by

$$\psi_{100} = \frac{1}{\sqrt{\pi a_0^3}} e^{-r/a_0} \tag{2.60}$$

The parameter a_0 appearing in the functions is called the *Bohr radius* and is given by

$$a_0 = \frac{4\pi\epsilon_0 \hbar^2}{m_0 e^2} = 0.53 \text{ Å} \tag{2.61}$$

2.4. Particles in an attractive potential: bound states

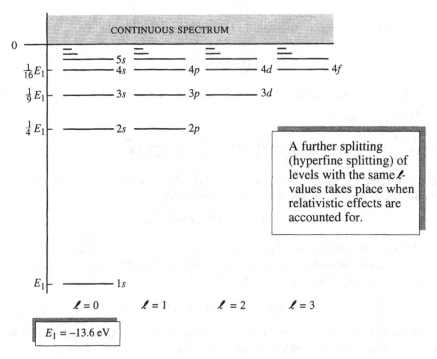

Figure 2.9: A schematic description of the energy levels of electrons in a hydrogen atom.

n	ℓ	m	$R(r)$	$F(\theta)$	$G(\phi)$
1	0	0	$\dfrac{2}{a_0^{3/2}} e^{-r/a_0}$	$\dfrac{1}{\sqrt{2}}$	$\dfrac{1}{\sqrt{2\pi}}$
2	0	0	$\dfrac{1}{(2a_0)^{3/2}} \left(2 - \dfrac{r}{a_0}\right) e^{-r/2a_0}$	$\dfrac{1}{\sqrt{2}}$	$\dfrac{1}{\sqrt{2\pi}}$
2	1	0	$\dfrac{1}{\sqrt{3}(2a_0)^{3/2}} \dfrac{r}{a_0} e^{-r/2a_0}$	$\sqrt{\dfrac{3}{2}} \cos\theta$	$\dfrac{1}{\sqrt{2\pi}}$
2	1	±1	$\dfrac{1}{\sqrt{3}(2a_0)^{3/2}} \dfrac{r}{a_0} e^{-r/2a_0}$	$\dfrac{\sqrt{3}}{2} \sin\theta$	$\dfrac{1}{\sqrt{2\pi}} e^{\pm i\phi}$

Table 2.1: Some low-lying hydrogen atom wavefunctions.

It roughly represents the spread of the ground state. Table 2.1 gives the functional forms of some of the low-lying wavefunctions.

Before ending the discussion on the hydrogen atom problem, it is useful to point out that states with $\ell = 0, 1, 2, 3, \ldots$ are called $s, p, d, f, \ldots$ in atomic physics notation. Such a notation is used for not only the hydrogen atom case, but also for all atomic spectra. The form of these functions, as well as the electron probability function, is shown in Fig. 2.10 for some low-lying states.

2.4.2 Particle in a quantum well

The quantum well problem is an important one technologically for several reasons. Using semiconductor heterostructures it is possible to fabricate quantum well systems. These systems are used for high-performance devices, such as lasers and modulators. The quantum well problem can also be used to understand how defects create trap levels.

A quantum well potential profile is shown in Fig. 2.11. The well (i.e., region where potential energy is lower) is described by a well size $W = 2a$ as shown and a barrier height V_0. In general the potential could be confining in one dimension with uniform potential in the other two directions (quantum well), or it could be confining in two dimensions (quantum wire) or in all three dimensions (quantum dot). We assume that the potential has a form

$$V(r) = V(x) + V(y) + V(z) \tag{2.62}$$

so that the wavefunction is separable and of the form

$$\psi(r) = \psi(x)\psi(y)\psi(z) \tag{2.63}$$

We will discuss the problem of the square potential well, which has acquired a great deal of importance in applied physics due to the use of quantum wells in optoelectronic devices such as semiconductor lasers and modulators.

The simplest form of the quantum well is one where the potential is zero in the well and infinite outside. The equation to solve then is (the wavefunction is non-zero only in the well region)

$$-\frac{\hbar^2}{2m}\frac{d^2\psi}{dx^2} = E\psi \tag{2.64}$$

which has the general solutions

$$\begin{aligned}\psi(x) &= B\cos\frac{n\pi x}{2a}, \quad n \text{ odd} \\ &= A\sin\frac{n\pi x}{2a}, \quad n \text{ even}\end{aligned} \tag{2.65}$$

The energy is

$$E = \frac{\pi^2 \hbar^2 n^2}{8ma^2} \tag{2.66}$$

Note that the well size is $2a$.

2.4. Particles in an attractive potential: bound states

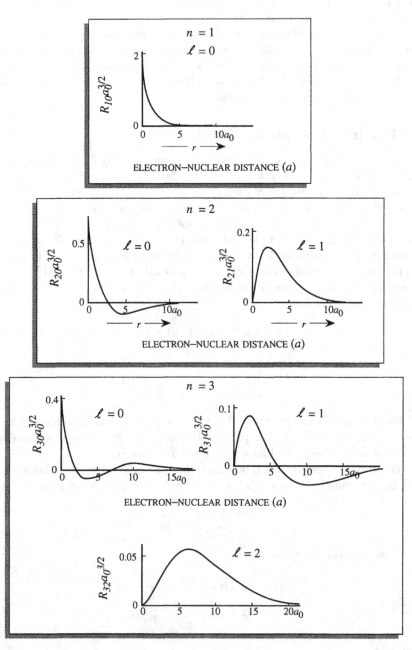

Figure 2.10: A plot of the probability density function as a function of the angle θ for the s, p, d electrons.

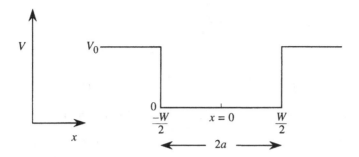

Figure 2.11: A quantum well of width 2a and infinite barrier height or barrier height V_0.

The normalized particle wavefunctions are

$$\psi(x) = \sqrt{\frac{2}{W}} \cos \frac{n\pi x}{W}, \quad n \text{ odd}$$

$$= \sqrt{\frac{2}{W}} \sin \frac{n\pi x}{W}, \quad n \text{ even} \quad (2.67)$$

If the potential barrier is not infinite, we cannot assume that the wavefunction goes to zero at the boundaries of the well. The equation for the barrier region is

$$\frac{-\hbar^2}{2m} \frac{d^2\psi}{dx^2} + V_0 \psi = E\psi \quad \text{for } |x| \geq a \quad (2.68)$$

where V_0 is the potential step.

The general bound-state ($E < V_0$) solution of the problem can now be taken to be of the form

$$\psi(x) = \begin{cases} Ae^{\beta x}, & x \leq -a \\ B \cos \alpha x + C \sin \alpha x, & -a \leq x \leq a \\ De^{-\beta x}, & x \geq a \end{cases} \quad (2.69)$$

where

$$\alpha = \sqrt{\frac{2mE}{\hbar^2}}$$

$$\beta = \sqrt{\frac{2m(V_0 - E)}{\hbar^2}} \quad (2.70)$$

We now impose the boundary conditions that at $x = \pm a$, ψ and $d\psi/dx$ are continuous. This corresponds to saying that the electron probability and the electron current do not suffer a discontinuity at the boundaries.

Matching the wavefunction and its derivative at the boundaries, we have the conditions

$$B \cos \frac{\alpha W}{2} - C \sin \frac{\alpha W}{2} = Ae^{-\beta W/2}$$

2.4. Particles in an attractive potential: bound states

$$\alpha B \sin \frac{\alpha W}{2} + \alpha C \cos \frac{\alpha W}{2} = \beta A e^{-\beta W/2}$$

$$B \cos \frac{\alpha W}{2} + C \sin \frac{\alpha W}{2} = D e^{-\beta W/2}$$

$$-\alpha B \sin \frac{\alpha W}{2} + \alpha C \cos \frac{\alpha W}{2} = -\beta D e^{-\beta W/2} \quad (2.71)$$

From these equations we get two pairs of conditions on the solutions

$$2B \cos \frac{\alpha W}{2} = (A + D)e^{-\beta W/2}$$

$$2\alpha B \sin \frac{\alpha W}{2} = \beta(A + D)e^{-\beta W/2} \quad (2.72)$$

and

$$2C \sin \frac{\alpha W}{2} = (D - A)e^{-\beta W/2}$$

$$2\alpha C \cos \frac{\alpha W}{2} = -\beta(D - A)e^{-\beta W/2} \quad (2.73)$$

These pairs give us two separate conditions for the solutions, obtained by dividing one equation by the other within each pair. The conditions for the allowed energy levels are the transcendental equations

$$\frac{\alpha W}{2} \tan \frac{\alpha W}{2} = \frac{\beta W}{2} \quad (2.74)$$

and

$$\frac{\alpha W}{2} \cot \frac{\alpha W}{2} = -\frac{\beta W}{2} \quad (2.75)$$

Eq. 2.74 results in states that have even parity (i.e., with $A = D$), while Eq. 2.75 gives states with odd parity (with $A = -D$).

Eqs. 2.74 and 2.75 can be solved by numerical techniques. One useful approach is a graphical technique shown in Fig. 2.12a. We start out by plotting curves in the $\beta W/2 - \alpha W/2$ plane which satisfy Eqs. 2.74 or 2.75. Note that several α-values satisfy the equations for a given value of β. Next, we note that we have the equality (from Eq. 2.70)

$$\left(\frac{\alpha W}{2}\right)^2 + \left(\frac{\beta W}{2}\right)^2 = \frac{mV_0 W^2}{2\hbar^2} \equiv R(d)^2 \quad (2.76)$$

We therefore draw a circle with radius $R(d)$. For given values of V_0 and W, there is one such circle. The intersection of this circle with the first set of curves gives the desired solutions. There may be several solutions for a given well thickness. As the well thickness increases, the number of allowed states also increases, as shown in Fig. 2.12a. To find the highest allowed state for a given well thickness, we note that $(\alpha W/2) \tan(\alpha W/2)$ and $(\alpha W/2) \cot(\alpha W/2)$ intersect the $\alpha W/2$ axis at values of $n\pi/2$. Thus the well width at which a state N is just allowed is given by

$$R(W_c) = \frac{N\pi}{2}$$

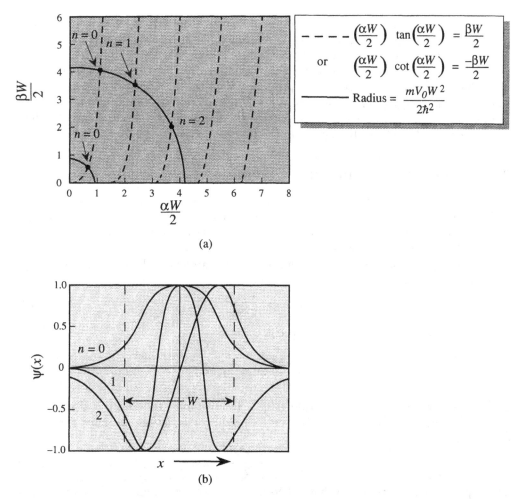

Figure 2.12: (a) The graphical approach to solving for the allowed modes in a finite quantum well. The figure shows two different cases of well size and potential energy combinations. (b) Typical solutions for the particle wavefunctions.

2.4. Particles in an attractive potential: bound states

The maximum mode number for a given well width is then the integer part of (the lowest state is given by $n = 0$)

$$\left(\frac{2mV_0W^2}{\pi^2\hbar^2}\right)^{1/2} + 1 \tag{2.77}$$

Another way (more accurate) to solve Eqs. 2.74 and 2.75 is to write a computer program based on the following steps: (i) Assume an electron energy starting from the lowest potential energy in the well. (ii) Calculate the left-hand and right-hand sides of the equation and see if they are equal (to some small error). (iii) Vary the electron energy until the equation(s) are satisfied. Typical shapes of the wavefunctions corresponding to the various energy levels are shown in Fig. 2.12b. Note that, in the case of the finite barrier potential, the wavefunctions penetrate the barrier region and the energy values are lower than the values obtained from the infinite barrier model.

2.4.3 Harmonic oscillator problem

As another example of a particle in an attractive potential we will discuss the harmonic oscillator problem. The mathematics of this problem is similar to that of many important problems. It also helps set up and understand "second quantization"–a development in which classical waves are represented by particles or quanta.

The potential energy for the harmonic oscillator is of the form shown in Fig. 2.13 and is given by

$$V(x) = \frac{1}{2}Kx^2 \tag{2.78}$$

The energy operator of the harmonic oscillator is

$$H = \frac{-\hbar^2}{2m}\nabla^2 + \frac{1}{2}Kx^2 \tag{2.79}$$

where K is the force constant describing the attractive force ($= -Kx$). If ψ represents the wavefunction of the time-independent Schrödinger equation, we have for a one-dimensional case

$$\frac{-\hbar^2}{2m}\frac{d^2\psi}{dx^2} + \frac{Kx^2\psi}{2} = E\psi \tag{2.80}$$

This equation can be recast so that it becomes a well-known differential equation with solutions known as Hermite polynomials, H_n. We will not discuss the mathematical details of how this equation is solved.

The general solutions are

$$\psi_n(x) = N_n H_n(\alpha x)e^{-\frac{1}{2}\alpha^2 x^2} \tag{2.81}$$

where

$$\alpha^4 = \frac{mK}{\hbar^2}$$

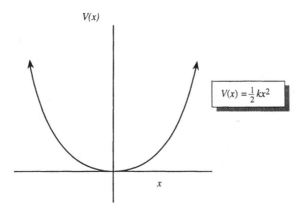

Figure 2.13: The potential energy profile for a harmonic oscillator.

The factor N_n is chosen so that

$$\int_{-\infty}^{\infty} |\psi_n(x)|^2 \, dx = \frac{|N_n|^2}{\alpha} \int_{-\infty}^{\infty} H_n^2(\xi) e^{-\xi^2} \, d\xi = 1$$

The function in the integral is integrable, and it is seen that the normalization factor is

$$N_n = \left(\frac{\alpha}{\pi^{1/2} 2^n n!} \right)^{1/2} \quad (2.82)$$

The first few Hermite polynomials are

$$\begin{aligned} H_0(\xi) &= 1 \\ H_1(\xi) &= 2\xi \\ H_2(\xi) &= 4\xi^2 - 2 \end{aligned} \quad (2.83)$$

The eigenenergy corresponding to the wavefunction $\psi_n(x)$ is given by the relation

$$E_n = \left(n + \frac{1}{2} \right) \hbar \omega_c \qquad n = 0, 1, 2, \ldots \quad (2.84)$$

Here the quantity ω_c is given by the classical frequency of the harmonic oscillator

$$\omega_c = \left(\frac{K}{m} \right)^{1/2} \quad (2.85)$$

In a classical harmonic oscillator, the energy of the oscillator is continuous and can be increased by increasing the amplitude of vibration. In the quantum treatment, the energy increases in steps of $\hbar \omega_c$. The number n in Eq. 2.84 is called the *occupation number* and tells us how many "quanta" of energy are in the oscillator. An important point is that the lowest energy of the quantum oscillator is not zero but $\frac{1}{2} \hbar \omega_c$. This lowest energy is called the *zero point energy* or, depending upon the context, in some problems, the *vacuum energy*. The Harmonic oscillator solutions are very useful in understanding lattice vibrations in crystals as well as in the quantum treatment of classical waves.

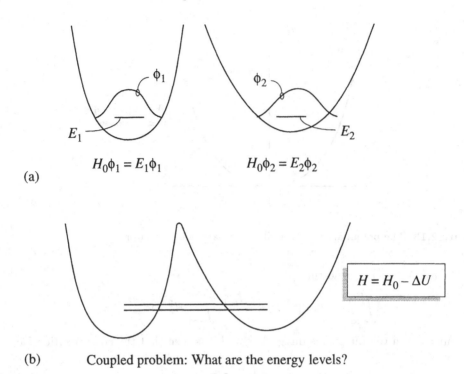

Figure 2.14: (a) A schematic of two uncoupled wells separated by a large potential barrier. (b) A schematic of two coupled potential wells.

2.5 FROM ATOMS TO MOLECULES: COUPLED WELLS

We have seen that in an attractive potential the electron (or a general particle) has discrete bound levels separated by energy gaps. We also know that, when some atoms are brought together, they attract each other to form a molecule. In this section we will examine how electronic levels in a quantum well change when another quantum well is brought close to it. This is the so-called coupled quantum well problem.

Let us consider two quantum wells separated by a large distance so that there is no *coupling* between them. Thus electrons occupying the bound states of each quantum well have no overlap with electrons in the other well, as shown in Fig. 2.14a. Now let us assume that the wells are brought closer to each other so that electrons in each well can sense the presence of the neighboring well, as shown in Fig. 2.14b. We are interested in finding out what happens to the allowed electronic levels in the coupled quantum wells problem.

We assume that we know the solutions for the uncoupled wells. When the quantum wells are spaced far apart and are uncoupled, let us say they each have states with energies E_1 and E_2, respectively, and wavefunctions ϕ_1 and ϕ_2, as shown in Fig. 2.14a. There may be other states in the quantum wells, but we will assume that their energies lie far from E_1 and E_2. We will see later that states far removed from the ones under consideration have a smaller effect on the solutions. We are interested in the solutions of the problem where the wells are brought closer so that there is some coupling between the wells. We write the Hamiltonian of the coupled problem as

$$H = H_0 - \Delta U \tag{2.86}$$

where ΔU is the correction due to the coupling of the well. The potential energy is reduced in the region between the wells, compared to the uncoupled system, so that ΔU is positive.

In the absence of any coupling, we have

$$H_0 \phi_1 = E_1 \phi_1$$
$$H_0 \phi_2 = E_2 \phi_2 \tag{2.87}$$

When the wells are coupled, the functions ϕ_1 and ϕ_2 are no longer the solutions. However, the new solutions can be expressed in terms of the uncoupled solutions to a good approximation. Let us write the solution for the coupled problem as

$$\psi = a_1 \phi_1 + a_2 \phi_2 \tag{2.88}$$

where a_1 and a_2 are unknown parameters which we will solve. If the system is uncoupled we have the two solutions

$$\psi_1 = \phi_1; \quad a_1 = 1, \; a_2 = 0$$
$$\psi_2 = \phi_2; \quad a_1 = 0, \; a_2 = 1$$

Coming to the coupled problem, the equation to be solved has the form

$$H(a_1 \phi_1 + a_2 \phi_2) = E(a_1 \phi_1 + a_2 \phi_2) \tag{2.89}$$

where H represents the full Hamiltonian of the coupled problem. We now multiply this equation from the left by ϕ_1^* and integrate over space to get

$$a_1 \int \phi_1^* (H_0 - \Delta U) \phi_1 d^3 r + a_2 \int \phi_1^* (H_0 - \Delta U) \phi_2 d^3 r$$
$$= E a_1 \int \phi_1^* \phi_1 d^3 r + E a_2 \int \phi_1^* \phi_2 d^3 r$$

Using the following equations

$$\int \phi_1^* (H_0 - \Delta U) \phi_1 d^3 r = E_1 \int \phi_1^* \phi_1 d^3 r - \int \phi_1^* \Delta U \phi_1 d^3 r \cong E_1$$
$$\int \phi_1^* \phi_1 d^3 r = 1$$
$$\int \phi_1^* \phi_2 d^3 r = 0$$

we get

$$a_1 (E_1 - E) + a_2 H_{12} = 0 \tag{2.90}$$

2.5. From atoms to molecules: coupled wells

where we have defined

$$H_{12} = -\int \phi_1^* \Delta U \phi_2 d^3r$$

The quantity H_{12} is called the matrix element of the Hamiltonian between the two states ϕ_1 and ϕ_2. In case the quantum wells are separated by a large distance the matrix element is zero. It increases as the coupling increases.

If we repeat the process described above but multiply Eq. 2.89 by ϕ_2^* instead of ϕ_1^*, we get

$$H_{21}a_1 + a_2(E_2 - E) = 0 \tag{2.91}$$

where

$$H_{21} = -\int \phi_2^* \Delta U \phi_1 d^3r$$

Assuming that the energy operator is real (energy is conserved) we have

$$H_{21} = H_{12}$$

The two coupled equations (Eqs. 2.90 and 2.91) can be written as a matrix vector product

$$\begin{vmatrix} E - E_1 & H_{12} \\ H_{21} & E - E_2 \end{vmatrix} \begin{vmatrix} a_1 \\ a_2 \end{vmatrix} = 0$$

To get non-trivial solutions of this equation, the determinant of the matrix must vanish. This gives us a quadratic equation with the following solutions

$$E = \frac{E_1 + E_2}{2} \pm \sqrt{\frac{(E_1 - E_2)^2}{4} + H_{12}^2} \tag{2.92}$$

The coefficients a_1 and a_2 can now be solved for and are

$$\frac{a_1}{a_2} = \frac{H_{12}}{E - E_1} \tag{2.93}$$

or

$$\frac{a_2}{a_1} = \frac{E - E_2}{H_{12}} \tag{2.94}$$

The simple equation we have derived has very useful implications for understanding many interesting and important physical systems.

Coupling of identical quantum wells

Let us examine in some detail the case where the two coupled quantum wells are identical and have the same initial energies and states, as shown in Fig. 2.15. It is important to keep in mind that when we talk of "quantum wells" we are simply referring to any problem with bound states. Thus the quantum wells may be atoms and molecules as well as potential wells created by use of semiconductors. Let us write

$$\begin{aligned} E_1 &= E_2 = E_0 \\ H_{12} &= H_{21} = -A \end{aligned} \tag{2.95}$$

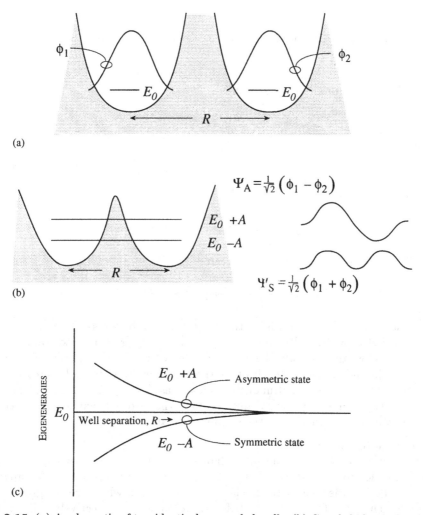

Figure 2.15: (a) A schematic of two identical uncoupled wells. (b) Coupled identical wells with energy levels and eigenfunctions. (c) Dependence of the symmetric and asymmetric eigenenergies on well separation or coupling.

2.5. From atoms to molecules: coupled wells

The quantity A represents the coupling coefficient between the wells.

From the derivation given above, the energy eigenvalues of the coupled system is now

$$E_S = E_0 - A$$
$$E_A = E_0 + A \tag{2.96}$$

For the state with energy E_S the coefficients of the state are, from Eq. 2.93,

$$a_1 = a_2 \tag{2.97}$$

while for the state with energy E_A we have

$$a_1 = -a_2 \tag{2.98}$$

If we normalize the state using

$$a_1^2 + a_2^2 = 1$$

we get the following solutions:
Symmetric state:

$$\Psi_S = \frac{1}{\sqrt{2}}[\phi_1 + \phi_2]; \quad E_S = E_0 - A \tag{2.99}$$

Asymmetric state:

$$\Psi_A = \frac{1}{\sqrt{2}}[\phi_1 - \phi_2]; \quad E_A = E_0 + A \tag{2.100}$$

The states are shown schematically in Fig. 2.15b. We see that, as a result of the coupling between the wells, the degenerate states E_0 are split into two states –one with energy below the uncoupled state and one with a higher energy. Note that as the wells are brought closer to each other the coupling strength will increase and the symmetric state energy will continuously decrease, as shown in Fig. 2.15c. If the electon in the coupled system occupy the symmetric state, the system behaves as if the coupling creates an attactive interaction in the symmetric state.

Hydrogen molecule ion

As an example of the coupled quantum well problem, let us consider two hydrogen nuclei with a single electron. As shown in Fig. 2.16, when the two nuclei are far apart, the electron can be in either one of the nuclei. Let these states, which are the ground states of the hydrogen atom, be denoted as before by ϕ_1 and ϕ_2, respectively. The energy of these states is just

$$E_0 = E_H = -13.6 \text{ eV}$$

As the nuclei are brought closer together, the electron on one atom feels the potential due to the attractive potential of the other nucleus. This gives a matrix element, which is a function of the inter-nucleus separation, R

$$H_{12} = -A(R)$$

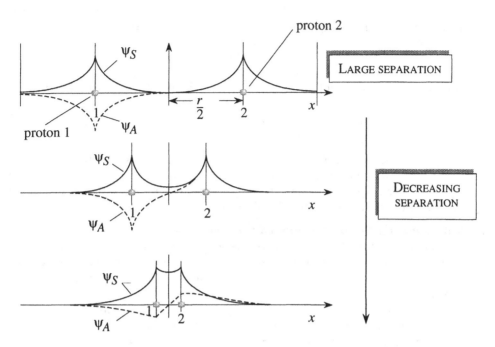

Figure 2.16: A schematic of the symmetric and antisymmetric wavefunction constructed from the ground state u_{100}. In the symmetric state, the electron is closer to the nuclei.

As noted above, the original state will now split into a lower energy symmetric state and a higher energy asymmetric state. These states are shown schematically in Fig. 2.16.

In Fig. 2.17a we show a plot of the change in the electronic energy as a function of the inter-nucleus separation. In the symmetric state, where the electronic function has a high probablility of occupying the space between the two nuclei, the system feels an attractive interaction, since the energy is reduced as the nuclei come closer. This is the reason the ion H_2^+ is stable. For the asymmetric state, where the electron is pushed away from the center of the two nuclei, there is an effective repulsive interaction, since the energy is larger than the energy when the nuclei are separated by a large distance.

To obtain the total energy of the H_2^+ ion as a function of inter-nucleus separation we need to add the repulsive interaction between the positively charged nuclei. This amounts to an energy

$$U_{\text{rep}} = \frac{1}{4\pi\epsilon_0} \frac{e^2}{R}$$

This energy is plotted in Fig. 2.17a. The total energy in the ion due to the presence of the two nuclei is now (in the ground state)

$$E(H_2^+) = E_0 - A(R) + \frac{1}{4\pi\epsilon_0} \frac{e^2}{R} \tag{2.101}$$

This total energy is plotted in Fig. 2.17b. We see that the energy minimizes at an

2.5. From atoms to molecules: coupled wells

inter-nucleus separation of R_{min}. Numerical calculations show that

$$R_{min} = 1.0 \text{ Å} \quad (2.102)$$

The binding energy of the ion is found to be (E_H is the magnitude of the ground state energy)

$$E_b = -0.2 E_H = -2.7 \text{ eV} \quad (2.103)$$

The example in this subsection shows how chemical bonds are formed by *sharing* of electrons between two nuclei. We have seen how the attractive interaction due to the coupling of the two bound states and the repulsive potential due to the nuclear charge play a role in setting the equilibrium bond distance.

Coupling between dissimilar quantum wells

Let us examine how the attractive and repulsive interactions due to coupling of wells are influenced when the two starting potential wells are dissimilar. This would occur if, for example, we had a hydrogen and a sodium nucleus coming together instead of two H nuclei, as considered in the previous subsection. Let the starting energies of the two potentials be E_1 and E_2, and let E_2 be larger than E_1. From the derivation given above, if the separation between E_2 and E_1 is much larger than the coupling coefficient A, we get (see Eq. 2.92) for the symmetric and asymmetric state

$$\begin{aligned} E_S &= E_1 - \frac{A^2}{E_2 - E_1} \\ E_A &= E_2 - \frac{A^2}{E_2 - E_1} \end{aligned} \quad (2.104)$$

If the value of $E_2 - E_1$ is much larger than the coupling coefficient A, we see that the effect of the coupling is very small. This is the reason bonding between dissimilar atoms is weak.

H_2 molecule

Another important and related problem is that of attraction between atoms. Let us consider the problem of how atoms attract each other to form a chemical bond. This is obviously an important question in chemistry, material science, and solid state physics. The problem of H_2^+ ion discussed previously sheds some light on the problem. However, in that problem we only had to consider one electron and two nuclei. What happens when there are two electrons?

Let us consider two H atoms initially far apart, as shown in Fig. 2.18a. In this uncoupled state each atom has an electron cloud around its nucleus, just as we expect in an isolated atom. The two states ϕ_1 and ϕ_2 are created through an exchange of the two electrons, as shown in Fig. 2.18a. For clarity we will call the electrons 1 and 2. As the atoms come closer to each other, there is an interaction between the atoms as each electron senses the attractive potential of the neighboring nucleus and the repulsive potential of the neighboring electron. The overall coupling is again represented by $A(R)$. Once again we have a symmetric state and an asymmetric state made from

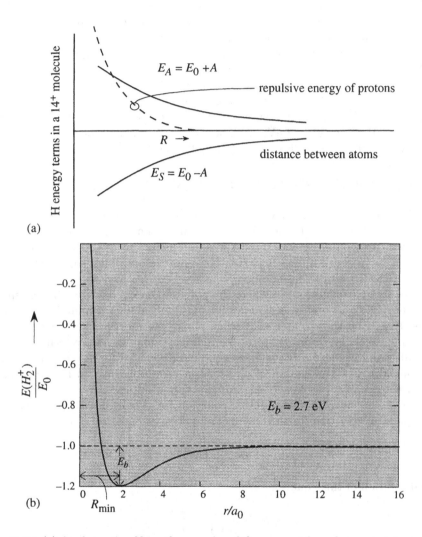

Figure 2.17: (a) A schematic of how the energies of the symmetric and asymmetric states vary with separation of the H-atom nuclei. Also shown is the repulsive energy arising from proton–proton repulsion; and (b) change in the energy of the H_2^+ ion as a function of inter-proton separation.

the original ϕ_1 and ϕ_2. In the terminology of chemical bonds, the symmetric state is called the bonding state and the asymmetric state is called the antibonding state. In the symmetric state the overall interaction is attractive and the energy of the system decreases in comparison with the energy it has when the atoms are well separated. In the asymmetric state there is a repulsive interaction. At very close spacing the repulsive potential of the charged particles dominates. The overall energy is shown in Fig. 2.18b. We see that in the symmetric state there is an equilibrium spacing between the atoms where the energy is minimum.

Detailed calculations show that the spacing between the atoms in equilibrium is 0.74 Å. This is the proton–proton spacing in the H_2 molecule. The binding energy of the molecule, i.e., the energy difference between the lowest energy state and the energy of two isolated H atoms is 4.52 eV, as shown in Fig. 2.18. Also shown are the binding energies (or dissociation energies) of several molecules.

2.6 ELECTRONS IN CRYSTALLINE SOLIDS

Most devices are made from crystalline materials (or crystalline materials with a very small density of defects). It is, therefore, important to understand the electronic properties of these materials. We will discuss a simple model to qualitatively examine the electronic levels in a crystalline material. However, before doing so, we will examine properties of electrons in free space.

The free particle problem where the potential energy term is zero (or spatially constant) provides very useful concepts that are applicable in a number of important problems. It will be shown later that when a particle moves in a perfectly periodic potential (in space) the solutions to the Schrödinger equation have a form very similar to the solutions in free space. Thus many concepts developed for the free-space particle can be applied to the description of electrons in crystalline media. In particular, the concept of density of states developed here is widely used.

Let us consider the Schrödinger equation for a free particle of mass m. The time-independent equation for the background potential equal to V_0 is

$$\frac{-\hbar^2}{2m}\left(\frac{\partial^2}{\partial x^2} + \frac{\partial^2}{\partial y^2} + \frac{\partial^2}{\partial z^2}\right)\psi(r) = (E - V_0)\psi(r) \tag{2.105}$$

A general solution of this equation is

$$\psi(\mathbf{r}) = \frac{1}{\sqrt{V}} e^{\pm i \mathbf{k} \cdot \mathbf{r}} \tag{2.106}$$

and the corresponding energy is

$$E = \frac{\hbar^2 k^2}{2m} + V_0 \tag{2.107}$$

where the factor $\frac{1}{\sqrt{V}}$ in the wavefunction occurs because we wish to have one particle per volume V or

$$\int_V d^3r \, |\psi(r)|^2 = 1 \tag{2.108}$$

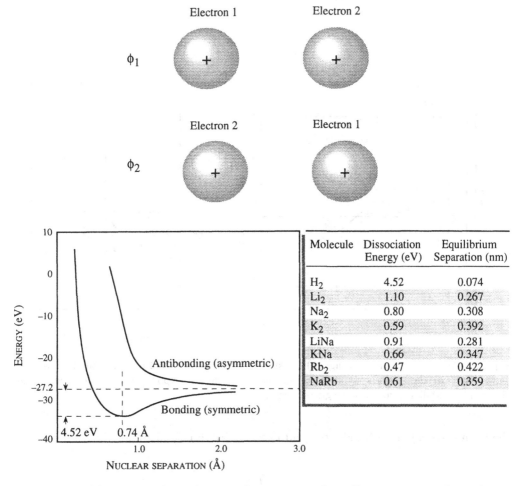

Figure 2.18: (a) A schematic of the two different states of two H atoms separated by a large distance; and (b) change in the symmetric (bonding) and asymmetric (antibonding) states as a function of nuclear spacing. Also shown are dissociative energies of several molecules.

We assume that the volume V is a cube of side L.

In classical mechanics the energy momentum relation for the free particle is $E = p^2/2m + V_0$, and p can be a *continuous variable*. The quantity $\hbar k$ seems to be replacing p in quantum mechanics. Due to the wave nature of the electron, in a finite volume, k is not continuous but discrete. To correlate with physical conditions we may want to describe, two kinds of boundary conditions are imposed on the wavefunction. In the first one the wavefunction is considered to go to zero at the boundaries of the volume, as shown in Fig. 2.19a. In this case, the wave solutions are standing waves of

2.6. Electrons in crystalline solids

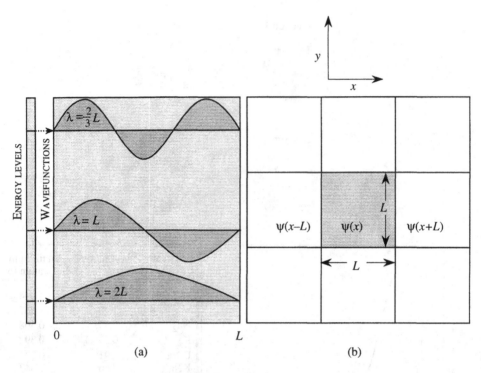

Figure 2.19: A schematic showing: (a) the stationary boundary conditions leading to standing waves and (b) the periodic boundary conditions leading to exponential solutions with the electron probability equal in all regions of space.

the form $\sin(k_x x)$ or $\cos(k_x x)$, etc., and k-values are restricted to positive values

$$k_x = \frac{\pi}{L}, \frac{2\pi}{L}, \frac{3\pi}{L} \ldots \tag{2.109}$$

Here we will use a different set of conditions. Periodic boundary conditions are shown in Fig. 2.19b. Even though we focus our attention on a finite volume V, the wave can be considered to spread in all space as we conceive the entire space was made up of identical cubes of sides L. Then

$$\begin{aligned}
\psi(x, y, z + L) &= \psi(x, y, z) \\
\psi(x, y + L, z) &= \psi(x, y, z) \\
\psi(x + L, y, z) &= \psi(x, y, z)
\end{aligned} \tag{2.110}$$

Because of the boundary conditions the allowed values of k are (n are integers – positive and negative)

$$k_x = \frac{2\pi n_x}{L}; \quad k_y = \frac{2\pi n_y}{L}; \quad k_z = \frac{2\pi n_z}{L} \tag{2.111}$$

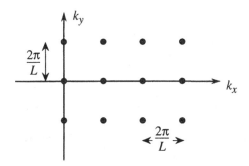

Figure 2.20: k-space volume of each electronic state. The separation between the various allowed components of the k-vector is $\frac{2\pi}{L}$.

If L is large, the spacing between the allowed k-values is very small. Also it is important to note that the results we obtain for properties of the particles in a *large volume are independent of whether we use the stationary or periodic boundary conditions*. It is useful to discuss the *volume in k-space that each electronic state occupies*. As can be seen from Fig. 2.20, this volume is (in three dimensions)

$$\left(\frac{2\pi}{L}\right)^3 = \frac{8\pi^3}{V} \qquad (2.112)$$

If Ω is a volume of k-space, the number of electronic states in this volume is

$$\boxed{\frac{\Omega V}{8\pi^3}} \qquad (2.113)$$

2.6.1 Electrons in a uniform potential

Density of states for a three-dimensional system

We will now use the discussion of the previous subsection to derive the extremely important concept of density of states. Although we will use the periodic boundary conditions to obtain the density of states, the stationary conditions lead to the same result, as long as the space under consideration is large, compared to the wavelength of the particle.

The concept of density of states is extremely powerful, and important physical properties in materials, such as optical absorption, transport, etc., are intimately dependent upon this concept. Density of states is the number of available electronic states *per unit volume per unit energy* around an energy E. If we denote the density of states by $N(E)$, the number of states in a unit volume in an energy interval dE around an energy E is $N(E)dE$. To calculate the density of states, we need to know the dimensionality of the system and the energy versus k relation that the particles obey. We will choose the particle of interest to be the electron, since in most applied problems we are dealing with electrons. Of course, the results derived can be applied to other particles

2.6. Electrons in crystalline solids

as well. For the free electron case we have the parabolic relation

$$E = \frac{\hbar^2 k^2}{2m_0} + V_0$$

The energies E and $E + dE$ are represented by surfaces of spheres with radii k and $k + dk$, as shown in Fig. 2.21. In a three-dimensional system, the k-space volume between vector k and $k + dk$ is (see Fig. 2.21a) $4\pi k^2 dk$. We have shown in Eq. 2.113 that the k-space volume per electron state is $(\frac{2\pi}{L})^3$. Therefore, the number of electron states in the region between k and $k + dk$ is

$$\frac{4\pi k^2 dk}{8\pi^3} V = \frac{k^2 dk}{2\pi^2} V$$

Denoting the energy and energy interval corresponding to k and dk as E and dE, we see that the number of electron states between E and $E + dE$ per unit volume is

$$N(E)\, dE = \frac{k^2 dk}{2\pi^2}$$

Using the E versus k relation for the free electron, we have

$$k^2 dk = \frac{\sqrt{2} m_0^{3/2} (E - V_0)^{1/2} dE}{\hbar^3}$$

and

$$N(E)\, dE = \frac{m_0^{3/2} (E - V_0)^{1/2} dE}{\sqrt{2}\pi^2 \hbar^3} \qquad (2.114)$$

Quantum mechanics tell us that particles such as electrons have an internal property associated with them known as spin. Electrons are called fermions and can have spin (which is an internal angular momentum) $\hbar/2$ or $-\hbar/2$. Accounting for spin, the density of states obtained is simply multiplied by 2

$$N(E) = \frac{\sqrt{2} m_0^{3/2} (E - V_0)^{1/2}}{\pi^2 \hbar^3} \qquad (2.115)$$

Density of states in sub-three-dimensional systems
Let us now consider a 2D system, a concept that has become a reality with the use of quantum wells. The two-dimensional density of states is defined as the number of available electronic states *per unit area per unit energy* around an energy E. Similar arguments as used in the derivation show that the density of states for a parabolic band (for energies greater than V_0) is (see Fig 2.21b)

$$N(E) = \frac{m_0}{\pi \hbar^2} \qquad (2.116)$$

The factor of 2, resulting from spin, has been included in this expression.

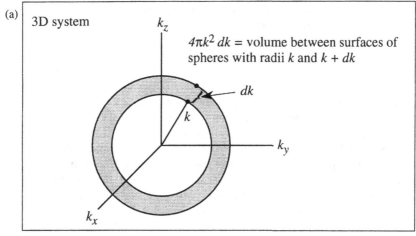
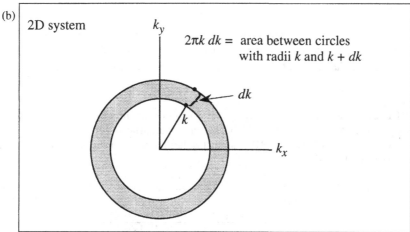
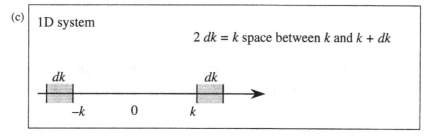

Figure 2.21: Geometry used to calculate density of states in three, two, and one dimensions. By finding the k-space volume in an energy interval between E and $E + dE$, we can find the number of allowed states.

2.6. Electrons in crystalline solids

Finally, we can consider a one-dimensional system often called a "quantum wire." The one-dimensional density of states is defined as the number of available electronic states *per unit length per unit energy* around an energy E. In a 1D system or a "quantum wire" the density of states is (including spin) (see Fig. 2.21c)

$$N(E) = \frac{\sqrt{2}m_0^{1/2}}{\pi \hbar}(E - V_0)^{-1/2} \tag{2.117}$$

Notice that as the dimensionality of the system changes, the energy dependence of the density of states also changes. As shown in Fig. 2.22, for a three-dimensional system we have $(E-V_0)^{1/2}$ dependence, for a two-dimensional system we have no energy dependence, and for a one-dimensional system we have $(E - V_0)^{-1/2}$ dependence.

We will see later in the next section that when a particle is in a periodic potential, its wavefunction is quite similar to the free particle wavefunction. Also, the particle responds to external forces as if it is a free particle *except that its energy-momentum relation is modified by the presence of the periodic potential*. In some cases it is possible to describe the particle energy by the relation

$$E = \frac{\hbar^2 k^2}{2m^*} + E_0 \tag{2.118}$$

where m^* is called the effective mass in the material. The expressions derived for the free electron density of states *can then be carried over to describe the density of states for a particle in a crystalline material (which has a periodic potential) by simply replacing m_0 by m^**.

EXAMPLE 2.4 Calculate the density of states of electrons in a 3D system and a 2D system at an energy of 0.1 eV. Assume that the background potential is zero.

The density of states in a 3D system (including the spin of the electron) is given by (E is the energy in Joules)

$$\begin{aligned} N(E) &= \frac{\sqrt{2}(m_0)^{3/2} E^{1/2}}{\pi^2 \hbar^3} \\ &= \frac{\sqrt{2}(0.91 \times 10^{-30} \text{ kg})(E^{1/2})}{\pi^2 (1.05 \times 10^{-34} \text{ J} \cdot \text{s})^3} \\ &= 1.07 \times 10^{56} E^{1/2} \text{ J}^{-1} \text{ m}^{-3} \end{aligned}$$

Expressing E in eV and the density of states in the commonly used units of eV^{-1} cm^{-3}, we get

$$\begin{aligned} N(E) &= 1.07 \times 10^{56} \times (1.6 \times 10^{-19})^{3/2}(1.0 \times 10^{-6})E^{1/2} \\ &= 6.8 \times 10^{21} E^{1/2} \text{ eV}^{-1} \text{ cm}^{-3} \end{aligned}$$

At $E = 0.1$ eV we get

$$N(E) = 2.15 \times 10^{21} \text{ eV}^{-1} \text{ cm}^{-3}$$

For a 2D system the density of states is independent of energy and is

$$N(E) = \frac{m_0}{\pi \hbar^2} = 4.21 \times 10^{14} \text{ eV}^{-1} \text{ cm}^{-2}$$

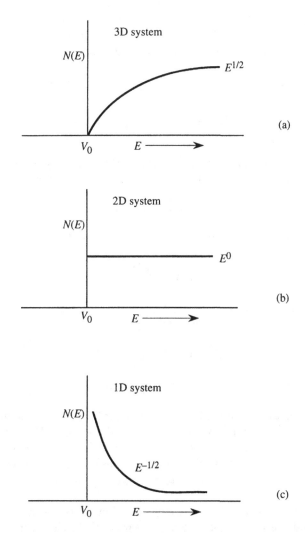

Figure 2.22: Variation in the energy dependence of the density of states in: (a) three-dimensional, (b) two-dimensional, and (c) one-dimensional systems. The energy dependence of the density of states is determined by the dimensionality of the system.

2.6.2 Particle in a periodic potential: Bloch theorem

How do electrons behave inside crystalline solids? The application of quantum mechanics to this problem was one of the first great successes of quantum theory. The resultant band theory eventually resolved the puzzle of electrical transport in metals, semiconductors, and insulators. The core of modern understanding of electronic and optical properties of solid-state materials is based on the band theory, which describes the properties of electrons in a periodic potential arising from the periodic arrangement of atoms in a crystal.

The description of the electron in the periodic material has to be via the Schrödinger equation

$$\left[\frac{-\hbar^2}{2m_0}\nabla^2 + U(\mathbf{r})\right]\psi(\mathbf{r}) = E\psi(\mathbf{r}) \tag{2.119}$$

where $U(\mathbf{r})$ is the background potential seen by the electrons. Due to the crystalline nature of the material, the potential $U(\mathbf{r})$ has the same periodicity, R, as the lattice

$$U(\mathbf{r}) = U(\mathbf{r} + \mathbf{R}) \tag{2.120}$$

If the background potential is zero, the electronic function in a volume V is

$$\psi(\mathbf{r}) = \frac{e^{i\mathbf{k}\cdot\mathbf{r}}}{\sqrt{V}}$$

and the electron momentum and energy are

$$\begin{aligned} \mathbf{p} &= \hbar\mathbf{k} \\ E &= \frac{\hbar^2 k^2}{2m_0} \end{aligned}$$

The wavefunction is spread in the entire sample and has equal probability ($\psi^*\psi$) at every point in space. Let us examine the periodic crystal. We expect the *electron probability to be same in all unit cells of the crystal because each cell is identical*. If the potential was random, this would not be the case, as shown schematically in Fig. 2.23a. This expectation is, indeed, correct and is put in a mathematical form by Bloch's theorem.

Bloch's theorem states that the eigenfunctions of the Schrödinger equation for a periodic potential are the product of a plane wave $e^{i\mathbf{k}\cdot\mathbf{r}}$ and a function $u_\mathbf{k}(\mathbf{r})$, which has the *same periodicity as the periodic potential*. Thus

$$\psi_\mathbf{k}(\mathbf{r}) = e^{i\mathbf{k}\cdot\mathbf{r}} u_\mathbf{k}(\mathbf{r}) \tag{2.121}$$

is the form of the electronic function. The periodic part $u_\mathbf{k}(\mathbf{r})$ has the same periodicity as the crystal, i.e.

$$u_\mathbf{k}(\mathbf{r}) = u_\mathbf{k}(\mathbf{r} + \mathbf{R}) \tag{2.122}$$

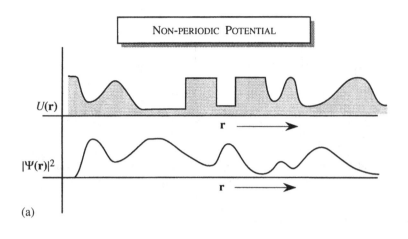
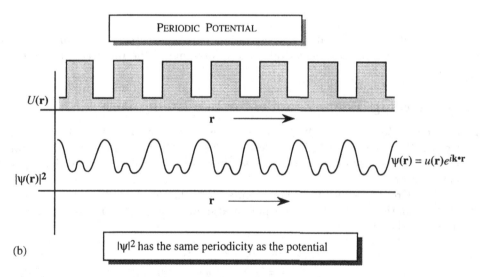

Figure 2.23: (a) Potential and electron probability value of a typical electronic wavefunction in a random material. (b) The effect of a periodic background potential on an electronic wavefunction. In the case of the periodic potential, $|\psi|^2$ has the same spatial periodicity as the potential. This puts a special constraint on $\psi(\mathbf{r})$ according to Bloch's theorem.

2.6. Electrons in crystalline solids

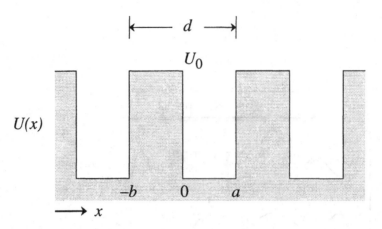

Figure 2.24: The periodic potential used to study the bandstructure in the Kronig–Penney model. The potential varies between 0 and U_0 as shown and has a periodicity of d.

The wavefunction has the property

$$\begin{aligned}\psi_{\mathbf{k}}(\mathbf{r}+\mathbf{R}) &= e^{i\mathbf{k}\cdot(\mathbf{r}+\mathbf{R})}u_{\mathbf{k}}(\mathbf{r}+\mathbf{R}) = e^{i\mathbf{k}\cdot\mathbf{r}}u_{\mathbf{k}}(\mathbf{r})e^{i\mathbf{k}\cdot\mathbf{R}} \\ &= e^{i\mathbf{k}\cdot\mathbf{R}}\psi_{\mathbf{k}}(\mathbf{r})\end{aligned} \quad (2.123)$$

The wavefunction is illustrated in Fig. 2.23b. Before discussing the solutions of the periodic potential problem, let us take a look at some of the important properties of crystalline materials.

2.6.3 Kronig–Penney model for bandstructure

A useful model for understanding how electrons behave inside crystalline materials is the Kronig–Penney model. Although not a realistic potential for crystals, it allows us to calculate the energy of the electrons as a function of the parameter k that appears in Bloch's theorem.

The Kronig–Penney model represents the background periodic potential seen by the electrons in the crystal as a simple potential shown in Fig. 2.24. The one-dimensional potential has the form

$$\begin{aligned}U(x) &= 0 \quad 0 \leq x \leq a \\ &= U_0 \quad -b \leq x \leq 0\end{aligned} \quad (2.124)$$

The potential is repeated periodically as shown in Fig. 2.24 with a periodicity distance $d\ (= a+b)$. Since the potential is periodic, the electron wavefunction satisfies Bloch's theorem and we may write

$$\psi(x+d) = e^{i\phi}\psi(x) \quad (2.125)$$

where the phase ϕ is written as

$$\phi = k_x d$$

In the region $-b < x < a$, the electron function has the form

$$\psi(x) = \begin{cases} Ae^{i\beta x} + Be^{-i\beta x}, & \text{if } -b < x < 0 \\ De^{i\alpha x} + Fe^{-i\alpha x}, & \text{if } 0 < x < a \end{cases} \quad (2.126)$$

where

$$\beta = \sqrt{\frac{2m_0(E - U_0)}{\hbar^2}}$$

$$\alpha = \sqrt{\frac{2m_0 E}{\hbar^2}} \quad (2.127)$$

Then, in the following period, $a < x < a + d$, from Eq. 2.123

$$\psi(x) = e^{i\phi} \begin{cases} Ae^{i\beta(x-d)} + Be^{-i\beta(x-d)}, & \text{if } a < x < d \\ De^{i\alpha(x-d)} + Fe^{-i\alpha(x-d)}, & \text{if } d < x < a + d \end{cases} \quad (2.128)$$

From the continuity conditions for the wavefunction and its derivative at $x = 0$ and at $x = a$, the following system of equations is obtained

$$\begin{aligned} A + B &= D + F \\ \beta(A - B) &= \alpha(D - F) \\ e^{i\phi}(Ae^{i\beta b} + Be^{-i\beta b}) &= De^{i\alpha a} + Fe^{-i\alpha a} \\ \beta e^{i\phi}(Ae^{i\beta b} - Be^{-i\beta b}) &= \alpha(De^{i\alpha a} - Fe^{-i\alpha a}) \end{aligned} \quad (2.129)$$

Non-trivial solutions for the variables A, B, D, F are obtained only if the determinant of their coefficients vanishes, which gives the condition

$$\begin{aligned} \cos\phi &= \cos a\alpha \cosh b\delta - \frac{\alpha^2 - \delta^2}{2\alpha\delta} \sin a\alpha \sinh b\delta, \quad \text{if } 0 < E < U_0 \\ &= \cos a\alpha \cos b\beta - \frac{\alpha^2 + \beta^2}{2\alpha\beta} \sin a\alpha \sin b\beta, \quad \text{if } E > U_0 \end{aligned} \quad (2.130)$$

where

$$\delta = \sqrt{\frac{2m_0(U_0 - E)}{\hbar^2}} \quad (2.131)$$

The energy E, which appears in Eq. 2.130 through α, β, and δ, is physically allowed only if

$$-1 \leq \cos\phi \leq +1$$

Consider the case where $E < U_0$. We denote the right-hand side of Eq. 2.130 by $f(E)$

$$\begin{aligned} f(E) &= \cos\left(a\sqrt{\frac{2m_0 E}{\hbar^2}}\right) \cosh\left(b\sqrt{\frac{2m_0(U_0 - E)}{\hbar^2}}\right) \\ &+ \frac{U_0 - 2E}{2\sqrt{E(U_0 - E)}} \sin\left(a\sqrt{\frac{2m_0 E}{\hbar^2}}\right) \sinh\left(b\sqrt{\frac{2m_0(U_0 - E)}{\hbar^2}}\right) \end{aligned} \quad (2.132)$$

2.6. Electrons in crystalline solids

This function must lie between -1 and $+1$ since it is equal to $\cos\phi$ ($=\cos k_x d$). We wish to find the relationship between E and ϕ or E and k_x. In general, we have to write a computer program in which we evaluate $f(E)$ starting from $E = 0$, and verifies if $f(E)$ lies between -1 and 1. If it does, we get the value of ϕ for each allowed value of E. The approach is shown graphically in Fig. 2.25a. As we can see, $f(E)$ remains between the ± 1 bounds only for certain regions of energies. These "allowed energies" form the allowed bands and are separated by "bandgaps." We can obtain the E versus k relation or the bandstructure of the electron in the periodic structure, as shown in Fig. 2.25b. In the figure, the energies between E_2 and E_1 form the first allowed band, the energies between E_4 and E_3 form the second bandgap, etc.

We note that the $\phi = k_x d$ term on the left-hand side of Eq. 2.130 appears as a cosine. As a result, if $k_x d$ corresponds to a certain allowed electron energy, then $k_x d + 2n\pi$ is also allowed. This simply reflects a periodicity that is present in the problem. It is customary to show the E–k relation for the smallest k-values. The smallest k-values lie in a region $\pm\pi/d$ for the simple problem discussed here. In more complex periodic structures the smallest k-values lie in a more complicated k-space. The term Brillouin zone is used to denote the smallest unity cell of k-space. If a k-value is chosen beyond the Brillouin zone values, the energy values are simply repeated. The concept of allowed bands of energy separated by bandgaps is central to the understanding of crystalline materials. Near the bandedges it is usually possible to define the electron E–k relation as

$$E = \frac{\hbar^2(k-k_o)^2}{2m^*} \qquad (2.133)$$

where k_o is the k-value at the bandedge and m^* is the effective mass. The concept of an effective mass is extremely useful, since it represents the response of the electron–crystal system to the outside world.

Significance of the k-vector
In our discussion of free electrons the quantity $\hbar k$ represents the momentum of the electron. We have now seen that when electrons are in crystalline systems their properties are described by a wavevector **k**. What is the significance of **k**?

For free electrons moving in space two important laws are used to describe their properties: (i) Newton's second law of motion tells us how the electron's trajectory evolves in the presence of an external force; (ii) the law of conservation of momentum allows us to determine the electron's trajectory when there is a collision. As noted in Section 2.3, the Ehrenfest theorem tells us that these laws are applicable to particles in quantum mechanics as well. We are obviously interested in finding out what the analogous laws are when an electron is inside a crystal and not in free space.

An extremely important implication of the Bloch theorem is that in the perfectly periodic background potential that the crystal presents, *the electron propagates without scattering*. The electronic state ($\sim \exp(i\mathbf{k}\cdot\mathbf{r})$) is an extended wave which *occupies the entire crystal*. To complete our understanding, we need to derive an equation of motion for the electrons which tells us how electrons will respond to external forces.

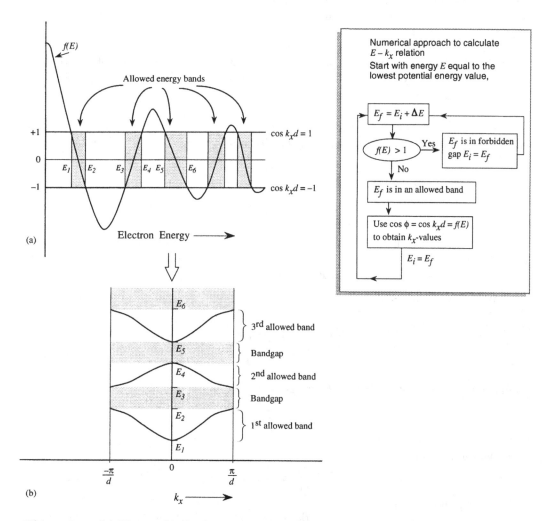

Figure 2.25: (a) The graphical solution to obtain the allowed energy levels. The function $f(E)$ is plotted as a function of E. Only energies for which $f(E)$ lies between $+1$ and -1 are allowed. (b) The allowed and forbidden bands are plotted in the E versus k relation using the results from (a). The inset shows a flow chart of how we can obtain the E-k_x relation.

2.6. Electrons in crystalline solids

The equation of motion

$$\frac{d\mathbf{p}}{dt} = \mathbf{F}_{ext} + \mathbf{F}_{int} \tag{2.134}$$

is not very useful for a meaningful description of the electron because it includes the internal forces on the electron. We need a description which does *not* include the evaluation of the internal forces.

As in classical wave theory, associated with any wave phenomena is the wave group velocity that represents the propagation of wave energy. In the case of a particle wave the group velocity represents the particle velocity. We can define the group velocity of this wavepacket as

$$\mathbf{v}_g = \frac{d\omega}{d\mathbf{k}} \tag{2.135}$$

where ω is the frequency associated with the electron of energy E; i.e., $\omega = E/\hbar$:

$$\begin{aligned} \mathbf{v}_g &= \frac{1}{\hbar} \frac{dE}{d\mathbf{k}} \\ &= \frac{1}{\hbar} \nabla_{\mathbf{k}} E(\mathbf{k}) \end{aligned}$$

If we have an electric field $\mathbf{F}$ present, the work done on the electron during a time interval δt is

$$\delta E = -e\mathbf{F} \cdot \mathbf{v}_g \delta t \tag{2.136}$$

We may also write, in general

$$\begin{aligned} \delta E &= \left(\frac{dE}{d\mathbf{k}}\right) \delta \mathbf{k} \\ &= \hbar \mathbf{v}_g \cdot \delta \mathbf{k} \end{aligned} \tag{2.137}$$

Comparing the two equations for δE, we get

$$\delta \mathbf{k} = -\frac{e\mathbf{F}}{\hbar} \delta t$$

giving us the relation

$$\hbar \frac{d\mathbf{k}}{dt} = -e\mathbf{F} \tag{2.138}$$

In general, we may write

$$\hbar \frac{d\mathbf{k}}{dt} = \mathbf{F}_{ext} \tag{2.139}$$

Eq. 2.139 looks identical to Newton's second law of motion

$$\frac{d\mathbf{p}}{dt} = \mathbf{F}_{ext}$$

in free space if we associate the quantity $\hbar \mathbf{k}$ with the momentum of the electron in the crystal. The term $\hbar \mathbf{k}$ responds to the *external forces as if it is the momentum of the electron*, although, as can be seen by comparing the true Newtons equation of motion, it

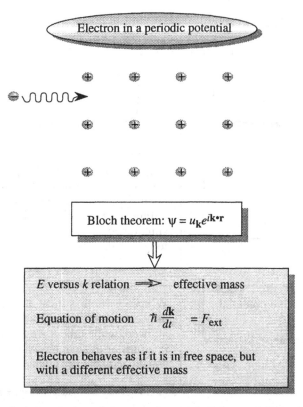

Figure 2.26: A physical description of electrons in a periodic potential. As shown the electrons can be treated as if they are in free space except that their energy–momentum relation is modified because of the potential. Near the bandedges the electrons respond to the outside world as if they have an effective mass m^*. The effective mass can have a *positive or negative value*.

is clear that $\hbar \mathbf{k}$ contains the effects of the internal crystal potentials and is therefore not the true electron momentum. The quantity $\hbar \mathbf{k}$ is called the crystal momentum. Once the E versus k relation is established, we can, for all practical purposes, forget about the background potential $U(\mathbf{r})$ and treat the electrons as if they are free and obey the effective Newtons equation of motion. This physical picture is summarized in Fig. 2.26.

2.7 SUMMARY

The topics covered in this chapter are summarized in Tables 2.2 to 2.5.

TOPICS STUDIED	KEY FINDINGS
Particle waves and wavepackets Uncertainty relations	In quantum mechanics the particle probability is described by a wavefunction. The wave nature of the particle results in uncertainty in defining the particle's position and momentum simultaneously or its energy and lifetime (how long it will have that energy) simultaneously.
Particles in attractive potential wells	When a particle is in an attractive potential only certain energy values (energy levels) are allowed, while others are forbidden. Electrons in atoms or in quantum wells have such "quantized" energy levels.

Table 2.2: Summary table.

2.8 PROBLEMS

2.1 The surface of the sun has a temperature of 6000 K. Calculate the wavelength at which the sun emits its peak radiancy of the photon corresponding to this peak wavelength.

2.2 Calculate the peak wavelength emitted by a healthy human. Can we see this radiation?

2.3 Estimate the power radiated by a typical human body.

2.4 Calculate the cutoff wavelength for the photoelectric effect for an aluminum sample.

2.5 A 1 W light source produces radiation in a uniform spatial distribution. The wavelength of light is 5500 Å. Calculate the number of photons per second striking an area 20 cm × 10 cm located 1 m from the source.

2.6 A metal surface is found to have a photoelectric cutoff wavelength of 3250 Å. Calculate the stopping potential if the surface is illuminated with light of wavelength 2000 Å.

2.7 It is found that when a sample of sodium is illuminated with light of a wavelength 3200 Å the stopping potential is 1.6 V. When it is illuminated with light of a wavelength

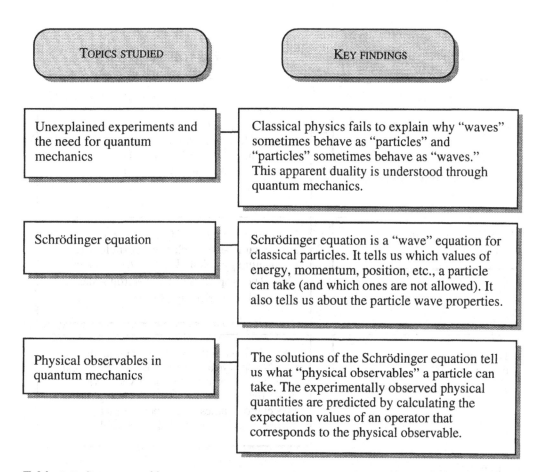

Table 2.3: Summary table.

2.8. Problems

Topics studied	Key findings
Coupled quantum wells Molecular levels	When two attractive potential wells are brought close to each other bonding and antibonding states are created. These are responsible for molecular attraction.
Electrons in a uniform potential	In a uniform potential V_0, the electron energies are the same as for a classical particle with momentum $p = \hbar \mathbf{k}$ and energy $E = V_0 + \hbar^2 k^2 /2m_0$. The properties of the electron states can be described by density of states.
Density of states and dimensionality	The solutions of the Schrödinger equation tell us what "physical observables" a particle can take. The experimentally observed physical quantities are predicted by calculating the expectation values of an operator that corresponds to the physical observable.

Table 2.4: Summary table.

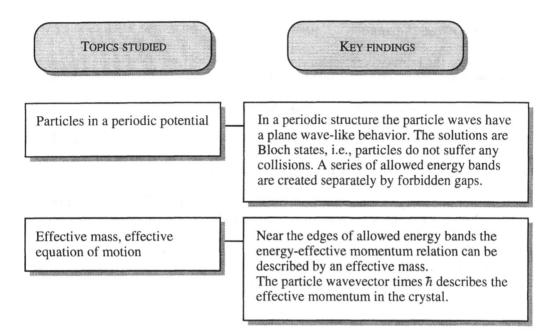

Table 2.5: Summary table.

2800 Å, the stopping potential is 2.15 V. Calculate the work function of sodium.

2.8 Use the energy levels calculated in the Bohr model for hydrogen to discuss what levels are involved in producing the Balmer and Paschen series.

2.9 Discuss how the Lyman series would be produced using the Bohr model for a hydrogen atom.

2.10 Calculate the speed of an electron in the $n = 3$ level of the hydrogen atom.

2.11 Calculate the radii and speed of an electron in the $n = 1$ and $n = 3$ levels in a hydrogen-like atom with a nuclear charge Ze. Assume that $Z = 2$ and $Z = 3$.

2.12 Compare the wavelengths of a 1 eV photon and a 1 eV electron.

2.13 Consider electrons with kinetic energy of 10 eV moving in (i) a medium where the smallest distance over which the potential energy changes is 1.0 μ m; (ii) a medium where this distance is 1.0 Å. Which problem will display "quantum behavior?"

2.14 Consider an electron wavefunction given by the general form

$$\psi = A\sin\frac{n\pi x}{L}; n = 1, 2, 3$$

Calculate the factor A (normalization factor) if the wavefunction is to be normalized between $x = 0$ and $x = L$.

2.15 Consider an electron wavefunction describing an electron state extending from $-W/2$ to $W/2$

$$\psi(x) = \sqrt{\frac{2}{W}}\cos\frac{n\pi x}{W}; n = 1, 3, 5$$

2.8. Problems

Calculate the energy, momentum and position expectation value of an electron in state $n = 1$ and $n = 3$.

2.16 Estimate the kinetic energy of an electron confined to a size 1 Å (twice the Bohr radius) using the uncertainty principle. How does this compare with the kinetic energy of an electron in the hydrogen atom?

2.17 In the $n = 2$ state of hydrogen, find the electron's velocity, kinetic energy, and potential energy.

2.18 Consider an electron in the $n = 5$ state of the hydrogen atom. Calculate the wavelengths of photons emitted as the electron relaxes to the ground state. Assume that only those transitions in which n changes by unity can be made.

2.19 Calculate the ionization energy of: (a) the $n = 3$ level of hydrogen. (b) The $n = 2$ level of He$^+$ (singly ionized helium). (c) The $n = 4$ level of Li^{++} (doubly ionized lithium). You can use the hydrogen atom model for He$^+$ and Li^{++} with appropriate changes in the nuclear charge.

2.20 The lifetimes of the levels in a hydrogen atom are of the order of 10^{-8} s. Find the energy uncertainty of the first excited state.

2.21 Consider a one-dimensional quantum well described by

$$V_0 = 2 \text{ eV}; W = 10 \text{ Å}$$

Calculate the energies of an electron (E_1 and E_2) in this well. Also plot the wavefunction ψ_1.

2.22 Consider a three-dimensional quantum well with a barrier height of 1.0 eV and $W_x = W_y = W_z = 10$ Å. Calculate the ground state energy and the first excited state energy.

2.23 Consider a pendulum of length 10 cm. Regard this as a quantum oscillator and calculate the energy difference between successive levels. Is it possible to observe these quantized levels?

2.24 Consider a carbon atom of atomic mass 12 vibrating in a material. Assume that the oscillator can be described by a force constant of 0.8 Nm^{-1}. The oscillator is in the $n = 10$ state. Calculate the energy of the oscillator. If this was a classical oscillator with the same energy, what would be the amplitude of vibration?

2.25 The ground state of an oscillator is 10 meV. How much energy is needed to excite this oscillator to the first excited state? If this oscillator is in thermal equilibrium at 300 K, calculate the energy of the system.

2.26 It can be shown that for two identical quantum wells with well size W, barrier height V_0, and separation W_b, the coupling matrix A defined in the text is

$$A = \frac{E_0}{2\pi} \exp\left(-\frac{\sqrt{2m(V_0 - E_0)}W_b}{\hbar}\right)$$

where E_0 is the energy position in the uncoupled well measured from the bottom of the well.

Calculate the splitting of the E_0 levels as a function of W_b as W_b changes from 50 Å to 10 Å if $V_0 = 1.0$ eV and $W = 10$ Å.

2.27 Consider two 20 Å quantum wells with a separation W_b and $V_0 = 1.0$ eV. Using the previous problem for the coupling coefficient, design a coupled well system (i.e., find W_b) so that the ground state energy splitting is 10 meV. What is the splitting of the first excited state?

2.28 Consider an electron in a Kronig–Penney model with $a = b = 2$ Å and $U_0 = 3.0$ eV. Calculate the positions (i.e., the starting energy and ending energy) of the lowest allowed band.

2.29 Consider electrons in a Kronig–Penney model with the following parameters

$$a = b = 3 \text{ Å}$$
$$U_0 = 10.0 \text{ eV}$$

Calculate the effective mass of electrons near the start of the first allowed band. You can find this by fitting the E–k results to an equation

$$E = E_1 + \frac{\hbar^2 k^2}{2m^*}$$

where E_1 is the start of the first allowed band.

2.30 Consider the previous problem. Calculate the effective mass in the first allowed band at (i) the bottom of the band, (ii) at the top of the band, and (iii) at the middle of the band.

2.31 Consider electrons in a Kronig–Penney potential with the following parameters

$$a = 5 \text{ Å}$$
$$b = 1 \text{ Å}$$
$$U_0 = 4 \text{ eV}$$

Calculate the positions of the bandedges of the first two allowed energy bands. Also calculate the effective mass of electrons near the start of the second energy band. Redo the problem if a and b decrease by 0.2 Å.

2.32 Consider the following periodic potential

$$a = 5 \text{ Å}; b = 2 \text{ Å}; U_0 = 5 \text{ eV}$$

Calculate the positons of the bandedges $E_1, E_2,$ and E_3. If there is a change in a and b of ± 0.1 Å, how do the positions of E_1, E_2, E_3 change? Consider the $+$ and $-$ signs as two different cases.

2.33 Consider three periodic structures with the following potential profiles:

$$a = 5 \text{ Å}; b = 2 \text{ Å}; U_0 = 4 \text{ eV}$$
$$a = 5 \text{ Å}; b = 5 \text{ Å}; U_0 = 4 \text{ eV}$$
$$a = 5 \text{ Å}; b = 10 \text{ Å}; U_0 = 4 \text{ eV}$$

Calculate the width of the first allowed band for each case and the effective mass of electrons at the bottom of the first band. Also find the width of the first bandgap $(E_3 - E_2)$ and the width of the second allowed band.

2.9 FURTHER READING

- **Review of classical physics**
    - R. Resnick, D. Halliday, and K.S. Krane, *Physics*, J. Wiley, New York (1992).
    - R.A. Serway, *Physics for Scientists and Engineers*, Saunders, Philadelphia (1990).
    - H.D. Young, *University Physics*, Addison-Wesley, Reading, MA (1992).

- **Quantum mechanics**
    - K. Krane, *Modern Physics*, J. Wiley, New York (1996).
    - P.A. Lindsay, *Introduction to Quantum Mechanics for Electrical Engineers*, 3rd edition, McGraw-Hill, New York (1967).

- **Electrons in crystalline materials**
    - J.R. Taylor, C.D. Zafiratos, M.A. Dubson, *Modern Physics for Scientists and Engineers*, Pearson Prentice Hall (2004).
    - C. Kittel, *Introduction to Solid State Physics*, J. Wiley, New York (1986).
    - R. E. Hummel, *Electronic Properties of Materials–An Introduction for Engineers*, Springer Verlag (1985)

Chapter 3

ELECTRONIC LEVELS IN SOLIDS

3.1 INTRODUCTION

In the previous chapter we have seen a number of important outcomes from the application of quantum mechanics to electronic states in solids. The key outcomes are: (i) Depending upon the physical system (i.e., the potential energy profile) certain energy levels are allowed and other energies are forbidden. This means an electron in the system can only occupy the allowed energy levels. (ii) In some systems the allowed energies form a continuous band extending over a range of energies. (iii) In allowed bands near the edge of the bands it is possible to describe electrons by an effective mass and an effective equation of motion that looks similar to Newton's equation. In this chapter we will examine several categories of solids and see how the outcomes listed above impact their electronic properties.

The solution of the Schrödinger equation for a particular system is just the first step in being able to understand and manipulate the behavior of the system. The second step in the problem is to obtain information on the distribution of particles in the allowed states (energy levels). Finally, we use quantum mechanics to understand the response of the particles to an external disturbance. In the next section we will discuss how electrons are distributed in allowed states.

3.2 OCCUPATION OF STATES: DISTRIBUTION FUNCTION

Let us say we have solved the Schrödinger equation for electrons or Maxwell's equation for photons and we have a certain number of particles. How will the particles distribute among the allowed states? To answer this question we need to use statistical physics; in particular, quantum statistical physics. According to quantum mechanics particles

3.2. Occupation of states: distribution function

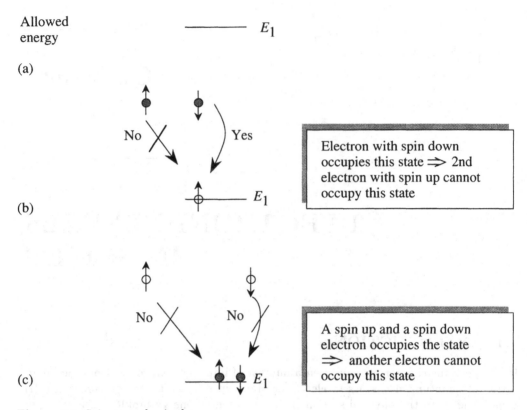

Figure 3.1: Spin up and spin down.

(this term includes classical particles and classical waves which are represented by particles) have an intrinsic angular momentum called spin. The spin of particles is a unique property (such as charge, mass, etc.) and can take a value of $0, 1/2\hbar, \hbar, 3/2\hbar$, etc. Particles which have integral spins (in units of $\hbar$) are called bosons, while those that have half-integral spins are called fermions.

According to quantum mechanics if there is an allowed state (energy level) any number of bosons can occupy that state. However, fermions obey the Pauli exclusion principle, according to which, at the most, one particle can occupy an allowed state. What this means for electrons is this. Let us say we solve the Schrödinger equation and obtain an allowed state E_1. One electron with spin up $(+\hbar/2)$ and one electron with spin down $(-\hbar/2)$ can be placed in this state. The possibilities are shown in Fig. 3.1.

According to thermodynamics, a system with a large number of particles can be described by macroscopic properties such as temperature, pressure, volume, etc. Under equilibrium conditions (no exchange of net energy with other systems) the system is described by a distribution function, which gives us the occupation number for any energy level. To find this occupation we have to minimize the free energy F of the system subject to any constraints from quantum mechanics (such as the Pauli exclusion principle). The following distribution functions are obtained:

- For fermions such as electrons in normal situations (the use of superconducting materials will be discussed later).

$$f(E) = \frac{1}{\exp\left[\frac{E - E_F}{k_B T}\right] + 1} \quad (3.1)$$

Here $f(E)$ is the occupation function; E_F is the Fermi energy (which, as we will discuss later, depends upon particle density).

It is useful to note that in classical physics the occupation function for electrons is

$$f(E) = \frac{1}{\exp\left(\frac{E - E_F}{k_B T}\right)} \quad (3.2)$$

Note that if $E - E_F \gg k_B T$; i.e., $f(E) \ll 1$, the classical function approaches the quantum Fermi distribution function.

- Massless bosons (like photons)

$$f(E) = \frac{1}{\exp\left(\frac{E}{k_B T}\right) - 1} \quad (3.3)$$

- Bosons with mass (this applies to electron pairs that occur in superconductors)

$$f(E) = \frac{1}{\exp\left(\frac{E - \mu}{k_B T}\right) - 1} \quad (3.4)$$

where μ is an energy determined from the particle density.

- There is one other distribution function that proves to be useful in solid state devices. As we have noted earlier, when solving the Schrödinger equation we can get more than one solution with the same energy. This is the degeneracy g_d of a state. Consider a case where a state has a degeneracy g_i and can, in principle, be occupied by g_d electrons. However, when one electron is placed in the allowed state, the next one cannot be placed because of the Coulombic attraction. This happens for some states, such as those states associated with donors or acceptors, traps, etc. Thus, even though Pauli exclusion principle would allow two (or more) electrons to reside on the state, the repulsion would not. In such cases the occupation function can be shown to be

$$f(E) = \frac{1}{\frac{1}{g_d} \exp\left(\frac{E - E_F}{k_B T}\right) + 1} \quad (3.5)$$

In Fig. 3.2 we show a schematic of the Fermi function for electrons and its dependence on temperature. It is important to note the following: (i) At $E - E_F$, $f(E) = 0.5$ regardless of the temperature. (ii) At zero temperature, the Fermi function becomes a step function with $f(E < E_F) = 1.0$ and $f(E) > E_F = 0.0$. (iii) Just because $f(E)$ is non-zero does not mean an electron will be present at that energy. The Schrödinger equation solutions must also allow the energy.

3.2. Occupation of states: distribution function

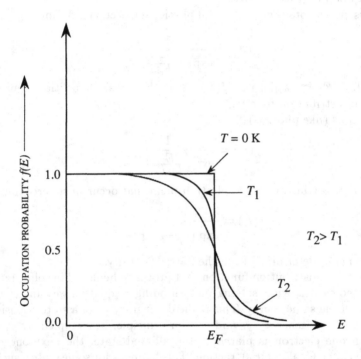

Figure 3.2: Schematic of the Fermi function for electrons and other fermions. In general the position of E_F is dependent on temperature.

3.3 METALS, INSULATORS, AND SUPERCONDUCTORS

In Chapter 2 we discussed how the allowed energy states of electrons in a crystalline material are described by a series of allowed bands separated by forbidden bandgaps. Two important situations arise when we examine the electron occupation of allowed bands. In one case we have a situation where an allowed band is completely filled with electrons, while the next allowed band is separated in energy by a gap E_g and is completely empty at 0 K. In a second case, the highest occupied band is only half full (or partially full). These cases are shown in Fig. 3.3.

It is important to note that, when an allowed band is completely filled with electrons, the electrons in the band cannot conduct any current. This important concept is central to the special properties of metals and insulators. Being fermions the electrons cannot carry any net current in a filled band since an electron can only move into an empty state. We can imagine a net cancellation of the motion of electrons moving one way and those moving the other. Because of this effect, when we have a material in which a band is completely filled, while the next allowed band is separated in energy and empty, the material has, in principle, infinite resistivity and is called an *insulator* or a *semiconductor*. The material in which a band is only half full with electrons has a very low resistivity and is called a *metal*.

The band that is normally filled with electrons at 0 K in semiconductors is called the valence band, while the upper unfilled band is called the conduction band. The energy difference between the vacuum level and the highest occupied electronic state in a metal is called the metal work function. The energy between the vacuum level and the bottom of the conduction band is called the electron affinity. This is shown schematically in Fig. 3.3.

The electrical conductivity of a material is proportional to the density of electrons that can participate in current flow. The metals have a very high conductivity because of the very large number of electrons that can participate in current transport. It is, however, difficult to alter the conductivity of metals in any simple manner as a result of this. In contrast, semiconductors have zero conductivity at 0 K and quite low conductivity at finite temperatures, but it is possible to alter their conductivity by orders of magnitude. This is the key reason why semiconductors can be used for active devices.

3.3.1 Holes in semiconductors

We have defined semiconductors as materials in which the valence band is full of electrons and the conduction band is empty at 0 K. At finite temperatures some of the electrons leave the valence band and occupy the conduction band. The valence band is then left with some unoccupied states. Let us consider the situation as shown in Fig. 3.4, where an electron with momentum $\mathbf{k}_e$ is missing from the valence band.

When all of the valence band states are occupied, the sum of the momentum over all is zero; i.e.

$$\sum \mathbf{k}_i = 0 = \sum_{\mathbf{k}_i \neq \mathbf{k}_e} \mathbf{k}_i + \mathbf{k}_e \qquad (3.6)$$

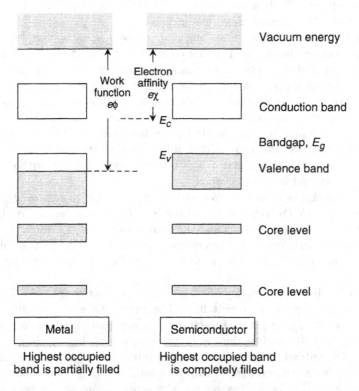

Figure 3.3: A schematic description of electron occupation of the bands in a metal and semiconductor (or insulator). In a metal, the highest occupied band at 0 K is partially filled with electrons. Also shown is the metal work function. In a semiconductor at 0 K, the highest occupied band is completely filled with electrons and the next band is completely empty. The separation between the two bands is the bandgap E_g. The electron affinity and work function are also shown.

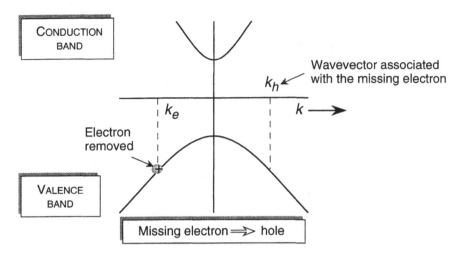

Figure 3.4: Illustration of the wavevector of the missing electron k_e. The wavevector is $-\mathbf{k}_e$, which is associated with the hole.

This result is just an indication that there are as many positive k states occupied as there are negative ones. Now, in the situation where the electron at wavevector $\mathbf{k}_e$ is missing, the total wavevector is

$$\sum_{k_i \neq k_e} \mathbf{k}_i = -\mathbf{k}_e \qquad (3.7)$$

The missing state is called a hole and the wavevector of the system $-\mathbf{k}_e$ is attributed to it. It is important to note that the electron is missing from the state $\mathbf{k}_e$ and the momentum associated with the hole is at $-\mathbf{k}_e$. The position of the hole is depicted as that of the missing electron. But in reality the hole wavevector $\mathbf{k}_h$ is $-\mathbf{k}_e$, as shown in Fig. 3.4 and we have

$$\mathbf{k}_h = -\mathbf{k}'_e \qquad (3.8)$$

If an electric field is applied, all the electrons move in the direction opposite to the electric field. This results in the unoccupied state moving in the field direction. *The hole thus responds as if it has a positive charge.* It therefore responds to external electric and magnetic fields $\mathbf{E}$ and $\mathbf{B}$, respectively, according to the equation of motion

$$\hbar \frac{d\mathbf{k}_h}{dt} = e\left[\mathbf{E} + \mathbf{v}_h \times \mathbf{B}\right] \qquad (3.9)$$

where $\hbar \mathbf{k}_h$ and $\mathbf{v}_h$ are the momentum and velocity of the hole.

Thus the equation of motion of holes is that of particles with a *positive* charge e. The mass of the hole has a positive value, although the electron mass in its valence band is negative. *When we discuss the conduction band properties of semiconductors or insulators we refer to electrons, but when we discuss the valence band properties, we refer to holes. This is because in the valence band only the missing electrons or holes lead to charge transport and current flow.*

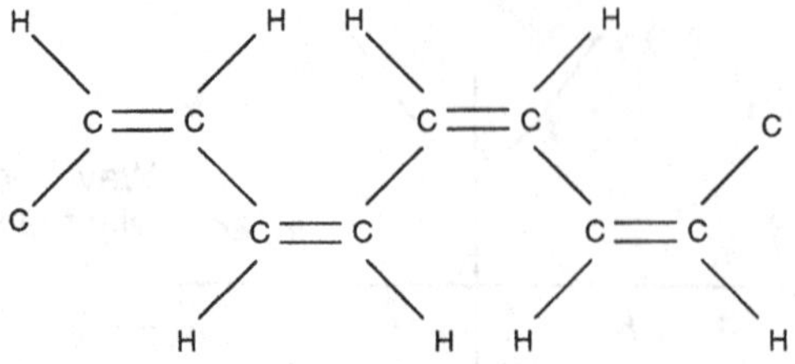

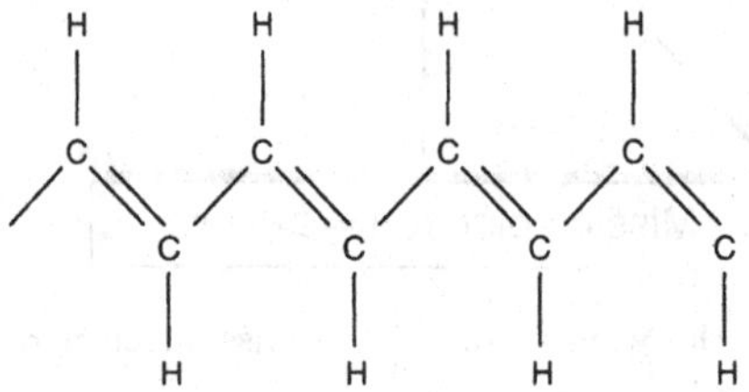

CIS-POLYACETYLENE

TRANS-POLYACETYLENE

Figure 3.5: Schematic of chains forming the polymer polyacetylene. Two forms of this polymer are shown.

3.3.2 Bands in organic and molecular semiconductors

At present, most of the information-processing devices are based on inorganic semiconductor technology. The most commonly used materials are Si and GaAs. However, there are many other material systems that have semiconducting properties and may be very useful if their technologies can be improved. Polymers (popularly known as plastics), which are a very familiar part of everyday life, have the potential of becoming important materials for electronics and optoelectronics. Most of the current uses of polymers rely on their properties, such as chemical inertness and durability. In electronics, polymers are primarily used as insulators. However, new kinds of materials and better understanding is now allowing scientists to develop polymers with properties such as controlled conductivity and light detection and emission. Given the ability of chemists to synthesize large-area polymers, if these properties can be harnessed polymers could become extremely important in future information-processing systems.

Polymers are formed from long chains of molecules. If there is a good fit between the molecules, the materials can crystallize upon drawing or cooling. In Fig. 3.5, we show chains of polyacetylene – an important polymer.

At present, the conductivity of polymers is not as easy to control as that of "traditional" semiconductors (Si, GaAs, etc.). As a result, they are not used for high-performance logic or memory applications. However, in many areas where cost is a primary concern and performance is not as critical, organic semiconductor devices are expected to play an important role.

Organic or molecular semiconductors, while having properties that are similar

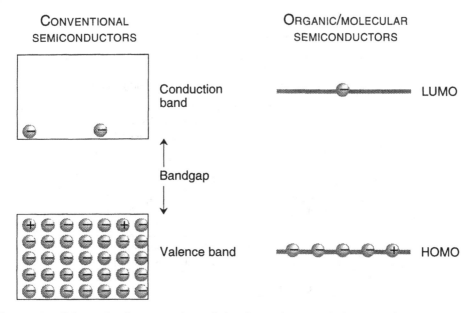

Figure 3.6: Schematic of a comparison of the electronic states in inorganic and organic semiconductors.

to inorganic semiconductors, are not entirely described by the band model for electrons. There are two important differences between inorganic (the traditional) semiconductors and organic semiconductors: (i) The interaction between the atoms within the molecules is usually weak, although intra-molecular interaction is quite strong. (ii) The crystalline quality of the solids is poor, compared to inorganic semiconductors. As a result, the allowed bands in organic semiconductors are very narrow and are better described as discrete levels. In Fig. 3.6 we show a schematic of the energy levels in organic semiconductors. The levels are produced through the bonding and antibonding states as discussed in Chapter 2, Section 2.5 for the coupled well problem. The highest occupied molecular orbital (HOMO) and the lowest unoccupied molecular orbital (LUMO) describes the electronic and optoelectronic properties of organic semiconductors. These levels can be thought of as the valence and conduction bands of the traditional semiconductors. As in traditional semiconductors, mobile carrier density and conductivity of organic semiconductors can be controlled by a bias. Also optical emission and absorption can occur. However, due to the strongly localized nature of the electronic levels in organic semiconductors, there are important quantitative and qualitative differences between these materials and the conventional semiconductors. These differences will be probed in later chapters.

3.3.3 Normal and superconducting states

It is found that certain metals lose their resistance when the temperature is lowered below a critical temperature. Recently, a new class of ceramic materials has been dis-

3.3. Metals, insulators, and superconductors

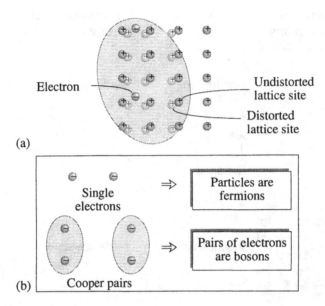

Figure 3.7: (a) A schematic showing how electrons distort the lattice occupied by positive ions to create an attractive potential. A bound pair can be formed in this attractive potential. (b) Single-electron states are antisymmetric under exchange, but if the electrons pair up the overall wavefunction is symmetric under exchange.

covered, which displays a very high critical temperature (and are thus known as high-T_c materials). While there are still aspects of high-T_c materials which are unexplained, quantum theory has been remarkably successful in explaining metallic superconductivity, which is one of the most fascinating and counter intuitive phenomenon.

The basis of the theory known as the *Bardeen–Cooper–Schrieffer* (BCS) theory is that in "normal" metals electrons behave as fermions, while in the superconducting state they behave as bosons. Of course, an individual electron is always a fermion, but they can form "pairs," known as *Cooper pairs*, and in this paired state they can act as if they are bosons. This distinction is of central importance in determining whether or not there is superconductivity.

Normally, electrons, being charged particles, repel each other and do not form bound pairs. However, inside a material the electrons interact with the ions on the crystal lattice and create an interaction between each other. We can physically think of this interaction via the simple picture in Fig. 3.7. An electron interacts with the positively charged background ions, creating a local potential disturbance. Another electron can then be attracted by this disturbance. The binding energy of the two electrons is extremely small (~ 1 meV), and at high temperatures the pairs dissociate and we have the usual unpaired electrons. However, at low temperatures, the electrons can pair up (forming Cooper pairs) and exist in the bound state.

Details of the BCS theory show that the lowest state of the system is one in which Cooper pairs are formed. The lowest state is separated from the next excited state

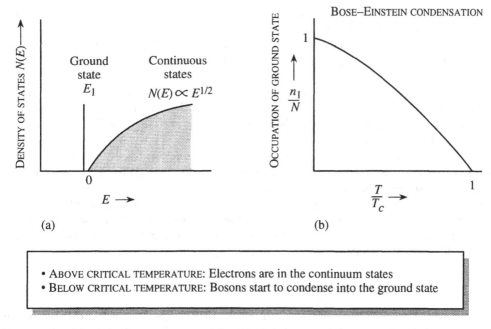

Figure 3.8: (a) A density of states model to describe a system of bosons. (b) The occupation of the lowest (discrete) state as a function of temperature.

by a gap, which is of the order of 1 meV (the gap energy depends upon the nature of the ions), as shown in Fig. 3.8a. In the ground state, since the electron pairs are bosons, a very large number of pairs can occupy the same state. There, the Pauli exclusion principle no longer applies. Thus, to carry current, the electron pairs do not have to move from an occupied state to an unoccupied one. Also, since the excited state (which forms the normal state) is separated by a gap, as long as the temperature is small the electrons do not suffer scattering (which is a source of resistance).

At low temperatures, the occupation of the ground state is high, as shown in Fig. 3.8b. As the temperature increases, the occupation decreases, until, at a critical temperature, T_c, there are no Cooper pairs. Thus superconductivity disappears above the critical temperature.

3.4 BANDSTRUCTURE OF SOME IMPORTANT SEMICONDUCTORS

In Chapter 2 we have discussed a simple model (the Kronig–Penney model) for electronic energies in periodic potentials. We see from that discussion that an energy-effective momentum or E-k diagram can be drawn for the solutions. In a three-dimensional periodic potential k is a three-dimensional vector and the bandstructure (E-k relation) is considerably more complex. To represent the bandstructure on a figure that is two-dimensional, we draw the E-k diagram in several (at least 2) panels where k goes

3.4. Bandstructure of some important semiconductors

from zero to its maximum value along the (100) direction or the (111) direction, etc. The maximum k-point is the one after which, as discussed in Section 2.6.3, the E values repeat. For the fcc lattice, the maximum k-value along the (100) direction is $2\pi/a(1,0,0)$. This point is called the X-point and there are five other equivalent points, due to the cubic symmetry of the lattice. Similarly, along the (111) direction, the maximum k-point is $\pi/a(1,1,1)$ and seven other similar points. This point is called the L-point. Thus we commonly display the E-k diagram with k going from the origin (called the Γ-point) to the X-point and from the origin to the L-point. In this section we will see several such plots for different semiconductors.

We will notice that in the valence band or conduction band there are a number of allowed energy–momentum relations. This is due to the nature of multiple atomic levels in the atoms making up the crystal.

We will now examine special features of some semiconductors. Of particular interest are the bandedge properties, since they dominate the transport and optical properties. In this context, it is important to appreciate the range of energies away from the bandedges which control various physical properties of devices.

3.4.1 Direct and indirect semiconductors: effective mass

The top of the valence band of most semiconductors occurs at $k = 0$; i.e., at effective momentum equal to zero. A typical bandstructure of a semiconductor near the top of the valence band is shown in Fig. 3.9. We notice the presence of three bands near the valence bandedge. These curves or bands are labeled I, II, and III in the figure and are called the heavy hole (HH), light hole (LH), and the split off hole bands.

The bottom of the conduction band in some semiconductors occurs at $k = 0$. Such semiconductors are called direct bandgap materials. Semiconductors, such as GaAs, InP, InGaAs, etc., are direct bandgap semiconductors. In other semiconductors, the bottom of the conduction band does not occur at the $k = 0$ point, but at certain other points. Such semiconductors are called indirect semiconductors. Examples are Si, Ge, AlAs, etc.

An important outcome of the alignment of the bandedges in the valence band and the conduction band is that direct gap materials have a strong interaction with light. Indirect gap materials have a relatively weak interaction with electrons. This is a result of the law of momentum conservation.

When the bandedges are at $k = 0$ it is possible to represent the bandstructure by a simple relation of the form

$$E(\mathbf{k}) = E_c + \frac{\hbar^2 k^2}{2m^*} \tag{3.10}$$

where E_c is the conduction bandedge, and the bandstructure is a simple parabola. The equation for the E-k relation looks very much like that of an electron in free space except that the *free electron mass, m_0, is replaced by a new quantity m^**.

Silicon
Silicon forms the backbone of modern electronics industry. The bandstructure of silicon

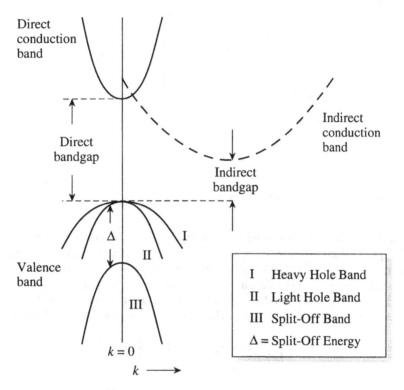

Figure 3.9: Schematic of the valence band, direct bandgap, and indirect bandgap conduction bands. The conduction band of the direct gap semiconductor is shown as the solid line, while the conduction band of the indirect semiconductor is shown as the dashed line. The curves I, II, III in the valence band are called heavy hole, light hole, and split-off hole states, respectively.

3.4. Bandstructure of some important semiconductors

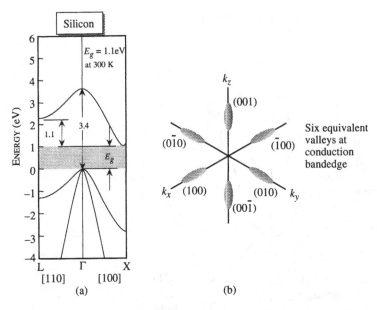

Figure 3.10: (a) Bandstructure of Si. (b) Constant energy ellipsoids for the Si conduction band. There are six equivalent valley in Si at the bandedge.

is shown in Fig. 3.10 and, as can be seen, it has an indirect bandgap. This fact greatly limits the applications of Si in optical devices, particularly for light-emitting devices. The bottom of the conduction band in Si is at point ($\sim (2\pi/a)(0.85, 0.0)$; i.e., close to the X-point. There are six degenerate X-points and, consequently, six conduction bandedge valleys. The near bandedge bandstructure can be represented by ellipsoids of energy with simple E vs. k relations of the form (for examples for the [100] valley)

$$E(\mathbf{k}) = \frac{\hbar^2 k_x^2}{2m_l^*} + \frac{\hbar^2 (k_y^2 + k_z^2)}{2m_t^*} \tag{3.11}$$

where we have two masses, the longitudinal and transverse. The constant energy surfaces of Si are ellipsoids according to Eq. 3.11. The six surfaces are shown in Fig. 3.10.

The longitudinal electron mass m_l^* is approximately $0.98\ m_0$, while the transverse mass is approximately $0.19\ m_0$.

The next valley in the conduction band is the L-point valley, which is about 1.1 eV above the bandedge. Above this is the Γ-point edge. The direct bandgap of Si is ~ 3.4 eV. This direct gap is quite important for optical transitions since, as we shall see later, the absorption coefficient for photons above this energy is very strong. It is important to note that, due to the six-fold degeneracy of the conduction bandedge, the electron transport in Si is quite poor. This is because of the very large density of states near the bandedge, leading to a high scattering rate.

The top of the valence band has the typical features seen in all semiconductor valence bands. One has the HH, LH degeneracy at the zone edge. The split-off (SO)

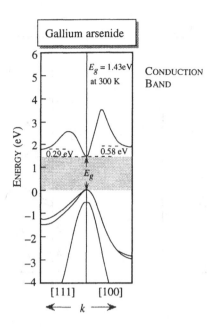

Figure 3.11: Bandstructure of GaAs. The bandgap at 0 K is 1.51 eV and at 300 K it is 1.43 eV. The bottom of the conduction band is at $k = (0,0,0)$, i.e., the Γ-point. The upper conduction band valleys are at the L-point.

band is also very close for Si, since the split-off energy is only 44 meV. This is one of the smallest split off energies of any semiconductors.

GaAs

The near bandedge bandstructure of GaAs is shown in Fig. 3.11. The bandgap is direct, which is the chief attraction of GaAs. The direct bandgap ensures excellent optical properties of GaAs, as well as superior electron transport in the conduction band. The bandstructure can be represented by the relation (referenced to E_c)

$$E = \frac{\hbar^2 k^2}{2m^*} \tag{3.12}$$

with $m^* = 0.067 m_0$. A better relationship is the non-parabolic approximation

$$E(1 + \alpha E) = \frac{\hbar^2 k^2}{2m^*} \tag{3.13}$$

with $\alpha = 0.67$ eV^{-1}.

For high electric field transport, it is important to note that the valleys above Γ-point are the L-valleys. There are eight L-points, but, since half of them are connected by a reciprocal lattice vector, there are four valleys. The separation $\Delta E_{\Gamma L}$ between the Γ- and L- minima is 0.29 eV. The L-valley has a much larger effective mass than the Γ-valley. For GaAs, $m_L^* \sim 0.25 m_0$. This difference in masses is extremely important

3.4. Bandstructure of some important semiconductors

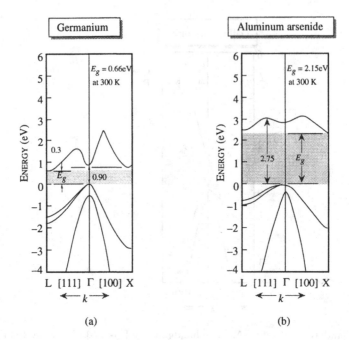

Figure 3.12: (a) Bandstructure of Ge. As for Si, Ge is an indirect semiconductor. The bottom of the conduction band occurs at the L-point. The hole properties of Ge are the best of any semiconductor, with extremely low hole masses. (b) Bandstructure of AlAs. AlAs is an important III–V semiconductor because of its excellent lattice constant, matching GaAs. The material has an indirect bandgap and is usually used in AlGaAs alloy for barrier materials in GaAs/AlGaAs heterostructures.

for high electric field transport and leads to negative differential resistance. Above the L-point in energy is the X-valley with $\Delta E_{\Gamma L} \sim 0.58$ eV. The mass of the electron in the X-valley is also quite large ($m_X^* \sim 0.6 m_0$). At high electric fields, electrons populate both the L- and X- valleys in addition to the Γ-valley, making these regions of bandstructure quite important.

The valence band of GaAs has the standard HH, LH, and SO bands. Due to the large spin–orbit splitting, for most purposes the SO band does not play any role in electronic or optoelectronic properties.

The bandstructures of Ge and AlAs, two other important semiconductors, are shown in Fig. 3.12, along with brief comments about their important properties.

		Experimental bandgap E_G (eV)		
Compound	Type of bandgap	0 K	300 K	Temperature dependence of bandgap $E_G(T)$ (eV)
AlP	Indirect	2.52	2.45	$2.52 - 3.18 \times 10^{-4} T^2/(T+588)$
AlAs	Indirect	2.239	2.163	$2.239 - 6.0 \times 10^{-4} T^2/(T+408)$
AlSb	Indirect	1.687	1.58	$1.687 - 4.97 \times 10^{-4} T^2/(T+213)$
GaP	Indirect	2.338	2.261	$2.338 - 5.771 \times 10^{-4} T^2/(T+372)$
GaAs	Direct	1.519	1.424	$1.519 - 5.405 \times 10^{-4} T^2/(T+204)$
GaSb	Direct	0.810	0.726	$0.810 - 3.78 \times 10^{-4} T^2/(T+94)$
InP	Direct	1.421	1.351	$1.421 - 3.63 \times 10^{-4} T^2/(T+162)$
InAs	Direct	0.420	0.360	$0.420 - 2.50 \times 10^{-4} T^2/(T+75)$
InSb	Direct	0.236	0.172	$0.236 - 2.99 \times 10^{-4} T^2/(T+140)$

Table 3.1: Bandgaps of binary III–V compounds (From Casey and Panish, 1978).

InN, GaN, and AlN
The III–V nitride family of GaN, InN, and AlN have become quite important due to progress in the ability to grow the semiconductor. The nitrides and their combinations, which have a wurtzite structure, can provide bandgaps ranging from ~1.0 eV to over 6.0 eV. This large range is very useful for short wavelength light emitters (for blue light emission and for high-resolution reading/writing applications in optoelectronics) and high power electronics.

It should be noted that it is difficult to obtain the bandgap of InN, since it is difficult to grow thick defect free layers due to substrate non-availability. Recent results have shown a bandgap closer to 0.9 eV.

Also important to note is that the bandgap of semiconductors generally decreases as temperature increases. The bandgap of GaAs, for example, is 1.51 eV at $T = 0K$ and 1.43 eV at room temperature. These changes have very important consequences for both electronic and optoelectronic devices. The temperature variation alters the laser frequency in solid state lasers, and alters the response of modulators and detectors. It also has effects on intrinsic carrier concentration in semiconductors. In Table 3.1 we show the temperature dependence of bandgaps of several semiconductors.

3.5 MOBILE CARRIERS

From our brief discussion of metals and semiconductors in Section 3.3, we see that in a metal current flows because of the electrons present in the highest (partially) filled band. This is shown schematically in Fig. 3.13a. The density of such electrons is very high ($\sim 10^{23}$ cm^{-3}). In a semiconductor, in contrast, no current flows if the valence band is filled with electrons and the conduction band is empty of electrons. However, if somehow empty states or holes are created in the valence band by removing electrons,

3.5. Mobile carriers

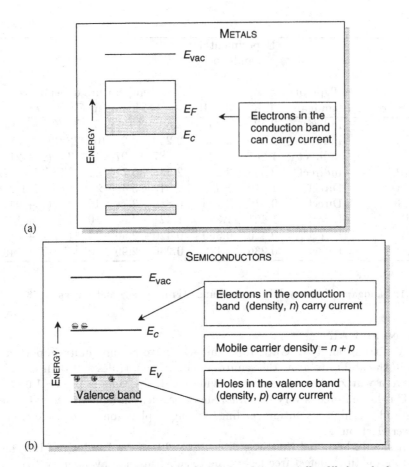

Figure 3.13: (a) In metals the highest occupied band is partially filled and electrons can carry current. (b) A schematic showing the valence band and conduction band in a typical semiconductor. In semiconductors only electrons in the conduction band and holes in the valence band can carry current.

current can flow through the holes. Similarly, if electrons are placed in the conduction band, these electrons can carry current. This is shown schematically in Fig. 3.13b. If the density of electrons in the conduction band is n and that of holes in the valence band is p, the total mobile carrier density is $n + p$.

In the next section we will calculate the density of mobile carriers in a metal and in a pure semiconductor.

3.5.1 Electrons in metals

In a metal, we have a series of filled bands and a partially filled band called the conduction band. The filled bands are inert as far as electrical and optical properties of metals are concerned. The conduction band of metals can be assumed to be described by the parabolic energy–momentum relation

$$E(k) = E_c + \frac{\hbar^2 k^2}{2m_0} \tag{3.14}$$

Note that we have used an effective mass equal to the free electrons mass. This is a reasonable approximation for metals. The large electron density in the band "screens" out the background potential and the electron effective mass is quite close to the free space value.

The electron density in the conduction band of a metal is related to the Fermi level by the relation

$$n = \int_{E_c}^{\infty} \frac{\sqrt{2}m_0^{3/2}}{\pi^2 \hbar^3} \frac{E^{1/2} dE}{\exp\left(\frac{E - E_F}{k_B T}\right) + 1} \tag{3.15}$$

This integral is particularly simple to evaluate as 0 K, since, at this temperature

$$\frac{1}{\exp\left(\frac{E - E_F}{k_B T}\right) + 1} = 1 \text{ if } E \leq E_F$$

$$= 0 \text{ otherwise}$$

Thus (choosing the conduction bandedge as the origin)

$$n = \int_0^{E_F} N(E) dE$$

We then have

$$n = \frac{\sqrt{2}m_0^{3/2}}{\pi^2 \hbar^3} \int_0^{E_F} E^{1/2} dE$$

$$= \frac{2\sqrt{2}m_0^{3/2}}{3\pi^2 \hbar^3} E_F^{3/2}$$

or

$$E_F = \frac{\hbar^2}{2m_0} \left(3\pi^2 n\right)^{2/3} \tag{3.16}$$

The expression is applicable to metals such as copper, gold, etc. In Table 3.2 we show the conduction band electron densities for several metals. The quantity E_F, which is the highest occupied energy state at 0 K, is called the Fermi energy. We can define a corresponding wavevector k_F, called the Fermi vector, and a velocity v_F, called the Fermi velocity as

$$k_F = (3\pi^2 n)^{1/3}$$

$$v_F = \left(\frac{\hbar}{m_0}\right)(3\pi^2 n)^{1/3} \tag{3.17}$$

3.5. Mobile carriers

Element	Valence	Density (gm/cm^3)	Conduction electron density (10^{22} cm^{-3})
Al	3	2.7	18.1
Ag	1	10.5	5.86
Au	1	19.3	5.90
Na	1	0.97	2.65
Fe	2	7.86	17.0
Zn	2	7.14	13.2
Mg	2	1.74	8.61
Ca	2	1.54	4.61
Cu	1	8.96	8.47
Cs	1	1.9	0.91
Sn	4	7.3	14.8

Table 3.2: Properties of some metals. In the case of elements that display several values of chemical valence, one of the values has been chosen arbitrarily.

It is important to note that even at 0 K, the velocity of the highest occupied state is v_F and not zero, as would be the case if we used classical statistics.

At finite temperatures, the Fermi level is approximately given by

$$E_F(T) = E_F(0)\left[1 - \frac{\pi^2}{12}\frac{(k_B T)^2}{(E_F(0))^2}\right] \tag{3.18}$$

where $E_F(T)$ and $E_F(0)$ are the Fermi levels at temperatures T and 0 K, respectively.

EXAMPLE 3.1 A particular metal has 10^{22} electrons per cubic centimeter. Calculate the Fermi energy and the Fermi velocity (at 0 K).

The Fermi energy is the highest occupied energy state at 0 K and is given by (measured from the conduction bandedge)

$$\begin{aligned}E_F &= \frac{\hbar^2}{2m_0}\left(3\pi^2 n\right)^{2/3}\\ &= \frac{(1.05\times 10^{-34})^2[3\pi^2(10^{28})]^{2/3}}{2(0.91\times 10^{-30})} = 2.75\times 10^{-19}\text{ J}\\ &= 1.72\text{ eV}\end{aligned}$$

The Fermi velocity is

$$\begin{aligned}v_F &= \frac{\hbar}{m_0}\left(3\pi^2 n\right)^{1/3}\\ &= \frac{(1.05\times 10^{-34}\text{ J.s})(3\pi^2\times 10^{28}\text{ m}^{-3})^{1/3}}{0.91\times 10^{-30}\text{ kg}} = 7.52\times 10^5\text{ m/s}\\ &= 7.52\times 10^7\text{ cm/s}\end{aligned}$$

Thus, the highest energy electron has a large energy and is moving with a very large speed. In a classical system the electron energy would be $\sim \frac{3}{2}k_B T$, which would be zero at 0 K. The electron velocity will also be zero at 0 K in classical physics.

3.5.2 Mobile carriers in pure semiconductors

In semiconductors, as discussed earlier, there are no mobile carriers at zero temperature. As temperature is raised, electrons from the valence band are thermally excited into the conduction band, and in equilibrium there is an electron density n and an equal hole density p, as shown in Fig. 3.14a. To calculate the electron and hole densities in a pure semiconductor (i.e., no defects are present) we first recall some important expressions for the density of states. The density of states has the form

$$N(E) = \frac{\sqrt{2}\,(m_{dos}^*)^{3/2}(E - E_c)^{1/2}}{\pi^2 \hbar^3} \tag{3.19}$$

where m_{dos}^* is the density of states mass and E_c is the conduction bandedge. A similar expression exists for the valence band except the energy term is replaced by $(E_v - E)^{1/2}$ and the density of states exist below the valence bandedge E_v. In Fig. 3.14 we show a schematic view of the density of states.

3.5. Mobile carriers

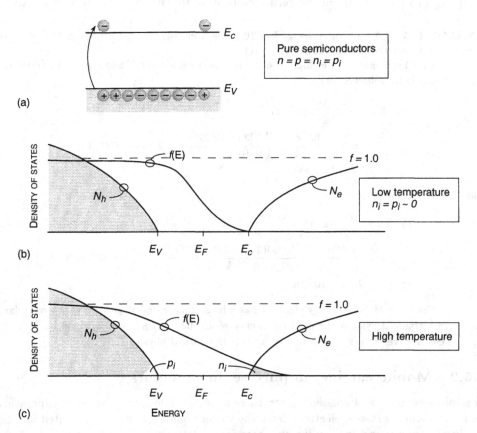

Figure 3.14: (a) A schematic showing that electron and hole densities are equal in a pure semiconductor. (b) Density of states and Fermi occupation function at low temperatures. (c) Density of states and Fermi function at high temperatures when n_i and p_i become large.

In direct gap semiconductors m^*_{dos} is just the effective mass for the conduction band. In indirect gap materials it is given by

$$m^*_{dos} = (m^*_1 m^*_2 m^*_3)^{1/3}$$

where $m^*_1 m^*_2 m^*_3$ are the effective masses along the three principle axes. For Si counting the six degenerate X-valleys we have

$$m^*_{dos} = 6^{2/3} \left(m_\ell m_t^2\right)^{1/3}$$

For the valence band we can write a simple expression for a density of states masses, which includes the HH and LH bands

$$m^*_{dos} = \left(m_{hh}^{*3/2} + m_{\ell h}^{*3/2}\right)^{2/3}$$

In pure semiconductors, electrons in the conduction come from the valence band and $n = p = n_i = p_i$, where n_i and p_i are the intrinsic carrier concentrations. In general the electron density in the conduction band is

$$n = \int_{E_c}^{\infty} N_e(E) f(E) dE$$

$$n = \frac{1}{2\pi^2} \left(\frac{2m_e^*}{\hbar^2}\right)^{3/2} \int_{E_c}^{\infty} \frac{(E-E_c)^{1/2} dE}{\exp\left(\frac{E-E_F}{k_B T}\right) + 1} \quad (3.20)$$

In Fig. 3.14b we show how a change of temperature alters the shape of the Fermi function and alters the electron and hole densities. For small values of n (non-degenerate statistics where we can ignore the unity in the Fermi function) we get

$$n = N_c \exp\left[(E_F - E_c)/k_B T\right] \quad (3.21)$$

where the effective density of states N_c is given by

$$N_c = 2 \left(\frac{m_e^* k_B T}{2\pi \hbar^2}\right)^{3/2}$$

A similar derivation for hole density gives

$$p = N_v \exp\left[(E_v - E_F)/k_B T\right] \quad (3.22)$$

where the effective density of states N_v is given by

$$N_v = 2 \left(\frac{m_h^* k_B T}{2\pi \hbar^2}\right)^{3/2}$$

We also obtain

$$np = 4 \left(\frac{k_B T}{2\pi \hbar^2}\right)^3 (m_e^* m_h^*)^{3/2} \exp\left(-E_g/k_B T\right) \quad (3.23)$$

3.5. Mobile carriers

MATERIAL	CONDUCTION BAND EFFECTIVE DENSITY (N_c)	VALENCE BAND EFFECTIVE DENSITY (N_v)	INTRINSIC CARRIER CONCENTRATION ($n_i = p_i$)
Si (300 K)	2.78×10^{19} cm^{-3}	9.84×10^{18} cm^{-3}	1.5×10^{10} cm^{-3}
Ge (300 K)	1.04×10^{19} cm^{-3}	6.0×10^{18} cm^{-3}	2.33×10^{13} cm^{-3}
GaAs (300 K)	4.45×10^{17} cm^{-3}	7.72×10^{18} cm^{-3}	1.84×10^{6} cm^{-3}

Table 3.3: Effective densities and intrinsic carrier concentrations of Si, Ge, and GaAs. The numbers for intrinsic carrier densities are the accepted values even though they are smaller than the values obtained by using the equations derived in the text.

We note that the product np is independent of the position of the Fermi level and is dependent only on the temperature and intrinsic properties of the semiconductor. This observation is called the law of mass action. If n increases, p must decrease, and vice versa. For the intrinsic case $n = n_i = p = p_i$, we have from the square root of the equation above

$$n_i = p_i = 2 \left(\frac{k_B T}{2\pi \hbar^2} \right)^{3/2} (m_e^* m_h^*)^{3/4} \exp(-E_g/2k_B T)$$

$$E_{Fi} = \frac{E_c + E_v}{2} + \frac{3}{4} k_B T \ln(m_h^*/m_e^*) \quad (3.24)$$

Thus the Fermi level of an intrinsic material lies close to the midgap. Note that in calculating the density of states masses m_h^* and m_e^*, the number of valleys and the sum of heavy and light hole states have to be included.

In Table 3.3 we show the effective densities and intrinsic carrier concentrations in Si, Ge, and GaAs. The values given are those accepted from experiments. These values are lower than the ones we get by using the equations derived in this section. The reason for this difference is due to inaccuracies in carrier masses and the approximate nature of the analytical expressions.

We note that the carrier concentration increases exponentially as the bandgap decreases. Results for the intrinsic carrier concentrations for Si, Ge, and GaAs are shown in Fig. 3.15. The strong temperature dependence and bandgap dependence of intrinsic carrier concentration can be seen from this figure. In electronic devices where current has to be modulated by some means, the concentration of intrinsic carriers is fixed by the temperature and therefore is detrimental to device performance. Once the intrinsic carrier concentration increases to $\sim 10^{15}$ cm^{-3}, the material becomes unsuitable for electronic devices, due to the high leakage current arising from the intrinsic carriers. A growing interest in high-bandgap semiconductors, such as diamond (C), SiC, etc., is partly due to the potential applications of these materials for high-temperature devices where, due to their larger gap, the intrinsic carrier concentration remains low up to very high temperatures.

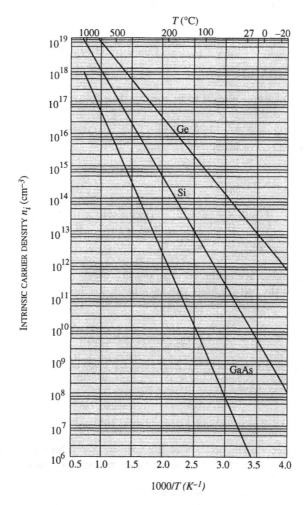

Figure 3.15: Intrinsic carrier densities of Ge, Si, and GaAs as a function of reciprocal temperature.

3.5. Mobile carriers

EXAMPLE 3.2 Calculate the effective density of states for the conduction and valence bands of GaAs and Si at 300 K. Let us start with the GaAs conduction-band case. The effective density of states is

$$N_c = 2 \left(\frac{m_e^* k_B T}{2\pi \hbar^2} \right)^{3/2}$$

Note that at 300 K, $k_B T = 26$ meV $= 4 \times 10^{-21}$ J.

$$N_c = 2 \left(\frac{0.067 \times 0.91 \times 10^{-30} \text{ (kg)} \times 4.16 \times 10^{-21} \text{ (J)}}{2 \times 3.1416 \times (1.05 \times 10^{-34} \text{ (Js)})^2} \right)^{3/2} \text{ m}^{-3}$$

$$= 4.45 \times 10^{23} \text{ m}^{-3} = 4.45 \times 10^{17} \text{ cm}^{-3}$$

In silicon, the density of states mass is to be used in the effective density of states. This is given by

$$m_{dos}^* = 6^{2/3} (0.98 \times 0.19 \times 0.19)^{1/3} \ m_0 = 1.08 \ m_0$$

The effective density of states becomes

$$N_c = 2 \left(\frac{m_{dos}^* k_B T}{2\pi \hbar^2} \right)^{3/2}$$

$$= 2 \left(\frac{1.06 \times 0.91 \times 10^{-30} \text{ (kg)} \times 4.16 \times 10^{-21} \text{ (J)}}{2 \times 3.1416 \times (1.05 \times 10^{-34} \text{ (Js)})^2} \right)^{3/2} \text{ m}^{-3}$$

$$= 2.78 \times 10^{25} \text{ m}^{-3} = 2.78 \times 10^{19} \text{ cm}^{-3}$$

We can see the large difference in the effective density between Si and GaAs.

In the case of the valence band, we have the heavy hole and light hole bands, both of which contribute to the effective density. The effective density is

$$N_v = 2 \left(m_{hh}^{3/2} + m_{\ell h}^{3/2} \right) \left(\frac{k_B T}{2\pi \hbar^2} \right)^{3/2}$$

For GaAs we use $m_{hh} = 0.45 m_0$, $m_{\ell h} = 0.08 m_0$ and for Si we use $m_{hh} = 0.5 m_0$, $m_{\ell h} = 0.15 m_0$, to get

$$N_v(\text{GaAs}) = 7.72 \times 10^{18} \text{ cm}^{-3}$$
$$N_v(\text{Si}) = 9.84 \times 10^{18} \text{ cm}^{-3}$$

EXAMPLE 3.3 Calculate the position of the intrinsic Fermi level in Si at 300 K.

The density of states effective mass of the combined six valleys of silicon is

$$m_{dos}^* = (6)^{2/3} \left(m_\ell^* \ m_t^{*2} \right)^{1/3} = 1.08 \ m_0$$

The density of states mass for the valence band is 0.55 m_0. The intrinsic Fermi level is given by (referring to the valence bandedge energy as zero)

$$E_{Fi} = \frac{E_g}{2} + \frac{3}{4} k_B T \ln \left(\frac{m_h^*}{m_e^*} \right) = \frac{E_g}{2} + \frac{3}{4} (0.026) \ln \left(\frac{0.55}{1.08} \right)$$

$$= \frac{E_g}{2} - (0.0132 \text{ eV})$$

The Fermi level is then 13.2 meV below the center of the mid-bandgap.

EXAMPLE 3.4 Calculate the intrinsic carrier concentration in InAs at 300 K and 600 K.

The bandgap of InAs is 0.35 eV and the electron mass is $0.027 m_0$. The hole density of states mass is $0.4 m_0$. The intrinsic concentration at 300 K is

$$n_i = p_i = 2\left(\frac{k_B T}{2\pi \hbar^2}\right)^{3/2} (m_e^* m_h^*)^{3/4} \exp\left(\frac{-E_g}{2k_B T}\right)$$

$$= 2\left(\frac{(0.026)(1.6 \times 10^{-19})}{2 \times 3.1416 \times (1.05 \times 10^{-34})^2}\right)^{3/2}$$

$$\left(0.027 \times 0.4 \times (0.91 \times 10^{-30})^2\right)^{3/4} \exp\left(-\frac{0.35}{0.052}\right)$$

$$= 1.025 \times 10^{21} \text{ m}^{-3} = 1.025 \times 10^{15} \text{cm}^{-3}$$

The concentration at 600 K becomes

$$n_i(600 \text{ K}) = 2.89 \times 10^{15} \text{cm}^{-3}$$

3.6 DOPING OF SEMICONDUCTORS

The intrinsic carrier density discussed in the previous section is usually undesirable in semiconductor devices. It leads to noise, leakage current, and limits the high-temperature operation of electronic and optoelectronic devices. Semiconductor devices operate at temperatures where the intrinsic carrier density is small ($\lesssim 10^{14}$ cm^{-3}). To introduce electrons (holes) in a semiconductor the material is doped with dopants. The electrons (holes) created by the dopants are used in device design.

There are two kinds of dopants–donors which can donate an electron to the conduction band and acceptors which can accept an electron from the valence band and thus create a hole. To solve the donor (or acceptor) problem, we consider a donor atom on a crystal lattice site. The donor atom could be a pentavalent atom in Si or a Si atom on a Ga site in GaAs. Focusing on the pentavalent atom in Si, four of the valence electrons of the donor atom behave as they would in a Si atom; the remaining fifth electron now sees a positively charged ion to which it is attracted, as shown in Fig. 3.16. The ion has a charge of unity and the attraction is simply Coulombic suppressed by the dielectric constant of the material. The problem is now that of the hydrogen atom case, except that the electron mass is the effective mass at the bandedge. The attractive potential is

$$U(r) = \frac{-e^2}{4\pi \epsilon r} \tag{3.25}$$

where ϵ is the dielectric constant of the semiconductor; i.e., the product of ϵ_0 and the relative dielectric constant. In this simplification the properties of the dopant atom can be described by a simple hydrogen-like model, where the *electron mass is simply the effective mass at the bandedge*. This approximation is called the effective mass approximation for impurities.

We have seen that electrons in the crystal can be represented by an effective mass near the bandedge. We get the effective mass equation for the donor level which

3.6. Doping of semiconductors

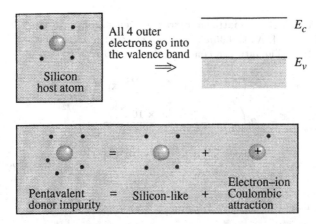

Figure 3.16: A schematic showing the approach we can take to understand donors in semiconductors. The donor problem is treated as the host atom problem, together with a Coulombic interaction term. The silicon atom has four "free" electrons per atom. All four electrons occupy the valence band at 0 K. The dopant has five electrons out of which four contribute to the valence band, while the fifth one can be used for increasing electrons in the conduction band.

has an energy for E_d of

$$\left[\frac{-\hbar^2}{2m_e^*}\nabla^2 - \frac{e^2}{4\pi\epsilon r}\right] F_c(r) = (E_d - E_c)F_c(r) \quad (3.26)$$

where m_e^* is the conduction bandedge mass and $E_d - E_c$ is the impurity energy with respect to the conduction bandedge E_c levels.

This equation is now essentially the same as that of an electron in the hydrogen atom problem. The only difference is that the electron mass is m^* and the Coulombic potential is reduced by ϵ_0/ϵ.

The energy solutions for this problem are

$$E_d = E_c - \frac{e^4 m_e^*}{2(4\pi\epsilon)^2 \hbar^2} \frac{1}{n^2}, \quad n = 1, 2, \ldots \quad (3.27)$$

A series of energy levels are produced, with the ground state energy level being at

$$\begin{aligned} E_d &= E_c - \frac{e^4 m_e^*}{2(4\pi\epsilon)^2 \hbar^2} \\ &= E_c - 13.6 \left(\frac{m^*}{m_o}\right)\left(\frac{\epsilon_o}{\epsilon}\right)^2 \text{ eV} \end{aligned} \quad (3.28)$$

Note that in the hydrogen atom problem the electron levels are measured from the vacuum energy level which is taken as $E = 0$. In the donor problem, the energy level is measured from the bandedge. Fig. 3.17 shows the energy level associated with a donor impurity.

The wavefunction of the ground state is as in the hydrogen atom problem

$$F_c(r) = \frac{1}{\sqrt{\pi a^3}} e^{-r/a} \tag{3.29}$$

where a is the donor Bohr radius and is given by

$$a = \frac{(4\pi\epsilon)\hbar^2}{m_e^* e^2} = 0.53 \left(\frac{\epsilon/\epsilon_0}{m_e^*/m_0} \right) \text{ Å} \tag{3.30}$$

For most semiconductors the donor energies are a few meVs below the conduction bandedge and the Bohr radius is ~ 100 Å.

Another important class of intentional impurities is the acceptors. Just as donors are defect levels, which are neutral when an electron occupies the defect level and positively charged when unoccupied, the acceptors are neutral when empty and negatively charged when occupied by an electron. The acceptor levels are produced when impurities, which have a similar core potential as the atoms in the host lattice, but have one less electron in the outermost shell, are introduced into the crystal. Thus group III elements can form acceptors in Si or Ge, while Si could be an acceptor if it replaces As in GaAs.

As shown in Fig. 3.18 the acceptor impurity potential could now be considered to be equivalent to a host atom potential, together with the Coulombic potential of a negatively charged particle. The "hole" (i.e., the absence of an electron in the valence band) can then bind to the acceptor potential. The effective mass equation can again be used, since only the top of the valence band contributes to the acceptor level. The valence band problem is considerably more complex and requires the solution of multiband effective mass theory. However, the acceptor level can be reasonably predicted by using the heavy hole mass. Due to the heavier hole masses, the Bohr radius for the acceptor levels is usually a factor of 2 to 3 smaller than that for donors.

Population of dopant levels

We have discussed above how the presence of a dopant impurity creates a bound level E_d (or E_a) near the conduction (or valence) bandedge. If the extra electron associated with the donor occupies the donor level, it does not contribute to the mobile carrier density. The purpose of doping is to create a mobile electron or hole. When the electron associated with a donor (or a hole associated with an acceptor) is in the conduction (or valence) band, the dopant is said to be ionized. To calculate the ionization probability, we refer back to our discussion of the occupation probability of discrete levels.

Consider a semiconductor containing both donors and acceptors. At finite temperatures, the electrons will be redistributed, but their numbers will be conserved and will satisfy the following equality resulting from charge neutrality

$$(n - n_i) + n_d = N_d \tag{3.31}$$
$$(p - p_i) + p_a = N_a \tag{3.32}$$

which gives

$$n + n_d = N_d - N_a + p + p_a \tag{3.33}$$

3.6. Doping of semiconductors

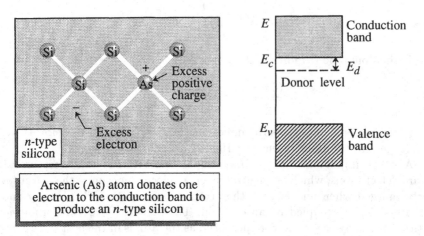

Figure 3.17: A schematic of doping of Si with arsenic. A donor level is produced below the conduction bandedge.

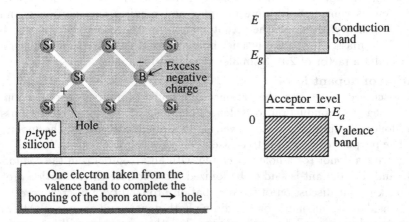

Figure 3.18: Boron has only three valence electrons. It can complete its tetrahedral bonds only by taking an electron from an Si–Si bond, leaving behind a hole in the silicon valence band. The positive hole is then available for conduction. The boron atom is called an acceptor because when ionized it accepts an electron from the valence band.

where

$$n = \text{total free electrons in the conduction band}$$
$$n_d = \text{electrons bound to the donors}$$
$$p = \text{total free holes in the valence band}$$
$$p_a = \text{holes bound to the acceptors}$$

The number density of electrons attached to the donors has been derived in Section 3.6 and is given by

$$\frac{n_d}{N_d} = \frac{1}{\frac{1}{2}\exp\left(\frac{E_d - E_F}{k_B T}\right) + 1} \qquad (3.34)$$

The factor $\frac{1}{2}$ essentially arises from the fact that there are two states an electron can occupy at a donor site corresponding to the two spin-states.

The probability of a hole being trapped to an acceptor level is given by

$$\frac{p_a}{N_a} = \frac{1}{\frac{1}{4}\exp\left(\frac{E_F - E_a}{k_B T}\right) + 1} \qquad (3.35)$$

The factor of $\frac{1}{4}$ comes about because of the presence of the two bands, light hole, heavy hole, and the two spin-states.

To find the fraction of donors or acceptors that are ionized, we have to use a computer program in which the position of the Fermi level is adjusted so that the charge neutrality condition given Eq. 3.33 is satisfied. Once E_F is known, we can calculate the electron or hole densities in the conduction and valence bands. For doped systems, it is useful to use the Joyce–Dixon approximation, which gives the relation between the Fermi level and the free carrier concentration. This approximation is more accurate than the Boltzmann approximation. According to the Joyce–Dixon approximation, we have

$$E_F = E_c + k_B T \left[\ln\frac{n}{N_c} + \frac{1}{\sqrt{8}}\frac{n}{N_c}\right] = E_v - k_B T \left[\ln\frac{p}{N_v} + \frac{1}{\sqrt{8}}\frac{p}{N_v}\right] \qquad (3.36)$$

This relation can be used to obtain the Fermi level if n is specified. Or else it can be used to obtain n if E_F is known by solving for n iteratively. *If the term $(n/\sqrt{8}\, N_c)$ is ignored, the result corresponds to the Boltzmann approximation.*

In general, if we have a doped semiconductor and we examine its mobile carrier density dependence upon temperature, there are three regimes. Let us consider an n-type semiconductor. At low temperatures, the electrons coming from the donors are attached to the donors and occupy the impurity levels E_d. Thus there is no contribution to the mobile carrier density from the dopants. This regime is called the *carrier freeze-out* regime. At higher temperatures, the dopants ionize until most of them are ionized. Then over a temperature regime, the mobile carrier is essentially equal to the dopant density and independent of temperature. This is the saturation regime and semiconductor devices are operated in this regime. At very high temperatures, the intrinsic carrier density overwhelms the dopant density and the material acts as an intrinsic material. The three regimes are shown in Fig. 3.19.

3.7. Tailoring electronic properties

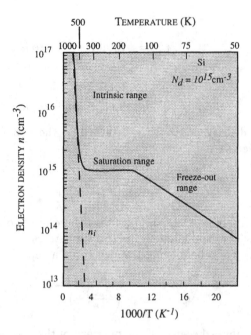

Figure 3.19: Electron density as a function of temperature for a Si sample with donor impurity concentration of 10^{15} cm^{-3}.

3.7 TAILORING ELECTRONIC PROPERTIES

Quite often we encounter applications where we need materials with bandgaps or other electronic properties which are not available in single crystal materials. It is possible to tailor electronic properties by using alloys and quantum wells.

3.7.1 Electronic properties of alloys

Alloys are made from combinations of two or more materials and can be exploited to create new bandgaps or lattice constants. In Section 1.2.2 (see Eq. 1.10) we have discussed how the lattice constant of alloys changes with composition. To the first order the electronic properties are also given by a similar relation. Consider an alloy $A_x B_{1-x}$ made from materials A with bandstructure given by $E_A(k)$ and B with bandstructure given by $E_B(k)$. The bandstructure of the alloy is then given by

$$E_{all}(k) = x E_A(k) + (1 - x) E_B(k) \qquad (3.37)$$

Note that the energy averaging is done at the same k value. If we make an alloy from a direct and an indirect material, one does not simply average the bandgaps to get the alloy bandgap. Instead the bandgaps at the same k values are averaged and the bandgap is then given by the lowest energy difference between the conduction and valence energies.

Based on the equation above the effective mass of the alloy is to be averaged as

$$\frac{1}{m^*_{all}} = \frac{x}{m^*_A} + \frac{(1-x)}{m^*_B} \qquad (3.38)$$

It is important to note that alloys have inherent disorder since they have random arrangements of atoms. This leads to disorder related scattering discussed in the next chapter.

3.7.2 Electronic properties of quantum wells

Quantum wells offer a very useful approach to bandstructure tailoring. In Chapter 2, Section 2.4 we have discussed electronic properties in quantum wells. In quantum wells electrons behave as if they are in a 2-dimensional space and acquire properties that are especially useful for many electronic and optoelectronic applications.

Using epitaxial crystal growth techniques it is possible to grow atomically abrupt semiconductor heterostructures. When two semiconductors with different bandgaps (and chemical compositions) form an interface, an important question that arises is: How does the conduction band (valence band) on one material line up with the other materials bands? This information is usually obtained through experiments. There are three possible scenarios as shown in Fig. 3.20. In type I structures the layer bandgap material "surrounds" the bandgap of the small gap material. In quantum wells made from such materials, both electrons and holes are confined in the same physical quantum well. Most optoelectronic devices (laser diodes, modulators, detectors, etc.,) are based on type I lineup. In type II lineup the conduction band of material A is below that of the material B, but the valence band of A is above that of B as shown. In quantum wells made from such materials the electrons and holes are confined in spatially different quantum wells. These structures are useful for applications in the long wavelength regime, since their "effective" bandgap can be very small. Finally, in type III heterostructures, both the conduction and valence band edges of material A are above the conduction band edge of material B.

In Fig. 3.21 we show a schematic of a type I quantum well made from a smaller bandgap material B sandwiched between a large bandgap material A. To understand the electronic properties of the quantum well we use the effective mass approach and the discussion of Section 2.4.

The electronic structure in a quantum well has been discussed in Section 2.4.2. For completeness, we will repeat the equations here noting that the electron mass is given by the effective mass, rather than free electron mass. The reader should review Section 2.4.2.

The confinement of electrons and holes by quantum wells alters the electronic properties of the system. This has important consequences for optical properties and optoelectronic devices. The Schrödinger equation for the electron states in the quantum well can be written in a simple approximation as

$$\left[-\frac{\hbar^2}{2m^*}\nabla^2 + V(z) \right] \Psi = E\Psi$$

3.7. Tailoring electronic properties

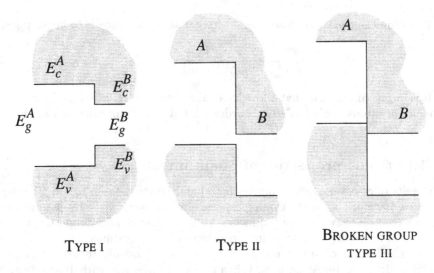

Figure 3.20: Various possible bandedge lineups in heterostructures.

where m^* is the effective mass of the electron. The wavefunction Ψ can be separated into its z and ρ (in the x-y plane) dependence and the problem is much simplified

$$\Psi(s,y,z) = e^{ik_x \cdot x} \cdot e^{ik_y \cdot y} f(z)$$

where $f(z)$ satisfies

$$\left[-\frac{\hbar^2}{2m^*}\frac{\partial^2}{\partial z^2} + V(z)\right] f(z) = E_n f(z) \qquad (3.39)$$

Assuming an infinite barrier approximation, the values of $f(z)$ are (W is the well size)

$$\begin{aligned} f(z) &= \cos\frac{\pi n z}{W}, \text{ if } n \text{ is even} \\ &= \sin\frac{\pi n z}{W}, \text{ if } n \text{ is odd} \end{aligned} \qquad (3.40)$$

with energies

$$E_n = \frac{\pi^2 \hbar^2 n^2}{2m^* W^2} \qquad (3.41)$$

The energy of the electron bands are then

$$E = E_n + \frac{\hbar^2 k_\parallel^2}{2m^*} \qquad (3.42)$$

The two-dimensional quantum well structure thus creates electron energies that can be described by *subbands* ($n = 1, 2, 3 \cdots$). The subbands for the conduction band and valence band are shown schematically in Fig. 3.22.

If the barrier potential V_c is not infinite, the wavefunction decays exponentially into the barrier region, and is a sine or cosine function in the well. By matching the

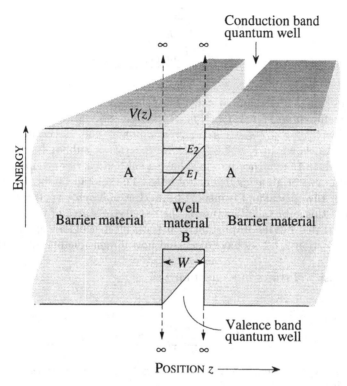

Figure 3.21: A schematic of a quantum well formed for the electron and holes in a heterostructure.

3.7. Tailoring electronic properties

wavefunction and its derivative at the boundaries one can show that the energy and the wavefunctions are given by the solution to the transcendental equations (see Section 2.4.2)

$$\alpha \tan \frac{\alpha W}{2} = \beta$$
$$\alpha \cot \frac{\alpha W}{2} = -\beta \qquad (3.43)$$

where

$$\alpha = \sqrt{\frac{2m^* E}{\hbar^2}}$$
$$\beta = \sqrt{\frac{2m^*(V_c - E)}{\hbar^2}}$$

These equations can be solved numerically. The solutions give the energy levels E_1, E_2, E_3 ... and the wavefunctions, $f_1(z), f_2(z), f_3(z), \cdots$.

Each level E_1, E_2, etc., is actually a subband due to the electron energy in the x–y plane. As shown in Fig. 3.22 we have a series of subbands in the conduction and valence band. In the valence band we have a subband series originating from heavy holes and another one originating from light holes. The subband structure has important consequences for the optical and transport properties of heterostructures. An important manifestation of this subband structure is the density of states of the electronic bands. The density of states figures importantly in both electrical and optical properties of any system. In Section 2.4.2 we have discussed how dimensionality alters the density of states.

The density of states in a quantum well is

- Conduction band

$$N(E) = \sum_i \frac{m^*}{\pi \hbar^2} \theta(E - E_i) \qquad (3.44)$$

where θ is the heavyside step function (unity if $E > E_i$; zero otherwise) and E_i are the subband energy levels.
- Valence band

$$N(E) = \sum_i \sum_{j=1}^{2} \frac{m_j^*}{\pi \hbar^2} \theta(E_{ij} - E) \qquad (3.45)$$

where i represents the subbands for the heavy hole ($j = 1$) and light holes ($j = 2$). The density of states is shown in Fig. 3.22 and has a staircase-like shape.

The differences between the density of states in a quantum well and a three-dimensional semiconductor is one of the important reasons why quantum wells are useful for optoelectronic devices. The key difference is that the density of states in a quantum well is large and finite at the effective bandedges (lowest conduction subband and highest valence subband). As a result the carrier distribution is highest at the bandedges.

The relationship between the electron or hole density (areal density for 2D systems) and the Fermi level is different from that in three-dimensional systems because the density of states function is different. The 2D electron density in a single subband starting at energy E_1^e is

$$n = \frac{m_e^*}{\pi \hbar^2} \int_{E_1^e}^{\infty} \frac{dE}{\exp\left(\frac{E - E_F}{k_B T}\right) + 1}$$

$$= \frac{m_e^* k_B T}{\pi \hbar^2} \left[\ell n \left\{ 1 + \exp\left(\frac{E_F - E_1^e}{k_B T}\right) \right\} \right]$$

or $\quad E_F = E_1^e + k_B T \ell n \left[\exp\left(\frac{n \pi \hbar^2}{m_e^* k_B T}\right) - 1 \right] \quad$ (3.46)

If more than one subband is occupied we can add their contribution similarly. For the hole density we have (considering both the HH and LH ground state subbands)

$$p = \frac{m_{hh}^*}{\pi \hbar^2} \int_{E_1^{hh}}^{-\infty} \frac{dE}{\exp\left(\frac{E_F - E}{k_B T}\right) + 1} + \frac{m_{\ell h}^*}{\pi \hbar^2} \int_{E_1^{\ell h}}^{-\infty} \frac{dE}{\exp\left(\frac{E_F - E}{k_B T}\right) + 1} \quad (3.47)$$

where m_{hh}^* and $m_{\ell h}^*$ are the in-plane density of states masses of the HH and LH subbands. We then have

$$p = \frac{m_{hh}^* k_B T}{\pi \hbar^2} \left[\ell n \left\{ 1 + \exp \frac{(E_1^{hh} - E_{Fp})}{k_B T} \right\} \right]$$

$$+ \frac{m_{\ell h}^* k_B T}{\pi \hbar^2} \left[\ell n \left\{ 1 + \exp \frac{(E_1^{\ell h} - E_{Fp})}{k_B T} \right\} \right] \quad (3.48)$$

If $\quad E_1^{hh} - E_1^{\ell h} > k_B T$

The occupation of the light hole subband can be ignored.

Quantum wells are used in a number of high performance devices such as transistors, lasers, modulators etc. The key reasons for their use are bandgap tailoring, confinement of electrons or holes and changes in the density of states.

3.8 LOCALIZED STATES IN SOLIDS

The band theory discussed in Chapter 2 and in this chapter is valid only for perfect crystals. As noted in Chapter 1, devices are now made from polycrystalline and amorphous materials as well. Even in good-quality crystals there are defects, which break the periodicity of the structure. Typical defects in crystalline materials are: (i) defects in the structure arise from missing atoms (vacancies), atoms at the wrong sites, unintended impurities, etc. (ii) We may also have dislocations at surfaces of a crystal the arrangement of atoms does not have the same periodicity as in the bulk. (iii) We could also have absorbed atoms or molecules at the surface; disordered solids such as amorphous or polycrystalline materials.

3.8. Localized states in solids

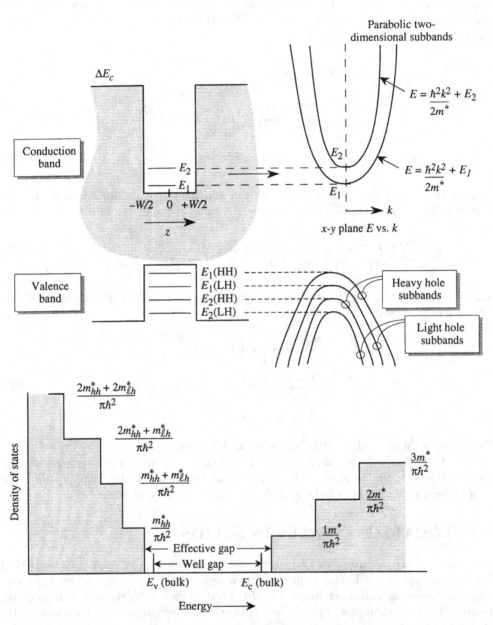

Figure 3.22: Schematic of density of states in a 3–, 2– and 1–dimensional system with parabolic energy momentum relations.

Defects and surface states

In Fig. 3.23 we show a schematic of a perfectly periodic material and one with a defect. A deep potential region indicates the region of defect. In the case of the periodic system we have seen the electrons see a bandedge and are described by simple a effective mass equations near the bandedge. There are no allowed states in the bandgap region. In the case of a defect the deep level causes new electronic states, which can have energies in the bandgap.

The key difference between electronic states in the perfect crystal and a non-perfect crystal is related to the wavefunction. In the periodic state, the electron state is extended over the entire system, as shown in Fig. 3.23a. This reflects the fact that the electron can propagate from one region to another. In the case of a defect a bandgap state may be created with an associated wavefunction that is spatially localized near the defect region, as shown in Fig. 3.23b. When an electron is occupying such a localized state its transport (mobility, diffusion) properties are seriously affected. Localized electrons cannot move across the material as easily.

In Fig. 3.24 we show a comparison of the density of states in a perfectly periodic and of a defect-containing material. In the case of the perfect material we have a well-defined bandgap, while in the presence of defects we have bandgap states. Electrons can be trapped into the bandgap states (hence these states are also called traps).

3.8.1 Disordered materials: extended and localized states

In Chapter 1 we have discussed non-crystalline materials, which have a disordered arrangement of atoms. Amorphous materials, such as amorphous Si, and polycrystalline materials, such as PZT, and other materials used for sensors fall into this category. Non-crystalline materials are used in a number of technologies, such as displays and sensors and their use is increasing. Since the non-crystalline materials lack periodicity, the electronic states are not described by Bloch-like plane wave states. As noted in Chapter 2, electrons in Bloch states have equal probability in all unit cells. In disordered systems this is not so, and electrons are localized in regions of potential fluctuations. In the discussion of levels produced by dopants or defects, we saw that the wavefunction associated with the point defect is not an extended or Bloch state of the form

$$\psi_{\text{ex}} = \frac{1}{\sqrt{V}} u_{\mathbf{k}}(r) e^{i \mathbf{k} \cdot \mathbf{r}} \tag{3.49}$$

but a localized state with a finite extent in space. The defect states have a general form $\psi_{\text{loc}}(r, r_0)$, representing the fact that they are localized around a point r_0 in space. Typical localized states may have an exponentially decaying behavior.

In amorphous semiconductors, electrons see a random background potential, in contrast to the periodic potential of a crystal. In the random potential, electrons can find local potential wells, where they can be trapped or localized. At low energies in a random potential (in an infinite structure), we get a continuum of localized states. As we go up to higher electronic energies, the electron wavefunction spreads over a larger volume, until eventually it becomes extended to the entire volume, as shown schematically in Fig. 3.25. However, the extended state is not the Bloch state of the crystal–it simply

3.8. Localized states in solids

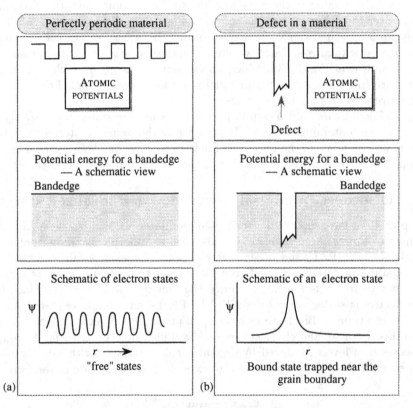

Figure 3.23: A schematic of the structural and electronic properties of (a) crystals and of (b) a polycrystalline material.

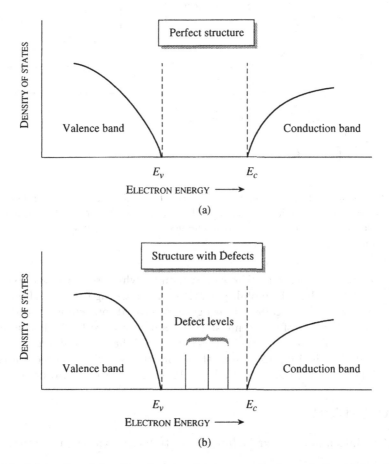

Figure 3.24: Schematic of density of states (a) in a perfectly periodic solid and (b) in a material with defects.

3.9. Summary

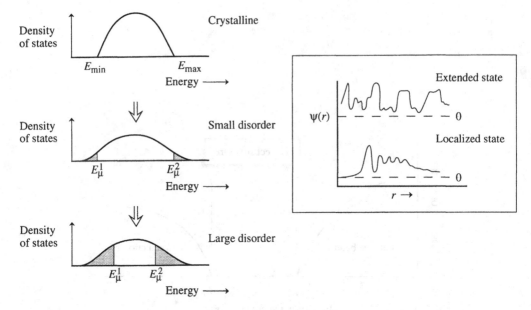

Figure 3.25: Density of states and the influence of disorder. The shaded region represents the region where the electronic states are localized in space. The mobility edges E_μ separate the region of localized and extended states. The inset shows a schematic of an extended and a localized state.

spreads over the entire sample. The energy points, which separate the extended and localized states, are called the mobility edge, the term arising from the observation that the dc conductivity of the localized states goes to zero (at low temperatures). The effect of disorder on the nature of electronic states was first studied by Anderson, and it was shown that as the disorder is increased the extent of the localized states increases, as shown schematically in Fig. 3.25. Near the bandedges we get localized states forming "bandtails." The width of these bandtails is related to the level of disorder in the system.

3.9 SUMMARY

Tables 3.4 to 3.6 summarize our findings of electronic properties in a variety of solids.

3.10 PROBLEMS

Section 3.2
3.1 Plot the Fermi function and the classical (Boltzmann) function as a function of temperature for (i) $T = 1$ K; (ii) $T = 77$ K; (iii) $T = 300$ K; (iv) $T = 1000$ K; and (v) $T = 10000$ K. Assume that $E_F = 0$ and the energy varies from -1.0 eV to 1.0 eV. Examine the energy–temperature regime, where the Fermi function and the classical function are essentially the same.

Topics Studied	Key Findings
Fermi particles and Bose particles	Nature provides us two classes of particles: (i) particles where the multiparticle wavefunction is *asymmetric* under exchange of particles–these are *fermions*; (ii) particles where the multiparticle wavefunction is *symmetric* under exchange of particles–these are *bosons*.
Pauli exclusion principle	Only one fermion can occupy a given allowed state. In bosons, there is no such restriction.
Classical and quantum statistics	In classical physics, identical particles are distinguishable, leading to Boltzmann statistics. In quantum mechanics, identical particles are indistinguishable. If the particles are fermions, we get, for the particle distribution function, the Fermi–Dirac statistics. If the particles are bosons, the resulting statistics are the Bose–Einstein statistics. Although electrons are fermions, in some materials, pairs of electrons can act as bosons, resulting in superconductivity.
Metals, insulators, and semiconductors	In materials where the highest occupied band is half (or partially) filled; conductivity is high. These are metals. In materials where the highest occupied band is completely filled (at low temperatures) conductivity is very poor and these are insulators or semiconductors.

Table 3.4: Summary table.

3.10. Problems

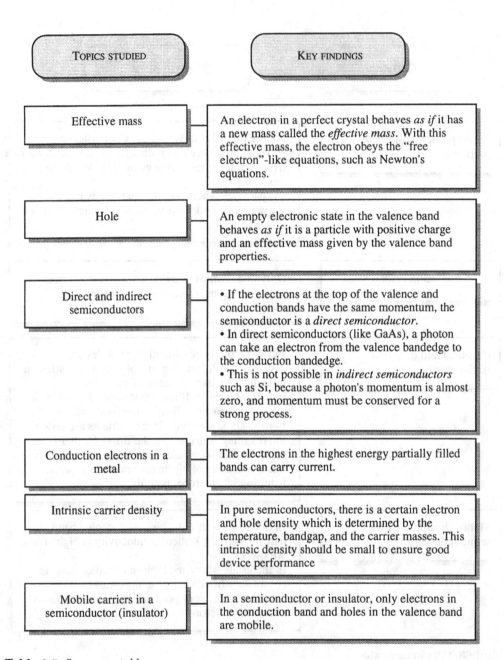

Table 3.5: Summary table.

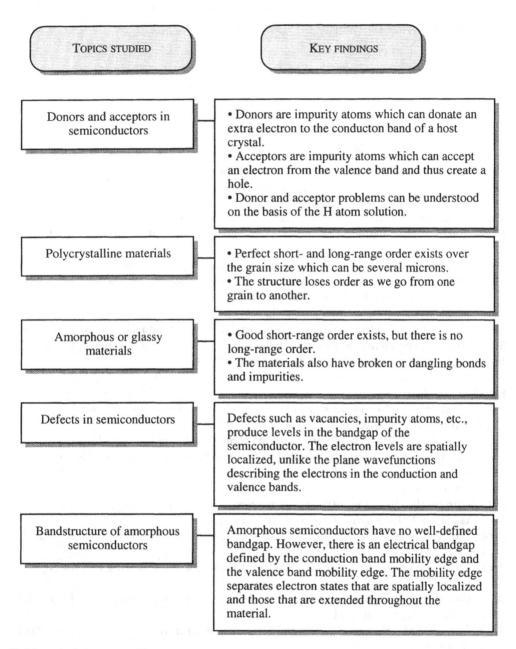

Table 3.6: Summary table.

3.10. Problems

Section 3.3–3.4

3.2 Plot the conduction band and valence band density of states in Si and GaAs from the bandedges to 0.5 eV into the bands. Use the units eV^{-1} cm^{-3}. Use the following data:

$$\begin{aligned}
\text{Si}: m_1^* &= m_\ell^* = 0.98\, m_0 \\
m_2^* &= m_3^* = m_t^* = 0.19\, m_0 \\
m_{hh}^* &= 0.49\, m_0 \\
m_{\ell h}^* &= 0.16\, m_0 \\
\text{GaAs}: m_e^* &= m_{dos}^* = 0.067\, m_0 \\
m_{hh}^* &= 0.45\, m_0 \\
m_{\ell h}^* &= 0.08\, m_0
\end{aligned}$$

3.3 The wavevector of a conduction band electron in GaAs is $k = (0.1, 0.1, 0.0)$ Å^{-1}. Calculate the energy of the electron measured from the conduction bandedge.

3.4 A conduction band electron in silicon is in the (100) valley and has a k-vector of $2\pi/a\, (1.0, 0.1, 0.1)$. Calculate the energy of the electron measured from the conduction bandedge. Here a is the lattice constant of silicon.

3.5 Calculate the energies of electrons in the GaAs and InAs conduction bands with k-vectors $(0.01, 0.01, 0.01)$ Å^{-1}. Refer the energies to the conduction bandedge values.

3.6 Calculate the lattice constant, bandgap, and electron effective mass of the alloy In$_x$Ga$_{1-x}$As as a function of composition from $x = 0$ to $x = 1$.

Section 3.5–3.6

3.7 Calculate the effective density of states at the conduction and valence bands of Si and GaAs at 77 K, 300 K, and 500 K.

3.8 Estimate the intrinsic carrier concentration of diamond at 700 K. You can assume that the carrier masses are similar to those in Si. Compare the results with those for GaAs and Si. The result illustrates one reason why diamond is useful for high-temperature electronics.

3.9 Estimate the change in intrinsic carrier concentration per K change in temperature for InAs, Si, and GaAs at near room temperature.

3.10 Calculate the position of the intrinsic Fermi level, measured from the midgap for GaAs and InAs.

3.11 Calculate the Fermi energy and Fermi velocity for the following metals: Ag, Au, Ca, Cs, Cu, Na.

3.12 Calculate the de Broglie wavelength of electrons at the Fermi energy for the following metals: Ag, Cu, Au, Al.

3.14 Calculate the change in the Fermi level as temperature changes from 0 to 300 K for Al and Cu.

Section 3.7

3.14 Using Vegard's law for the lattice constant of an alloy (i.e., the lattice constant is the weighted average) find the bandgaps of alloys made in InAs, InP, GaAs, GaP which can be lattice matched to InP.

3.15 For long-haul optical communication, the optical transmission losses in a fiber dictate that the optical beam must have a wavelength of either 1.3 μm or 1.55 μm. Which alloy combinations lattice matched to InP have a bandgap corresponding to these wavelengths?

3.16 Calculate the composition of $Hg_xCd_{1-x}Te$ which can be used for a night vision detector with bandgap corresponding to a photon energy of 0.1 eV. Bandgap of CdTe is 1.6 eV *and that of HgTe is* −0.3 eV at low temperatures around 4 K.

3.17 In the $In_{0.53}Ga_{0.47}As/InP$ system, 40% of the bandgap discontinuity is in the conduction band. Calculate the conduction and valence band discontinuities. Calculate the effective bandgap of a 100 Å quantum well. Use the infinite potential approximation and the finite potential approximation and compare the results.

3.18 Calculate the first and second subband energy levels for the conduction band in a $GaAs/Al_{0.3}Ga_{0.7}As$ quantum well as a function of well size. Assume that the barrier height is 0.18 eV.

3.19 Calculate the width of a GaAs/AlGaAs quantum well structure in which the effective bandgap is 1.6 eV. The effective bandgap is given by

$$E_g^{eff} = E_g(\text{GaAs}) + E_1^e + E_1^h$$

where E_g (GaAs) is the bandgap of GaAs (= 1.5 eV) and E_1^e and E_1^h are the ground state energies in the conduction and valence band quantum wells. Assume that $m_e^* = 0.067\ m_0$, $m_{hh}^* = 0.45\ m_0$. The barrier heights for the conduction and valence band well is 0.2 eV and 0.13 eV, respectively.

Section 3.8

3.20 Assume that a particular defect in silicon can be represented by a three-dimensional quantum well of depth 1.5 eV (with reference to the conduction bandedge). Calculate the position of the ground state of the trap level if the defect dimensions are 5 Å× 5 Å× 5 Å. The electron effective mass is $0.26\ m_0$.

3.21 A defect level in silicon produces a level at 0.5 eV below the conduction band. Estimate the potential depth of the defect if the defect dimension is 5 Å× 5 Å×5 Å. The electron mass is $0.25\ m_0$.

3.11 FURTHER READING

- **General bandstructure**

    - H.C. Casey Jr. and M.B. Panish, *Heterostructure Lasers*, Part A, "Fundamental Principles," Part B, "Materials and Operating Characteristics," Academic Press, New York (1978).

    - R.E. Hummel, *Electronic Properties of Materials–An Introduction for Engineers*, Springer Verlag, New York (1985).

3.11. Further reading

- Landolt-Bornstein, *Numerical Date and Functional Relationship in Science and Technology*, Vol. 22, Eds. O. Madelung, M. Schulz, and H. Weiss, Springer-Verlag, New York (1987).
- K. Seeger, *Semiconductor Physics: An Introduction*, Springer, Berlin (1985).
- H.F. Wolf, *Semiconductors*, Wiley-Interscience, New York (1971).

- **Bandstructure modification**

    - A.G. Milnes and D.L. Feucht, *Heterojunctions and Metal Semiconductor Junctions*, Academic Press, New York (1972).
    - For a simple discussion of electrons in quantum wells any book on basic quantum mechanics is adequate. An example is L. Schiff, *Quantum Mechanics*, McGraw-Hill, New York (1968).
    - J. Singh, *Electronic and Optoelectronic Properties of Semiconductor Structures*, Cambridge University Press (2003).

- **Intrinsic and extrinsic carriers**

    - J.S. Blakemore, *Electron. Commun.*, 29, 131 (1952).
    - J.S. Blakemore, *Semiconductor Statistics*, Pergamon Press, New York (1962) reprinted by Dover, New York (1988).
    - K. Seeger, *Semiconductor Physics: An Introduction*, Springer Verlag, Berlin (1985)

- **Organic semiconductors**

    - M. Pope and C.E. Swenberg, *Electronic Processes in Organic Crystals*, Clarendon, Oxford (1982).
    - R.S. Kohlman, J. Joo, and A.J. Epstein, *Physical Properties of Polymers Handbook*, eds. by J.E. Mark, AIP Press, Woodbury, New York (1996).

- **Non-crystalline solids**

    - N.F. Mott and E.A. Davis, *Electronic Processes in Non-Crystalline Materials*, Oxford University Press, Cleveland (1971).
    - J. Kanicki (editor) *Amorphous and Microcrystalline Semiconductor Devices: Materials and Device Physics*, Artech House, Boston (1992).

Chapter 4

CHARGE TRANSPORT IN MATERIALS

4.1 INTRODUCTION

In this chapter we will examine how electrical current flows occur in materials. Electrical current can be carried by the transport of electron, holes, and even ions. These charges respond to applied electric fields and also move if there are concentration gradients. The transport phenomena is central in a number of important devices. The charges in a solid can be loosely classified as fixed and mobile. When an external perturbation is applied (e.g., an electric field) the mobile charges can move from one point in space to another. In particular they can move from one contact on a device to another. The fixed charge, however, can only be disturbed slightly from its equilibrium position, but cannot move over the length of a device. Both fixed charges and mobile charges play an important role in technology, as shown in Fig. 4.1. Essentially all electronic devices such as field effect transistors, bipolar transistors, diodes, as well as optoelectronic devices, such as lasers and detectors depend upon free or mobile charges. Mobile charges are the electrons in the conduction band and holes in the valence band for semiconductors and insulators. In some insulators with high defect densities, ions can carry current also, but the conductivity is very small. As we have discussed in previous chapters, in metals the mobile charges are the electrons in the conduction band.

Fixed charges in materials also play an important role in devices, even though they cannot participate in current flow. Small movements in the position of the fixed charges are responsible for the dielectric response of solids. The fixed charges are also responsible for polarization effects, which are exploited for devices, such as sensors and detectors.

On the subject of carrier transport there are two key issues that need to be discussed: (i) How do mobile carriers respond to electric fields? (ii) How do mobile

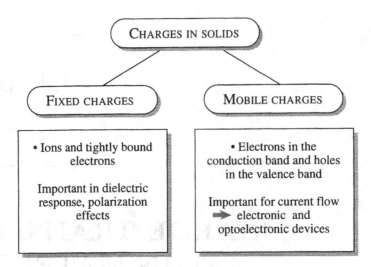

Figure 4.1: An overview of fixed and mobile charges in solids and their impact on physical phenomena.

carriers respond to changes in mobile carrier densities? The answers to these questions help us understand how electronic and optoelectronic devices operate.

4.2 AN OVERVIEW OF ELECTRONIC STATES

Before discussing issues in free carrier (or mobile carrier) transport we show an overview of the nature of electronic states in solids in Fig. 4.2. In Fig. 4.2a we show a schematic of the density states as a function of energy. In the case of the perfect crystals we see that in the conduction and valence bands the electronic states are "free," as discussed in Chapter 3. There are no allowed energy levels in the bandgap (density of states is zero in the bandgap, as shown).

In the case of a crystal with defects we still have the free states in the conduction and valence bands, but we also have defect-related allowed states in the bandgap region, as shown in Fig. 4.2b. These states are called trap states and electrons are not free to move under an electric field if they are occupying these states.

In Fig. 4.2c we show a schematic of the electronic states in a solid that is disordered; i.e., there is no underlying crystalline lattice. In this case we can divide the electronic states into "free" and localized categories, as shown. We still have a conduction band and valence band, but the bandgap region is no longer a true bandgap. There are allowed states in the bandgap region, but electrons are not mobile if they are in these states. These bandgap states are called localized states, while the free states in disordered materials are called extended states. Mobility edges E_μ^v and E_μ^c denote energy positions, where transitions occur form localized to extended states.

In Fig. 4.3 we show a schematic of how electrons (holes) move through a sample when an electric field is applied. In Fig. 4.3a we show the situation in a good-quality

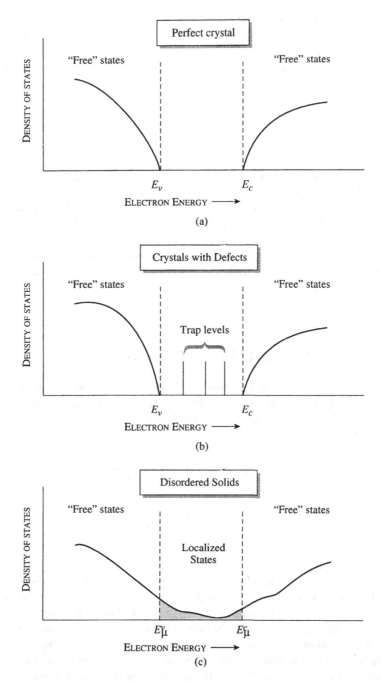

Figure 4.2: A schematic of the nature of electronic states in solids: (a) for a perfect crystal, (b) for a crystal with defects, and (c) for a disordered solid. In the disordered system, mobility edges E_μ^v, E_μ^c represent the transition from localized to free states.

crystalline material (most high-quality semiconductor devices are in this category). The electron moves under the electric field force, but suffers a number of scattering processes. The scattering occurs due to various imperfections, such as defects, vibrations of atoms (due to thermal energy), and other small imperfections. If we examine the distance versus time trajectory of a typical electron we observe that the electron shows a zigzag path (shown). On average the electron trajectory is described by

$$d = vt$$
$$v = \mu E \tag{4.1}$$

where d is the distance traveled in time t. The velocity v is proportional to the electric field applied through μ, the mobility.

In disordered solids or materials with a high defect density (Fig. 4.3b) there are interesting differences when one examines the trajectory of a mobile carrier. In this case the electron occasionally falls into a trap state or a localized state, where it is immobile, and the distance versus time trajectory shows that during these periods the distance does not change with time. Eventually the electron is able to escape from the trap state and resume its progress.

We will discuss models for transport for both crystalline and disordered materials.

4.3 TRANSPORT AND SCATTERING

In this section we will discuss how electron (hole) transport occurs in good-quality crystalline materials. By good quality we imply that there are negligible trap (or localized) states. In equilibrium the electron (hole) distribution in energy (or momentum) is given by the Fermi–Dirac distribution

$$f(E) = f^\circ(E) = \frac{1}{\exp\left(\frac{E - E_F}{k_B T}\right) + 1}$$

$$E = E_i + \frac{\hbar^2 k^2}{2m^*}$$

where E_i is the bandedge.

We can see that in the absence of any applied electric field, the occupation of a state with momentum $+\hbar\mathbf{k}$ is the same as that of a $-\hbar\mathbf{k}$ state. Thus there is net cancellation of momenta and there is no net current flow. The distribution function in momentum space is shown schematically in Fig. 4.4a. The question we would like to answer is the following: If an electric field is applied, what happens to the free electrons (holes)? When a field is applied the electron distribution will shift, as shown schematically in Fig. 4.4b, and there will be a net momentum of the electrons. If the crystal is rigid and perfect, according to the Bloch theorem the electron states are described by

$$\psi_k(r,t) = u_k \exp i(k \cdot r - \omega t) \tag{4.2}$$

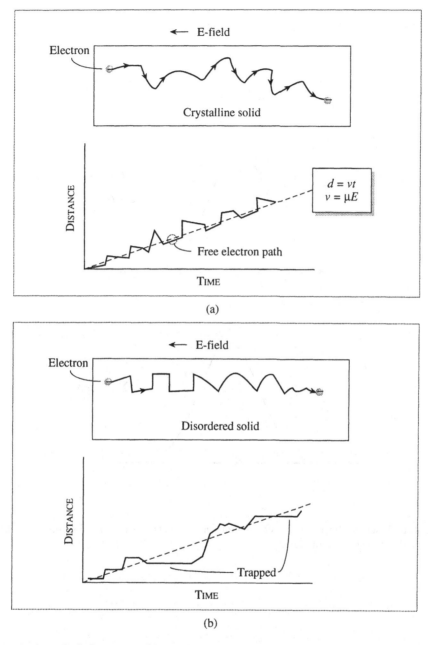

Figure 4.3: A typical electron trajectory in a sample and the distance versus time profile: (a) crystalline solid, (b) disordered solid with trap or localized states.

4.3. Transport and scattering

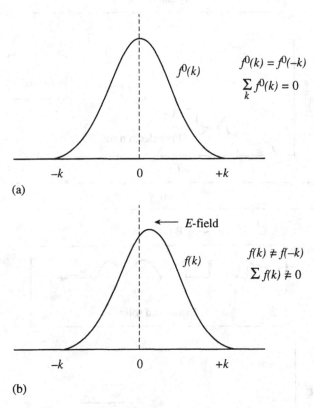

Figure 4.4: A schematic of the electron momentum distribution function in (a) equilibrium and (b) in the presence of an electric field.

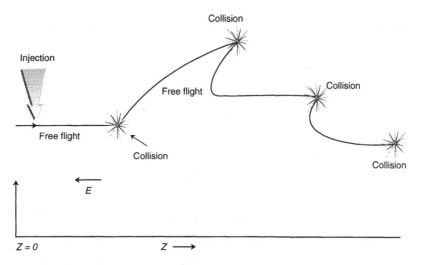

Figure 4.5: Schematic view of an electron as it moves under an electric field in a semiconductor. The electron suffers a scattering as it moves. In between scattering the electron moves according to the "free" electron equation of motion.

where $\omega = E/\hbar$ is the electron wave frequency. There is no scattering of the electron in the perfect system. Also, if an electric field E is applied, the electron behaves as a "free" space electron would, obeying the equation of the motion

$$\frac{\hbar d\mathbf{k}}{dt} = F_{\text{ext}} = -e\mathbf{E} \tag{4.3}$$

The electron would travel along a particular E–k band.

In a real material, there are always imperfections which cause scattering of electrons so that the equation of motion of electrons is not given by Eq. 4.3. A conceptual picture of electron transport can be developed where the electron moves in space for some time, then scatters and then again moves in space and again scatters. The process is shown schematically in Fig. 4.5. The average behavior of the ensemble of electrons will then represent the transport properties of the electron.

4.3.1 Scattering of electrons

The key to understanding the non-equilibrium properties of electrons is the understanding of the scattering process of the electrons. The scattering problem in solids is treated by using the perturbation theory in quantum mechanics. We are interested in solving the quantum mechanics problem formally represented by

$$H\Phi = E\Phi \tag{4.4}$$

where H is the full hamiltonian (potential energy + kinetic energy operator) of the problem and the electron states are denoted by Φ. This hamiltonian is, in our case, the

4.3. Transport and scattering

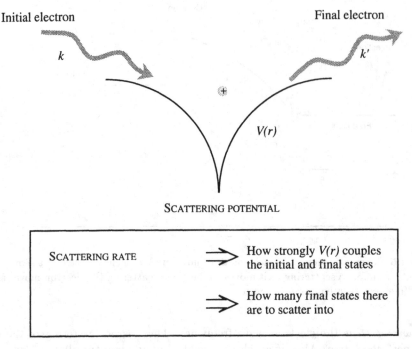

Figure 4.6: The scattering of an electron initially with momentum $\hbar \mathbf{k}$ from a scattering potential $V(r)$. The final momentum is $\hbar \mathbf{k}'$. The scattering process is assumed to be instantaneous. The scattering depends upon the coupling of the initial state to the final state by the scattering potential.

sum of the hamiltonian of the perfect crystal H_o and the energy V corresponding to the imperfection causing scattering. Thus

$$H = H_o + V \qquad (4.5)$$

We know how to solve the problem

$$H_o \psi = E \psi \qquad (4.6)$$

which just gives us the bandstructure of the semiconductor. In the perturbation theory, we use the approach that the effect of the perturbation V is to cause scattering of the electron from one perfect crystalline state to another. This theory works well if the perturbation is small. The effect of the scattering is shown schematically in Fig. 4.6. The rate of scattering for an electron initially in state i to a state f in the presence of a perturbation of the form

$$V(r,t) = V(r) \exp(i\omega t) \qquad (4.7)$$

is given by the Fermi golden rule

$$W_{if} = \frac{2\pi}{\hbar} \mid M_{ij} \mid^2 \delta(E_i \pm \hbar\omega - E_f) \qquad (4.8)$$

where the various quantities in the equation represent the following:

- $\frac{2\pi}{\hbar}$: this is a factor that appears from the details of the calculations.
- $|M_{ij}|^2$: The quantity is called the matrix element of the scattering and is given by

$$M_{ij} = \int \psi_f^* V(r) \psi_i d^3r \qquad (4.9)$$

The matrix element tells us how the potential couples the initial and the final state.
- $\delta(E_i \pm \hbar\omega - E_f)$: This δ-function is simply a representative of energy conservation. The process where

$$E_f = E_i + \hbar\omega \qquad (4.10)$$

is called absorption, while the process

$$E_f = E_i - \hbar\omega \qquad (4.11)$$

is called emission. Thus, both absorption or emission of energy can occur if the perturbation has a time dependence $\exp(i\omega t)$. If the potential is time independent, the scattering is elastic ($E_i = E_f$).

In principle, the evaluation of the scattering rates is fairly straightforward since it simply involves the calculations of some integrals. In practice, the problem is complicated by the fact that the scattering potential $V(r)$ is not well defined and models have to be constructed to represent a defect by a proper potential. Thus, while it may be easy to describe the physical nature of the defect, it is quite difficult to represent the potential perturbation that the electron sees due to this defect. We will now briefly review some important scattering sources in semiconductors.

Phonon scattering
In Chapter 1, we discussed the crystalline structure in which atoms were at fixed periodic positions. In reality, the atoms in the crystal are vibrating. These lattice vibrations are represented by "particles" in quantum mechanics and are called phonons. They satisfy an equation of motion similar to that of masses coupled to each other by springs. The properties of the lattice vibrations are represented by the relation between the vibration, amplitude, u, frequency, ω, and the wavevector q. The vibration of a particular atom is given by

$$u_i(q) = u_{oi} \exp i(q \cdot r - \omega t) \qquad (4.12)$$

which has the usual plane wave form that all solutions of periodic structures have. Recall that in a semiconductor there are two kinds of atoms in a basis. This results in a ω vs. k relation shown in Fig. 4.7. This relation is for GaAs typical of all compound semiconductors. We notice two kinds of lattice vibrations, denoted by acoustic and optical. Additionally, there are two transverse and one longitudinal modes of vibration for each kind of vibration. The acoustic branch can be characterized by vibrations where the two atoms in the basis of a unit cell (i.e., Ga and As for GaAs) vibrate with the same sign of the amplitude as shown in Fig. 4.7b. In optical vibrations, the two atoms

4.3. Transport and scattering

with opposing amplitudes are shown. A brief discussion of lattice vibrations or phonons is given in Appendix D.

While the dispersion relations represent the allowed lattice vibration modes, an important question is how many such modes are actually being excited at a given temperature. In quantum mechanics the modes are called phonons and the number of phonons with frequency ω are given by the Bose–Einstein distribution function

$$n_\omega = \frac{1}{\exp\left(\frac{\hbar\omega}{k_B T}\right) - 1} \tag{4.13}$$

The lattice vibration problem is mathematically similar to the harmonic oscillator problem. The quantum mechanics of the harmonic oscillator problem tells us that the energy in the mode frequency ω is then

$$E_\omega = (n_\omega + \frac{1}{2})\hbar\omega \tag{4.14}$$

Note that even if there are no phonons in a particular mode, there is a finite "zero point" energy $\frac{1}{2}\hbar\omega$ in the mode.

The vibrations of the atoms produce three kinds of potential disturbances that result in the scattering of electrons. A schematic of the potential disturbance created by the vibrating atoms is shown in Fig. 4.8. In a simple physical picture, we can imagine the lattice vibrations causing spatial and temporal fluctuations in the conduction and valence band energies. The electrons (holes) then scatter from these disturbances. The acoustic phonons produce a strain field in the crystal and the electrons see a disturbance which produces a potential of the form

$$V_{AP} = D \frac{\partial u}{\partial x} \tag{4.15}$$

where D is called a deformation potential (units are eV) and $\frac{\partial u}{\partial x}$ is the amplitude gradient of the atomic vibrations.

The optical phonons produce a potential disturbance, which is proportional to the atomic vibration amplitude, since in the optical vibrations the two atoms in the basis vibrate opposing each other

$$V_{op} = D_o u \tag{4.16}$$

where D_o (units are eV/cm) is the optical deformation potential.

In compound semiconductors, where the two atoms are in the basis of the crystal structure, an extremely important scattering potential arises additionally from optical phonons. Since the two atoms are different, there is an effective positive and negative charge e^* on each atom. When optical vibrations take place, the effective dipole in the unit cell vibrates, causing polarization fields from which the electron scatters. This scattering, called polar optical phonon scattering, has a scattering potential of the form

$$V_{po} \sim e^* u \tag{4.17}$$

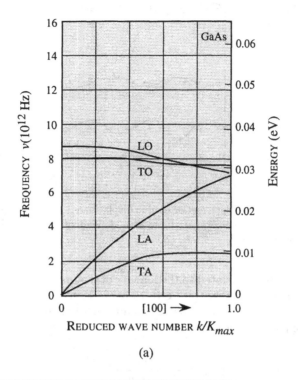

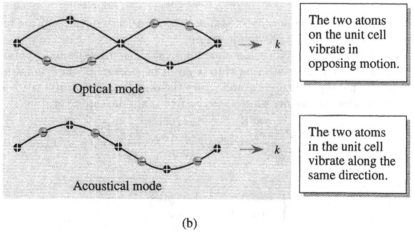

Figure 4.7: (a) Typical dispersion relations of a semiconductor (GaAs in this case). (b) The displacement of atoms in the optical and acoustic branches of the vibrations is shown. The motion of the atoms is shown for small k vibrations.

4.3. Transport and scattering

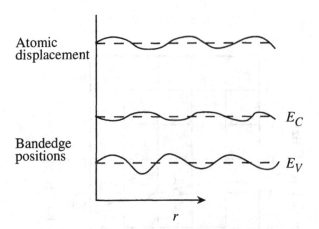

Figure 4.8: A schematic showing the effect of atomic displacement on bandedge energy levels in real space. The lattice vibrations cause spatial and time-dependent variations in the bandedges from which the electrons scatter.

The effective charge is related to the ionicity of the material. By using the Fermi golden rule we can calculate the scattering rates of electrons due to lattice vibrations. We will provide these rates and, in the solved examples, calculate some typical values.

The acoustic phonon scattering rate for an electron with energy E_k to any other state is given by

$$W_{\text{ac}}(E_k) = \frac{2\pi D^2 k_B T N(E_k)}{\hbar \rho v_s^2} \tag{4.18}$$

where $N(E_k)$ is the electron density of states, ρ is the density of the semiconductor, v_s is the sound velocity and T is the temperature.

In materials like GaAs, the dominant optical phonon scattering is polar optical phonon scattering, and the scattering rate is given by (assuming the bandstructure is defined by a non-parabolic band; ϵ_∞ and ϵ_s are the high frequency and static dielectric constants of the semiconductor, while ϵ_o is the free space dielectric constant)

$$W(k) = \frac{e^2 m^{*1/2} \omega_o}{4\pi\sqrt{2}\hbar} \left(\frac{\epsilon_o}{\epsilon_\infty} - \frac{\epsilon_o}{\epsilon_s} \right) \frac{1 + 2\alpha E'}{\gamma^{1/2}(E)} F_o(E, E')$$

$$\times \begin{cases} n(\omega_o) & \text{absorption} \\ n(\omega_o) + 1 & \text{emission} \end{cases} \tag{4.19}$$

where

$$\begin{aligned} E' &= E + \hbar\omega_o \text{ for absorption} \\ &= E - \hbar\omega_o \text{ for emission} \\ \gamma(E) &= E(1 + \alpha E) \end{aligned}$$

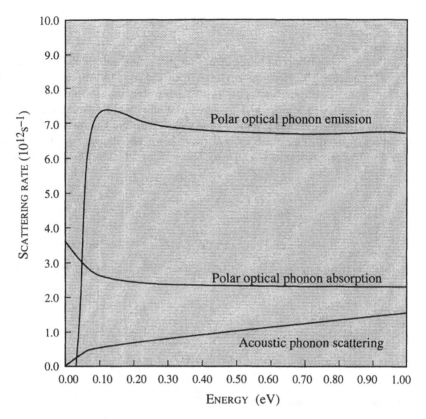

Figure 4.9: A comparison of the scattering rates due to acoustic and optical phonons for GaAs electrons at room temperature.

$$F_o(E, E') = C^{-1}\left(A\ell n \left|\frac{\gamma^{1/2}(E) + \gamma^{1/2}(E')}{\gamma^{1/2}(E) - \gamma^{1/2}(E')}\right| + B\right)$$

$$A = [2(1 + \alpha E)(1 + \alpha E') + \alpha\{\gamma(E) + \gamma(E')\}]^2$$

$$B = -2\alpha\gamma^{1/2}(E)\gamma^{1/2}(E')$$

$$= \times [4(1 + \alpha E)(1 + \alpha E') + \alpha\{\gamma(E) + \gamma(E')\}]$$

$$C = 4(1 + \alpha E)(1 + \alpha E')(1 + 2\alpha E)(1 + 2\alpha E')$$

It is important to examine typical values of scattering rates from these processes. The values for GaAs are shown in Fig. 4.9. *Note that the phonon emission process can start only after the electron has energy equal to the phonon energy.* The emission rates are about three times as strong as the absorption rates at room temperature (the ratio between $n(\omega_o) + 1$ and $n(\omega_o)$).

Optical phonon scattering is the most important scattering mechanism for high-field or high-temperature transport of electrons. It is also responsible for relaxation of hot electrons injected into a semiconductor. The hot carrier relaxation is a key process

4.3. Transport and scattering

in semiconductor laser performance and will be discussed later.

Ionized impurity scattering

In Section 3.6 we discussed doping in semiconductors. When a dopant ionizes to produce an extra "free" electron, the electron scatters from the ion. In Appendix E we discuss scattering from ionized impurities. The scattering potential is Coulombic in nature, except that the potential is suppressed by screening effects. The screening is due to the presence of the other free electrons, which form a cloud around the ion so the effect of the potential is short ranged. There are several models for the ionized impurity scattering potential. A good approximation is given by the screened Coulombic potential

$$V(r) = \frac{e^2}{\epsilon} \frac{e^{-\lambda r}}{r} \tag{4.20}$$

where

$$\lambda^2 = \frac{ne^2}{\epsilon k_B T} \tag{4.21}$$

with n the free electron density. The scattering rate then becomes, for an electron with energy E_k and momentum $\hbar k$

$$W(k) = 4\pi F \left(\frac{2k}{\lambda}\right)^2 \left[\frac{1}{1+(\lambda/2k)^2}\right]$$

$$F = \frac{1}{\hbar} \left(\frac{e^2}{\epsilon}\right)^2 \frac{N(E_k)}{32k^4} N_I \tag{4.22}$$

where N_I is the ionized impurity density.

Alloy scattering

Alloys are made from combinations of two or more materials. Since atoms on the lattice are arranged randomly there is random potential fluctuation which causes scattering. In Appendix E we discuss this scattering. The scattreing rate for an alloy $A_x B_{1-x}$ is found to be

$$\begin{aligned} W_{tot} &= \frac{2\pi}{\hbar} \left(\frac{3\pi^2}{16} V_0\right) U_{all}^2 \, N(E_{\mathbf{k}}) \left[x\,(1-x)^2 + (1-x)\,x^2\right] \\ &= \frac{3\pi^3}{8\hbar} V_0 \, U_{all}^2 \, N(E_{\mathbf{k}}) \, x\,(1-x) \end{aligned} \tag{4.23}$$

Here U_{all} is the potential difference between A type and B type potentials (see Appendix E), V_0 is the volume of the unit cell in the lattice and $N(E)$ is the density of states without counting spin degeneracy.

While the phonon and impurity scattering are the dominant scattering processes for most transport problems, electron–electron scattering, electron–hole scattering, and alloy potential scattering, etc., can also play an important role.

EXAMPLE 4.1 Calculate the polar optical phonon emission rate for an electron in GaAs with energy 0.2 eV. Use the following parameters:

$$w_o = 5.4 \times 10^{13} \, \text{rad/s}$$

$$\hbar w_o = 36 \text{ meV}$$
$$\frac{\epsilon_s}{\epsilon_o} = 13.2$$
$$\frac{\epsilon_\infty}{\epsilon_o} = 10.9$$
$$n(w_o) = 0.33$$
$$\alpha = 0.6 \text{ eV}^{-1}$$

The scattering rate is given by (see Eq. 4.19)

$$W(k) = \frac{(1.6 \times 10^{-19} \text{ C})^2 (0.067 \times 0.9 \times 10^{-30} \text{ kg})^{1/2} (5.4 \times 10^{13} \text{rad/s})(1.33)}{4\Pi(1.414)(1.05 \times 10^{-34} \text{ J.s})(8.84 \times 10^{-12} \text{ F/m})}$$
$$\times \left(\frac{1}{10.9} - \frac{1}{13.2}\right) \left(\frac{1 + 2(0.6 \text{ eV}^{-1})(0.164 \text{ eV})}{(0.224 \times 1.6 \times 10^{-19} \text{ J})^{1/2}}\right) F_o(E, E')$$
$$= 2.43 \times 10^{12} \ F_o(E, E') \text{ s}^{-1}$$

The value of $F_o(E, E')$ can be found to be 2.87. Thus, the emission rate is

$$W(k) = 7.0 \times 10^{12} \text{s}^{-1}$$

4.4 MACROSCOPIC TRANSPORT PROPERTIES

The scattering rates discussed in the previous section are only one of the ingredients of a transport theory. Note that the scattering rates are dependent upon the energy of the electron. What energy should be used to obtain transport properties? Clearly, an averaging must be carried out over the ensemble of the electrons in the semiconductor. However, this requires knowing the distribution function, which is only known at equilibrium. Since the scattering processes and the distribution function are inter-related at non-equilibrium, the problem is very complicated, and various numerical and computer simulation techniques are developed to solve the problem.

Two important approaches to understanding transport in semiconductors are the solution of the Boltzmann transport equation, using numerical methods and the Monte Carlo method using computer simulations. We will summarize the results of such theories by examining the drift velocity versus electric field relations in semiconductors.

4.4.1 Velocity–electric field relations in semiconductors

When an electron distribution is subjected to an electric field, the electrons tend to move in the field direction (opposite to the field **E**) and gain velocity from the field. However, because of imperfections in the crystal potential, they suffer scattering. A steady state is established in which the electrons have some net drift velocity in the field direction. The response of the electrons to the field can be represented by a velocity–field relation. We will briefly discuss the velocity-field relationships at low electric fields and moderately high electric fields.

Low field response: mobility

At low electric fields, the macroscopic transport properties of the material (mobility, conductivity) can be related to the microscopic properties (scattering rate or relaxation time) by simple arguments. We will not solve the Boltzmann transport equation, but we will use simple conceptual arguments to understand this relationship. In this approach we make the following assumptions:

(i) The electrons in the semiconductor do not interact with each other. This approximation is called the independent electron approximation.
(ii) Electrons suffer collisions from various scattering sources and the time τ_{sc} describes the mean time between successive collisions.
(iii) The electrons move according to the free electron equation

$$\frac{\hbar dk}{dt} = \mathbf{E} \qquad (4.24)$$

in between collisions. After a collision, the electrons lose all their excess energy (on the average) so that the electron gas is essentially at thermal equilibrium. This assumption is really valid only at very low electric fields.

According to these assumptions, immediately after a collision the electron velocity is the same as that given by the thermal equilibrium conditions. This average velocity is thus zero after collisions. The electron gains a velocity in between collisions; i.e., only for the time τ_{sc}.

This average velocity gain is then that of an electron with mass m^*, traveling in a field $\mathbf{E}$, for a time τ_{sc}

$$\mathbf{v}_{\text{avg}} = -\frac{e\mathbf{E}\tau_{sc}}{m^*} = \mathbf{v}_d \qquad (4.25)$$

where v_d is the drift velocity. The current density is now

$$\mathbf{J} = -ne\mathbf{v}_d = \frac{ne^2\tau_{sc}}{m^*}\mathbf{E} \qquad (4.26)$$

Comparing this with the Ohm's law result for conductivity σ

$$\mathbf{J} = \sigma\mathbf{E} \qquad (4.27)$$

we have

$$\sigma = \frac{ne^2\tau_{sc}}{m^*} \qquad (4.28)$$

The resistivity of the semiconductor is simply the inverse of the conductivity. From the definition of mobility μ, for electrons

$$\mathbf{v}_d = \mu\mathbf{E} \qquad (4.29)$$

we have

$$\mu = \frac{e\tau_{sc}}{m^*} \qquad (4.30)$$

	Bandgap (eV)		Mobility at 300 K (cm²/V-s)	
Semiconductor	300 K	0 K	Elec.	Holes
C	5.47	5.48	1800	1200
GaN	3.4	3.5	1400	350
Ge	0.66	0.74	3900	1900
Si	1.12	1.17	1500	450
α-SiC	3.00	3.30	400	50
GaSb	0.72	0.81	5000	850
GaAs	1.42	1.52	8500	400
GaP	2.26	2.34	110	75
InSb	0.17	0.23	80000	1250
InAs	0.36	0.42	33000	460
InP	1.35	1.42	4600	150
CdTe	1.48	1.61	1050	100
PbTe	0.31	0.19	6000	4000
$In_{0.53}Ga_{0.47}As$	0.8	0.88	11000	400

Table 4.1: Bandgaps along with electron and hole mobilities in several semiconductors. Properties of large bandgap materials (C, GaN, SiC) are continuously changing (mobility is improving), due to progress in crystal growth. Zero temperature bandgap is extrapolated.

If both electrons and holes are present, the conductivity of the material becomes

$$\sigma = ne\mu_n + pe\mu_p \tag{4.31}$$

where μ_n and μ_p are the electron and hole mobilities and n and p are their densities.

Notice that the mobility has an explicit $\frac{1}{m^*}$ dependence in it. Additionally τ_{sc} also decreases with m^*. Thus the mobility has a strong dependence on the carrier mass. In Table 4.1 we show the mobilities of several important semiconductors at room temperature. The results are shown for pure materials. If the semiconductors are doped, the mobility decreases. Note that Ge has the best hole mobility among all semiconductors.

In Appendix E we have derived mobility limited by ionized impurity and by alloy scattering.

The total relaxation time due to ionized impurity scattering is

$$\frac{1}{\langle\langle\tau\rangle\rangle} = N_i \frac{1}{128\sqrt{2\pi}} \left(\frac{Ze^2}{\epsilon}\right)^2 \frac{1}{m^{*1/2}(k_BT)^{3/2}} \\ \times \left[\ln\left(1+\left(\frac{8m^*k_BT}{\hbar^2\lambda^2}\right)^2\right) - \frac{1}{1+\left(\frac{\hbar^2\lambda^2}{8m^*k_BT}\right)^2}\right] \tag{4.32}$$

4.4. Macroscopic transport properties

The mobility limited from ionized impurity scattering is

$$\mu = \frac{e\langle\langle\tau\rangle\rangle}{m^*}$$

The mobility limited by ionized dopant has the special feature that it decreases with temperature ($\mu \sim T^{3/2}$). This temperature dependence is quite unique to ionized impurity scattering. One can understand this behavior physically by saying that at higher temperatures, the electrons are traveling faster and are less affected by the ionized impurities.

After doing the proper ensemble averaging the relaxation time for the alloy scattering is

$$\frac{1}{\langle\langle\tau\rangle\rangle} = \frac{3\pi^3}{8\hbar}V_0 U_{all}^2 x(1-x) \frac{m^{*3/2}(k_B T)^{1/2}}{\sqrt{2}\pi^2\hbar^3}\frac{1}{0.75} \tag{4.33}$$

according to which the mobility due to alloy scattering is

$$\mu_0 \propto T^{-1/2}$$

The temperature dependence of mobility is in contrast to the situation for the ionized impurity scattering. The value of U_{all} is usually in the range of 1.0 eV.

EXAMPLE 4.2 Consider a semiconductor with effective mass $m^* = 0.26\ m_0$. The optical phonon energy is 50 meV. The carrier scattering relaxation time is 10^{-13} sec at 300 K. Calculate the electric field at which the electron can emit optical phonons on the average.

In this problem we have to remember that an electron can emit an optical phonon only if its energy is equal to (or greater than) the phonon energy. According to the transport theory, the average energy of the electrons is (v_d is the drift velocity)

$$E = \frac{3}{2}k_B T + \frac{1}{2}m^* v_d^2$$

In our case, this has to be 50 meV at 300 K. Since $k_B T \sim 26$ meV at 300 K, we have

$$\frac{1}{2}m^* v_d^2 = 50 - 39 = 11\ \text{meV}$$

or

$$v_d^2 = \frac{2 \times (11 \times 10^{-3} \times 1.6 \times 10^{-19}\ \text{J})}{(0.91 \times 10^{-30} \times 0.26\ \text{kg})}$$

$$v_d = 1.22 \times 10^5\ \text{m/s}$$

Also we should note that the symbol E is being used for the electric field and energy

$$v_d = \frac{e\tau \mathbf{E}}{m^*}$$

Substituting for v_d, we get (for the average electrons) for the electric field

$$\mathbf{E} = \frac{(0.26 \times 0.91 \times 10^{-30}\ \text{kg})(1.22 \times 10^5\ \text{m/s})}{(4.8 \times 10^{-10}\ \text{esu})(10^{13}\ \text{s})}$$

$$= 18.04\ \text{kV/cm}$$

The results discussed correspond approximately to silicon. Of course, since the distribution function has a spread, electrons start emitting optical phonons at a field lower than the one calculated above for the average electron.

EXAMPLE 4.3 The mobility of electrons in pure GaAs at 300 K is 8500 cm^2/V·s. Calculate the relaxation time. If the GaAs sample is doped at $N_d = 10^{17}$ cm^{-3}, the mobility decreases to 5000 cm^2/V·s. Calculate the relaxation time due to ionized impurity scattering.

The relaxation time is related to the mobility by

$$\tau_{sc}^{(1)} = \frac{m^*\mu}{e} = \frac{(0.067 \times 0.91 \times 10^{-30} \text{ kg})(8500 \times 10^{-4} \text{ m}^2/\text{V}\cdot\text{s})}{1.6 \times 10^{-19} \text{ C}}$$

$$= 3.24 \times 10^{-13} \text{ s}$$

If the ionized impurities are present, the time is

$$\tau_{sc}^{(2)} = \frac{m^*\mu}{e} = 1.9 \times 10^{-13} \text{ s}$$

The total scattering rate is the sum of individual scattering rates. Since the scattering rate is inverse of scattering time we find that (this is called Mathieson's rule) the impurity-related time $\tau_{sc}^{(imp)}$ is given by

$$\frac{1}{\tau_{sc}^{(2)}} = \frac{1}{\tau_{sc}^{(1)}} + \frac{1}{\tau_{sc}^{(imp)}}$$

which gives

$$\tau_{sc}^{(imp)} = 4.6 \times 10^{-13} \text{ s}$$

EXAMPLE 4.4 The mobility of electrons in pure silicon at 300 K is 1500 cm^2/Vs. Calculate the time between scattering events using the conductivity effective mass.

The conductivity mass for indirect semiconductors, such as Si, is given by

$$m_\sigma^* = 3\left(\frac{2}{m_t^*} + \frac{1}{m_\ell^*}\right)^{-1}$$

$$= 3\left(\frac{2}{0.19m_o} + \frac{1}{0.98m_o}\right)^{-1} = 0.26 m_o$$

The scattering time is then

$$\tau_{sc} = \frac{\mu m_\sigma^*}{e} = \frac{(0.26 \times 0.91 \times 10^{-30})(1500 \times 10^{-4})}{1.6 \times 10^{-19}}$$

$$= 2.2 \times 10^{-13} \text{ s}$$

EXAMPLE 4.5 Consider two semiconductor samples, one Si and one GaAs. Both materials are doped n-type at $N_d = 10^{17}$ cm^{-3}. Assume 50 % of the donors are ionized at 300 K. Calculate the conductivity of the samples. Compare this conductivity to the conductivity of undoped samples.

You may assume the following values:

$$\mu_n(\text{Si}) = 1000 \text{ cm}^2/\text{V}\cdot\text{s}$$
$$\mu_p(\text{Si}) = 350 \text{ cm}^2/\text{V}\cdot\text{s}$$
$$\mu_n(\text{GaAs}) = 8000 \text{ cm}^2/\text{V}\cdot\text{s}$$
$$\mu_p(\text{GaAs}) = 400 \text{ cm}^2/\text{V}\cdot\text{s}$$

4.4. Macroscopic transport properties

In the doped semiconductors, the electron density is (50 % of 10^{17} cm^{-3})

$$n_{doped} = 5 \times 10^{16} \text{ cm}^{-3}$$

and hole density can be found from

$$p_{doped} = \frac{n_i^2}{n_{doped}}$$

For silicon we have

$$p_{doped} = \frac{2.25 \times 10^{20}}{5 \times 10^{16}} = 4.5 \times 10^3 \text{ cm}^{-3}$$

which is negligible for the conductivity calculation.
The conductivity is

$$\sigma_{doped} = ne\mu_n + pe\mu_p = 8 \text{ } (\Omega \text{ cm})^{-1}$$

In the case of undoped silicon we get ($n = n_i = p = 1.5 \times 10^{10}$ cm^{-3})

$$\sigma_{undoped} = n_i e\mu_n + p_i e\mu_p = 3.24 \times 10^{-6} \text{ } (\Omega \text{ cm})^{-1}$$

For GaAs we get

$$\sigma_{doped} = 5 \times 10^{16} \times 1.6 \times 10^{-19} \times 8000 = 64 \text{ } (\Omega \text{ cm})^{-1}$$

For undoped GaAs we get ($n_i = 1.84 \times 10^6$ cm^{-3})

$$\sigma_{undoped} = n_i e\mu_n + p_i e\mu_p = 2.47 \times 10^{-9} \text{ } (\Omega \text{ cm})^{-1}$$

You can see the very large difference in the conductivities of the doped and undoped samples. Also there is a large difference between GaAs and Si.

EXAMPLE 4.6 Consider a semiconductor in equilibrium in which the position of the Fermi level can be placed anywhere *within the bandgap*.

What is the maximum and minimum conductivity for Si and GaAs at 300 K? You can use the data given in the problem above.

The maximum carrier density occurs when the Fermi level coincides with the conduction bandedge if $N_c > N_v$ or with the valence bandedge if $N_v > N_c$. If $N_c > N_v$; the Boltzmann approximation gives

$$n_{max} = N_c$$

while if $N_v > N_c$ we get

$$p_{max} = N_v$$

This gives us for the maximum density: i) for Si, 2.78×10^{19} cm^{-3} ii) for GaAs, 7.72×10^{18} cm^{-3}. Based on these numbers we can calculate the maximum conducitvity:
For Si

$$\sigma_{max} = 2.78 \times 10^{19} \times 1.6 \times 10^{-19} \times 1000 = 4.45 \times 10^3 \text{ } (\Omega \text{ cm})^{-1}$$

For GaAs

$$\sigma_{max} = 7.72 \times 10^{18} \times 1.6 \times 10^{-19} \times 400 = 4.9 \times 10^2 \text{ } (\Omega \text{ cm})^{-1}$$

To find the minimum conductivity we need to find the minima of the expression

$$\sigma = ne\mu_n + pe\mu_p$$
$$= \frac{n_i^2}{p}e\mu_n + pe\mu_p$$

To find the minimum we take the derivative with respect to p and equate the result to zero. This gives

$$p = n_i\sqrt{\frac{\mu_n}{\mu_p}}$$

This then gives for the minimum conductivity

$$\sigma_{min} = n_i e[\mu_n\sqrt{\frac{\mu_p}{\mu_n}} + \mu_p\sqrt{\frac{\mu_n}{\mu_p}}]$$

For Si this gives upon plugging in numbers

$$\sigma_{min} = 2.8 \times 10^{-6} \; (\Omega \; cm)^{-1}$$

and for GaAs

$$\sigma_{min} = 1.05 \times 10^{-9} \; (\Omega \; cm)^{-1}$$

Note that these values are lower than the values we get in the the previous problem for the undoped cases. This example shows the tremendous variation in conductivity that can be obtained in a semiconductor.

High field transport: velocity–field relations

In most electronic devices a significant portion of the electronic transport occurs under strong electric fields. This is especially true of field effect transistors. At such high fields ($\sim 1- 100$ kV/cm) the electrons get "hot" and aquire a high average energy. The extra energy comes due to the strong electric fields. The drift velocities are also quite high. The description of electrons at such high electric fields is quite complex and requires either numerical techniques or computer simulations. We will only summarize the results.

At high electric field as the carriers gain energy from the field they suffer greater rates of scattering, i.e., τ_{sc} decreases. The mobility thus starts to decrease. It is usual to represent the response of the carriers to the electric field by velocity–field relations. In Fig. 4.10 we show such relations for several semiconductors. At very high fields the drift velocity becomes saturated; i.e., becomes independent of the electric field. The drift velocity for carriers in most materials saturates to a value of $\sim 10^7$ cm/s. The fact that the velocity saturates is very important in understanding current flow in semiconductor devices.

EXAMPLE 4.7 The mobility of electrons in a semiconductor decreases as the electric field is increased. This is because the scattering rate increases as electrons become hotter due to the applied field. Calculate the relaxation time of electrons in silicon at 1 kV/cm and 100 kV/cm at 300 K.

The velocity of the silicon electrons at 1 kV/cm and 100 kV/cm is approximately 1.4 $\times \; 10^6$ cm s and 1.0×10^7 cm/s, respectively, from the v-F curves given in Fig. 4.10. The mobilities are then

$$\mu(1 \; kV/cm) = \frac{v}{E} = 1400 \; cm^2/V \cdot s$$
$$\mu(100 \; kV/cm) = 100 \; cm^2/V \cdot s$$

4.4. Macroscopic transport properties

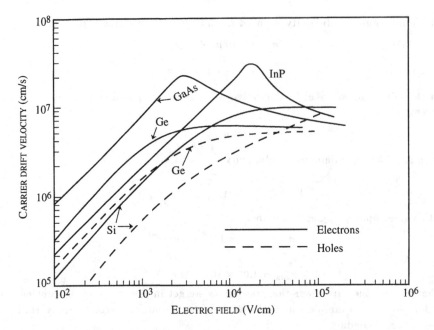

Figure 4.10: Velocity–field relations for several semiconductors at 300 K.

The corresponding relaxation times are

$$\tau_{sc}(1 \text{ kV/cm}) = \frac{(0.26 \times 0.91 \times 10^{-30} \text{ kg})(1400 \times 10^{-4} \text{ m}^2/\text{V})}{1.6 \times 10^{-19} \text{ C}} = 2.1 \times 10^{-13} \text{ s}$$

$$\tau_{sc}(100 \text{ kV/cm}) = \frac{(0.26 \times 0.91 \times 10^{-30})(100 \times 10^{-4})}{1.6 \times 10^{-19}} = 1.48 \times 10^{-14} \text{ s}$$

Thus the scattering rate has dramatically increased at the higher field.

EXAMPLE 4.8 The average electric field in a particular 2.0 μm GaAs device is 5 kV/cm. Calculate the transit time of an electron through the device (a) if the low field mobility value of 8000 cm²/V·s is used; (b) if the saturation velocity value of 10^7 cm/s is used.

If the low field mobility is used, the average velocity of the electron is

$$v = \mu E = (8000 \text{ cm}^2/\text{Vs}) \times (5 \times 10^3 \text{ V/cm}) = 4 \times 10^7 \text{ cm/s}$$

The transit time through the device becomes

$$\tau_{tr} = \frac{L}{v} = \frac{2.0 \times 10^{-4} \text{ cm}}{4 \times 10^7 \text{ cm/s}} = 5 \text{ ps}$$

The transit time, if the saturation velocity (which is the correct velocity value) is used, is

$$\tau_{tr} = \frac{L}{v} = \frac{2 \times 10^{-4}}{10^7} = 20 \text{ ps}$$

In our discussion on MESFET and MOSFET devices later in the text, we use a simple analytical model, and use the constant mobility model for electron velocity. As this example shows, this can cause an error in transit time.

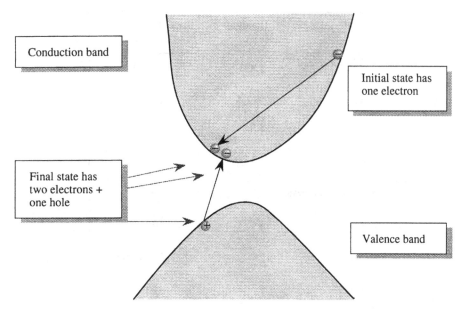

Figure 4.11: The impact ionization process where a high energy conduction-band electron scatters from a valence-band electron, producing two conduction-band electrons and a hole.

Very high field transport: breakdown phenomena

When the electric field becomes extremely high ($\gtrsim 100$ kV cm^{-1}), the semiconductor suffers a "breakdown" in which the current has a "runaway" behavior. The breakdown occurs due to carrier multiplication, which arises from the two sources discussed below. By carrier multiplication we mean that the number of electrons and holes that can participate in current flow increases. Of course, the total number of electrons is always conserved.

Avalanche breakdown

In the transport considered in the previous subsections, the electron (hole) remains in the same band during the transport. At very high electric fields, this does not hold true. In the impact ionization process shown schematically in Fig. 4.11, an electron, which is "very hot" (i.e., has a very high energy due to the applied field) scatters with an electron in the valence band via coulombic interaction, and knocks it into the conduction band. The initial electron must provide enough energy to bring the valence-band electron up into the conduction band. Thus the *initial electron should have energy slightly larger than the bandgap* (measured from the conduction-band minimum). In the final state we now have two electrons in the conduction band and one hole in the valence band. Thus the number of current carrying charges have multiplied, and the process is often called *avalanching*. Note that the same could happen to "hot holes" and thus could then trigger the avalanche.

4.4. Macroscopic transport properties

Material	Bandgap (eV)	Breakdown electric field (V/cm)
GaAs	1.43	4×10^5
Ge	0.664	10^5
InP	1.34	
Si	1.1	3×10^5
$In_{0.53}Ga_{0.47}As$	0.8	2×10^5
C	5.5	10^7
SiC	2.9	$2\text{-}3 \times 10^6$
SiO_2	9	$\sim 10^7$
Si_3N_4	5	$\sim 10^7$
GaN	3.4	$\sim 5 \times 10^6$

Table 4.2: Breakdown electric fields in some materials.

Once avalanching starts, the carrier density in a device changes as

$$\frac{dn(z)}{dz} = \alpha_{imp} n \qquad (4.34)$$

where n is the carrier density and α_{imp} represents the average rate of ionization per unit distance.

The coefficients α_{imp} for electrons and β_{imp} for holes depend upon the bandgap of the material in a very strong manner. This is because, as discussed above, the process can start only if the initial electron has a kinetic energy equal to a certain threshold (roughly equal to the bandgap). This is achieved for lower electric fields in narrow gap materials.

If the electric field is constant so that α_{imp} is constant, the number of times an initial electron will suffer impact ionization after traveling a distance x is

$$n(x) = \exp(\alpha_{imp} z) \qquad (4.35)$$

A critical breakdown field E_{crit} is defined where α_{imp} or β_{imp} approaches 10^4 cm^{-1}. When α_{imp} (β_{imp}) approaches 10^4 cm^{-1}, there is about one impact ionization when a carrier travels a distance of one micron. Values of the critical field are

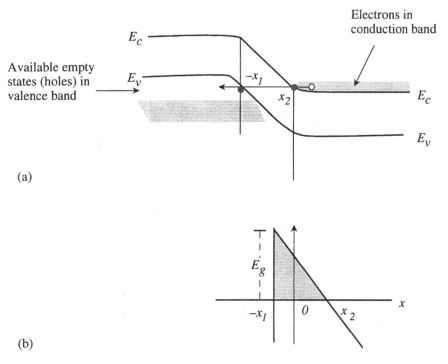

Figure 4.12: (a) A schematic showing the E–x and E–k diagram for a p–n junction. An electron in the conduction band can tunnel into an unoccupied state in the valence band or vice versa. (b) The potential profile seen by the electron during the tunneling process.

given for several semiconductors in Table 4.2. The avalanche process places an important limitation on the power output of devices. Once the process starts, the current rapidly increases due to carrier multiplication and the control over the device is lost. The push for high-power devices is one of the reasons for research in large gap semiconductor devices. It must be noted that in certain devices, such as avalanche photodetectors, the process is exploited for high gain detection. The process is also exploited in special microwave devices.

Band-to-band tunneling breakdown

In quantum mechanics electrons behave as waves and one of the outcomes of this is that electrons can tunnel through regions where classically they are forbidden. Thus they can penetrate regions where the potential energy is larger than their total energy. This process is described by the tunneling theory. This theory is invoked to understand another phenomenon responsible for high field breakdown. Consider a semiconductor under a strong field, as shown in Fig. 4.12a. At strong electric fields, the electrons in the valence band can tunnel into an unoccupied state in the conduction band. As the electron tunnels, it sees the potential profile shown in Fig. 4.12b.

4.5. Carrier transport by diffusion

The tunneling probability through the triangular barrier is given by

$$T = \exp\left(\frac{-4\sqrt{2m^*}\, E_g^{3/2}}{3e\hbar E}\right) \tag{4.36}$$

where E is the electric field in the semiconductor.

In narrow bandgap materials this band-to-band tunneling or Zener tunneling can be very important. It is the basis of the Zener diode, where the current is essentially zero until the band-to-band tunneling starts and the current increases very sharply. A tunneling probability of $\sim 10^{-6}$ is necessary to start the breakdown process.

EXAMPLE 4.9 Calculate the band-to-band tunneling probability in GaAs and InAs at an applied electric field of 2×10^5 V/cm.

The exponent for the tunneling probability is (m^*(GaAs) = 0.065 m_0; m^*(InAs) ~ 0.02 m_0; E_g(GaAs) = 1.5 eV; E_g(InAs) = 0.4 eV) for GaAs

$$= -\frac{4 \times (2 \times 0.065 \times 0.91 \times 10^{-30} \text{ kg})^{1/2}(1.5 \times 1.6 \times 10^{-19} \text{ J})^{3/2}}{3 \times (1.6 \times 10^{-19} \text{ C})(1.05 \times 10^{-34} \text{ Js})(2 \times 10^7 \text{ V/m})}$$

$$= -160$$

The tunneling probability is $\exp(-160) \cong 0$. For InAs the exponent turns out to be -12.5 and the tunneling probability is

$$T = \exp(-12.5) = 3.7 \times 10^{-6}$$

In InAs the band-to-band tunneling will start becoming very important if the field is $\sim 2 \times 10^5$ V/cm.

4.5 CARRIER TRANSPORT BY DIFFUSION

In the previous sections we have seen how the force eE applied by an electric field causes transport in materials. There is another important transport mechanism that does not involve such a direct force. This is the diffusion process. Whenever there is a gradient in the concentration of a species of mobile particles, the particles diffuse from the regions of high concentration to the regions of low concentration. This diffusion is due to the random motion of the particles.

In the case of electrons (or holes), as the particles move they suffer random collisions, as discussed in the previous section. The collision process can be described by the mean free path ℓ and the mean collision time τ_{sc}. The mean free path is the average distance the electron (hole) travels between successive collisions. These collisions are due to the various scattering processes that were discussed for the drift problem. In between the collisions the electrons move randomly, with equal probability of moving in any direction (there is no electric field). We are interested in finding out how the electrons move (diffuse) when there is a concentration gradient in space.

Consider a concentration profile $n(x,t)$ of electrons at time t, as shown in Fig. 4.13. We are going to calculate the electron flux $\phi(x,t)$ across a plane $x = x_o$ at any instant of time. We consider a region of space a mean free path ℓ to each side of x_o,

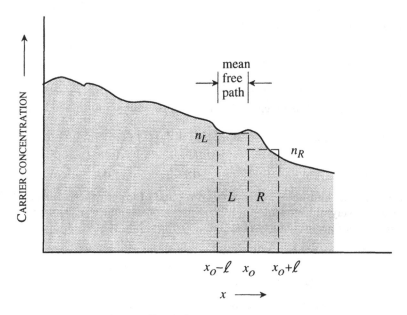

Figure 4.13: The concentration profile of electrons as a function of space. The terms n_L, n_R, L, and R are used to derive the diffusion law in the text. The distance ℓ is the mean free path for electrons; i.e., the distance they travel between collisions.

from which electrons can come across the $x = x_o$ boundary in time τ_{sc}. Electrons from regions further away will suffer collisions that will randomly change their direction. Since in the two regions labeled L and R in Fig. 4.13, the electrons move randomly, half of the electrons in region L will go across $x = x_o$ to the right and half in the region R will go across $x = x_o$ to the left in time τ_{sc}. The flux to the right is then

$$\phi_n(x,t) = \frac{(n_L - n_R)\ell}{2\tau_{sc}} \tag{4.37}$$

where n_L and n_R are the average carrier densities in the two regions. Since the two regions L and R are separated by the distance ℓ, we can write

$$n_L - n_R \cong -\frac{dn}{dx} \cdot \ell \tag{4.38}$$

Thus the net flux is

$$\phi_n(x,t) = -\frac{\ell^2}{2\tau_{sc}} \frac{dn(x,t)}{dx} = -D_n \frac{dn(x,t)}{dx} \tag{4.39}$$

where D_n is called the diffusion coefficient of the electron system and clearly depends upon the scattering processes that control ℓ and the τ_{sc}. Since the mean free path is essentially $v_{th}\tau_{sc}$, where v_{th} is the mean thermal speed, the diffusion coefficient depends upon the temperature as well. In a similar manner, the hole diffusion coefficient gives

4.5. Carrier transport by diffusion

the hole flux due to a hole density gradient

$$\phi_p(x,t) = -D_p \frac{dp(x,t)}{dx} \qquad (4.40)$$

Because of this electron and hole flux, a current can flow in the structure that, in the absence of an electric field, is given by (current is just charge multiplied by particle flux)

$$\begin{aligned}\mathbf{J}_{tot}(diff) &= \mathbf{J}_n(diff) + \mathbf{J}_p(diff) \\ &= eD_n \frac{dn(x,t)}{dx} - eD_p \frac{dp(x,t)}{dx}\end{aligned} \qquad (4.41)$$

While both electrons and holes move in the direction of lower concentration of electrons and holes respectively, the currents they carry are opposite, since electrons are negatively charged, while holes are positively charged.

EXAMPLE 4.10 In an n-type GaAs crystal at 300 K, the electron concentration varies along the x-direction as

$$n(x) = 10^{16} \exp\left(-\frac{x}{L}\right) \text{ cm}^{-3} \qquad x > 0$$

where L is 1.0 μm. Calculate the diffusion current density at $x = 0$ if the electron diffusion coefficient is 220 cm^2/s.

The diffusion current density at $x = 0$ is

$$\begin{aligned}J_n(diff) &= eD_n \frac{dn}{dx}\bigg|_{x=0} \\ &= \left(1.6 \times 10^{-19} \text{ C}\right)\left(220 \text{ cm}^2/\text{s}\right)\left(\frac{10^{16} \text{ cm}^{-3}}{10^{-4} \text{ cm}}\right) \\ &= 3.5\text{kA}/\text{cm}^2\end{aligned}$$

Note that in this problem the diffusion current of electrons changes with the position in space. Since the total current is constant in the absence of any source or sink of current, some other current must be present to compensate for the spatial change in electron diffusion current.

4.5.1 Transport by drift and diffusion: Einstein's relation

In many electronic devices the charge moves under the combined influence of electric fields and concentration gradients. The current is then given by

$$\begin{aligned}\mathbf{J}_n(x) &= e\mu_n n(x)\mathbf{E}(x) + eD_n \frac{dn(x)}{dx} \\ \mathbf{J}_p(x) &= e\mu_p p(x)\mathbf{E}(x) - eD_p \frac{dp(x)}{dx}\end{aligned} \qquad (4.42)$$

In our discussion of the diffusion coefficient we indicated that it is controlled by essentially the same scattering mechanisms that control the mobility. We will now establish an important relationship between mobility and diffusion coefficients. To do so, let us examine the effect of electric fields on the energy bands of the semiconductor.

We consider a case where a uniform electric field is applied, as shown in Fig. 4.14a. The potential energy associated with the field is shown in Fig. 4.14b. There is a positive potential on the left-hand side in relation to the right-hand side. For a uniform electric field the potential energy is

$$U(x) = U(0) - eEx \tag{4.43}$$

The potential energy profile is shown in Fig. 4.14b. We now discuss how the electron energy band profile is displayed. The electron energy band includes the effect of the negative charge of the electrons. The applied force is related to the potential energy by

$$\text{Force} = -\nabla U(x) \tag{4.44}$$

Thus, since the electron charge $-e$ is negative, the bands bend as shown in Fig. 4.14c according to the relation

$$E_c(x) = E_c(0) + eEx \tag{4.45}$$

Thus, if a positive potential is applied to the left of the material and a negative to the right, the energy bands will be lower on the left-hand side, as shown in Fig. 4.14c. The electrons drift downhill in the energy band picture and thus opposite to the field.

At equilibrium, the total electron and hole currents are individually zero and we have from Eq. 4.42 for the electrons

$$\mathbf{E}(x) = -\frac{D_n}{\mu_n} \frac{1}{n(x)} \frac{dn(x)}{dx} \tag{4.46}$$

To obtain the derivative of carrier concentration, we write $n(x)$ in terms of the intrinsic Fermi level, E_{Fi}, which serves as a reference level, and the Fermi level in the semiconductor, $E_F(x)$. If we assume that the electron distribution is given by the Boltzmann distribution we have

$$n(x) = n_i \exp\left\{-\left(\frac{E_{Fi} - E_F(x)}{k_B T}\right)\right\} \tag{4.47}$$

This gives

$$\frac{dn(x)}{dx} = \frac{n(x)}{k_B T}\left(-\frac{dE_{Fi}}{dx} + \frac{dE_F}{dx}\right) \tag{4.48}$$

At equilibrium, the Fermi level cannot vary spatially, otherwise the probability of finding electrons along a constant energy position will vary along the semiconductor. This would cause electrons at a given energy in a region where the probability is low to move to the same energy in a region where the probability is high. Since this is not allowed by definition of equilibrium conditions, i.e. no current is flowing, the Fermi level has to be uniform in space at equilibrium, or

$$\frac{dE_F}{dx} = 0 \tag{4.49}$$

We then have from Eqs. 4.43 and 4.45 (using $E(x) = \frac{1}{e}\frac{dE_{Fi}}{dx}$)

$$\frac{D_n}{\mu_n} = \frac{k_B T}{e} \tag{4.50}$$

4.5. Carrier transport by diffusion

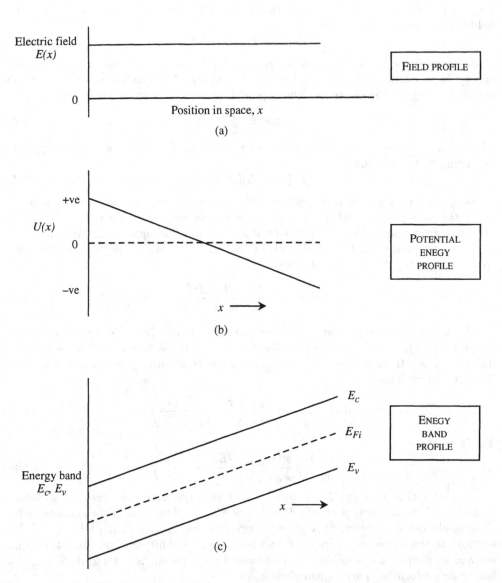

Figure 4.14: (a) Electric field profile in a semiconductor. (b) Plot of the potential energy associated with the electric field. (c) Electron energy band profile. The negative charge of the electron causes the energy band profile to have the opposite sign to the potential energy profile.

	D_n (cm²/s)	D_p (cm²/s)	μ_n (cm²/V·s)	μ_p (cm²/V·s)
Ge	100	50	3900	1900
Si	35	12.5	1350	480
GaAs	220	10	8500	400

Table 4.3: Low field mobility and diffusion coefficients for several semiconductors at room temperature. The Einstein relation is satisfied quite well.

which is the Einstein relation satisfied for electrons. A similar relation exists for the holes.

Table 4.3 lists the mobilities and diffusion coefficients for a few semiconductors at room temperature.

EXAMPLE 4.11 Use the velocity–field relations for electrons in silicon to obtain the diffusion coefficient at an electric field of 1 kV/cm and 10 kV/cm at 300 K.

According to the v-E relations given in Fig. 4.10, the velocity of electrons in silicon is $\sim 1.4 \times 10^6$ cm/s and $\sim 7 \times 10^6$ cm/s at 1 kV/cm and 10 kV/cm. Using the Einstein relation, we have for the diffusion coefficient

$$D = \frac{\mu k_B T}{e} = \frac{v k_B T}{eF}$$

This gives

$$D(1kV/\text{cm}^{-1}) = \frac{(1.4 \times 10^4 m/s)(0.026 \times 1.6 \times 10^{-19} \text{ J})}{(1.6 \times 10^{-19} \text{ C})(10^5 \text{ V/m}^{-1})}$$
$$= 3.64 \times 10^{-3} \text{ m}^2/\text{s} = 36.4 \text{ cm}^2/\text{s}$$
$$D(10kV/\text{cm}^{-1}) = \frac{(7 \times 10^4 \text{ m/s})(0.026 \times 1.6 \times 10^{-19} \text{ J})}{(1.6 \times 10^{-19} \text{ C})(10^6 \text{ Vm}^{-1})}$$
$$= 1.82 \times 10^{-3} \text{ m}^2/\text{s} = 18.2 \text{ cm}^2/\text{s}$$

The diffusion coefficient decreases with the field because of the higher scattering rate at higher fields.

4.6 IMPORTANT DEVICES BASED ON CONDUCTIVITY CHANGES

In this section we will provide a brief overview of the important devices that are based on electron transport in high-quality crystalline materials. High performance semiconductor devices, such as field effect transistors and bipolar transistors, are responsible for high-speed microprocessors, satellite communication systems, radar systems, and drivers for high-speed optical communication systems. We will mainly discuss the motivations driving advances in these devices – but not provide details of device design and operation.

4.6. Important devices based on conductivity changes

Electronic devices based on semiconductors fall into two categories–unipolar and bipolar. In unipolar devices only one type of charge flow occurs (i.e., either electrons or holes), while in bipolar devices both electrons and holes participate in current flow. Devices, such as field effect transistors (which include the metal oxide semiconductor field effect transistor (MOSFET)) and other variations are unipolar devices. Bipolar junction transistors (BJTs) or heterojunction bipolar transistors (HBTs) involve both electron and hole current flow.

We will briefly discuss how semiconductor devices function, but before that let us take a look at some of the driving forces behind semiconductor technology.

Semiconductor digital and analog devices operate by being able to alter rapidly the current flowing in the device by the application of a small input signal. The current in a device can be changed by changing the electron or hole density, or the area A through which current is flowing. The way the transistor operation occurs enables it to alter n and an effective channel opening area A quite easily, and at high speed.

As will be seen from the brief discussion below the needs for high performance devices involve (i) superior mobility to improve transit times, (ii) large bandgap to avoid high voltage breakdown; (iii) semiconductors, which can be combined with large bandedge discontinuities, etc. As a result, while Si and GaAs are the most important semiconductors many other materials are finding important applications.

4.6.1 Field effect transistor

The field effect transistor (FET) forms the backbone of the digital and analog microelectronic systems. From consumer electronic goods, such as stereo systems and microprocessors on automobiles, to satellite communication systems, the FETs provide a versatile device. This three-terminal device, consisting of source and drain ohmic contacts through which current flows and a gate that is isolated from the active channel, can be loosely compared to a water tap. The gate is equivalent to the faucet handle which shuts off the water supply by constricting the flow in the channel.

The most important FET structure is the Si-based Metal-Oxide-Semiconductor FET or MOSFET. One of the reasons for the great success of Si technology is that a high-quality oxide (an insulator) can be grown directly on it. This oxide layer, with a bandgap of 9.0 eV, effectively isolates the channel where the charge flows from the gate. A strong bias can then be applied to the gate without drawing any current into the gate.

In the MOSFET, the charge in the channel (i.e., at the oxide–semiconductor junction) is not due to any donors, but is produced by "inversion" created by the gate bias. For example, the n-MOS device is produced in p^- Si, as shown in Fig. 4.15a. In Fig. 4.15 we show how a gate bias can create band bending, so that there is an accumulation of holes (Fig. 4.15b) at the interface, and a depletion state, where there are very few mobile changes at the interface (Fig. 4.15c). This is the OFF state. The gate bias can also create an inversion state where there is a high density of electrons at the interface (Fig. 4.15d). This is the ON state. The variation of the gate bias can change the electron concentration and thus modulate the current flowing through the source and drain.

For semiconductors, such as GaAs, InP, InGaAs, etc., there is no oxide (or

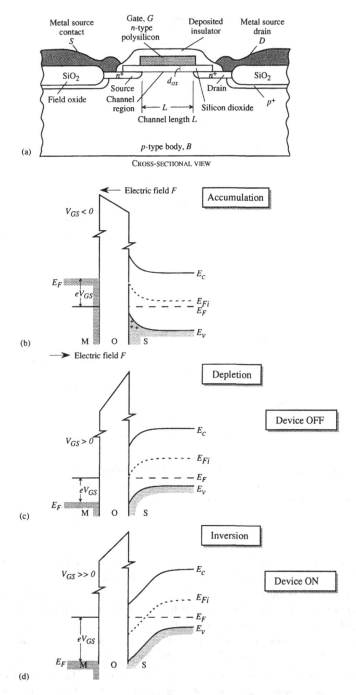

Figure 4.15: (a) A schematic of an n-MOS device. In figures (b), (c), and (d), the effect of the gate potential on the electron and hole density at the interfaces is shown. When the excess electrons are formed under inversion conditions, the device channel is conducting.

4.6. Important devices based on conductivity changes

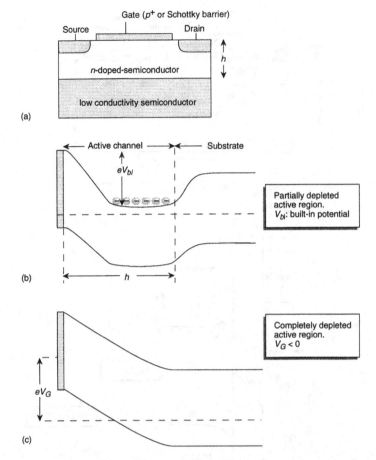

Figure 4.16: (a) A schematic of a JFET or MESFET showing the source, drain, and gate. The channel width of this device is h. (b) The band profile when the applied gate bias is zero as is the source-drain bias. (c) The band profile with a negative gate bias so that the channel is depleted.

other simple compound) which could act as an insulator to isolate the gate from the channel. This is a big disadvantage for these materials, since the simple and elegant MOS concept cannot be used. Instead we have to use a Schottky barrier to provide the gate to channel isolation, and the material has to be doped to provide the active charge in the channel. The gate potential modulates the charge in the channel by modulating the depletion width under the gate. This device is known as a Metal-Semiconductor FET or MESFET, and is shown schematically in Fig. 4.16.

In the MESFET the gate voltage controls the area of the channel opening where mobile carriers are present. As shown in Fig. 4.16 the gate bias increases or decreases the depletion region and thus controls the gate current flowing from the source to the drain. In the MESFET, the free charge moves through and scatters from the fixed donor (acceptor) impurities, resulting in mobilities that are lower than those in undoped materials. Also, the device operation suffers at low temperatures because of the carrier freeze-out effect discussed in Chapter 3. An ingenious way out of both of these problems is provided by the MODFET, using simple heterostructure concepts. In this device, as shown in Fig. 4.17, the dopants are placed in a larger bandgap material, which has a positive conduction band offset (for the n-MODFET). The electrons spill into the lower bandgap semiconductor, and the resulting charge separation causes band bending according to the Poisson equation, resulting in the formation of an electron channel. The key improvement over the MESFET is the improved mobility and superior low temperature performance due to reduced ionized impurity scattering and lack of carrier freeze-out. In fact, the MODFET has taken over all device functions in systems where very high performance is the key requirement.

What is the driving force for the FET structures? To answer this question we must examine a few simple concepts that describe device performance. The driving forces for the FET designer come from one of the following general considerations: faster switching time for digital applications; lower power dissipation for digital systems in the switching process; higher frequencies of operation for microwave applications; higher power output for microwave applications; and higher temperature performance.

We will briefly discuss the important motivation for device research in semiconductor electronics. As can be seen in Table 4.4 there are a number of important driving forces.

- **High frequency/high speed: scaling**

The most important approach in the march towards higher speed devices has been the scaling of device dimensions. The scaling approach has led to the famous Moore's law, according to which the complexity of a semiconductor chip doubles every 18 months.

Shrinking of FET gate length essentially reduces the transit time of electrons from the source to the drain. As an approximation the transit time t_{tr} is given by

$$t_{tr} \sim \frac{L}{v} \qquad (4.51)$$

where v is approximately the saturation velocity of the electrons. An obvious way to reduce the transit time is to reduce the device length. This has been a primary method to develop high-speed technologies. Device dimensions are scaled down through advances in technology.

To continue to scale devices, in addition to lithographic-tools-related challenges, an emerging challenge is becoming the tunneling-related leakage current at very small dimensions. In the Si MOSFET technology, scaling laws demand that the gate length L_G and oxide thickness are related by

$$L_G \sim 45 d_{ox} \qquad (4.52)$$

It is straightforward to see that once SiO_2 thickness becomes ~ 20 Å (i.e., L_G reaches sub 0.1 μm) tunneling current through the insulator becomes very large, and the device becomes leaky.

4.6. Important devices based on conductivity changes

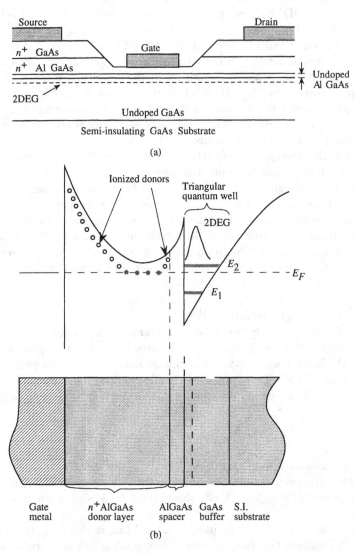

Figure 4.17: (a) A schematic of a GaAs/AlGaAs MODFET structure. (b) The transfer of electrons from the barrier region to the GaAs interface, forming a 2D electron gas. Since electrons are spatially separated from donors they do not suffer much scattering from ionized impurities.

MOSFETs	Smaller dimensions
	Use of SiGe systems
	Superior interface
	High dielectric constant insulators
MESFETs MODFETs	New, faster material systems
	Submicron gates
	Improved ohmic technology
	Large bandgap semiconductors

Table 4.4: Approaches for improving FET performance.

To avoid the gate-tunneling leakage current there is considerable research work in using high dielectric constant insulators. Several systems including oxides, such as $BzTiO_3$, $SrTiO_3$, etc., are being examined. A large dielectric constant allows the insulator to be thicker (and hence tunneling is suppressed), while maintaining gate control.

- **High speed/high frequency: new faster materials**

It is possible to increase device speed (i.e., decrease electron transit time) by using materials with small effective masses. A smaller effective mass not only ensures greater acceleration in a field, there is also less scattering. The push from Si to GaAs to InGaAs, etc., is based on this motivation. It has to be remembered, though, that silicon remains the most reliable and inexpensive technology. "Fast materials" are used only when speed is critical.

- **High power/high temperature devices**

Both high-power and high-temperature devices require large bandgap semiconductors. A large bandgap allows a large breakdown field and the low intrinsic carrier concentration necessary for operations under large bias and high temperatures. Large bandgap technologies currently in commercial use include SiC- and GaN-based devices. Diamond (C) is a large bandgap semiconductor with great potential, although many technology challenges have to be overcome to make C-based devices feasible.

4.6.2 Bipolar junction devices

Bipolar junction transistors (BJTs) and their superior versions, heterojunction BJTs (or HBTs), are important electronic devices used for analog and digital applications. As we have seen in the previous subsection, FETs operate by a gate signal manipulating the total channel charge. In the case of a bipolar transistor, a "potential barrier" is placed in the path of electrons (or holes). The raising or lowering of the barrier controls the current flow.

4.6. Important devices based on conductivity changes

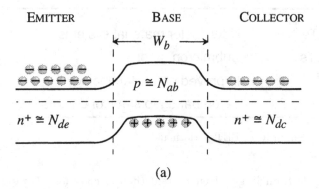

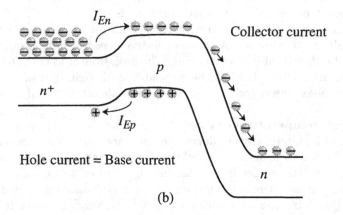

Figure 4.18: (a) Band profile of an unbiased n^+p–n BJT. (b) Band profile of a BJT biased for high gain.

DESIGN CONSIDERATIONS

Low base resistance ⇒ wide base and high doping

Low transit time ⇒ narrow base

High base-emitter valence band offset (for *n-p-n* devices)

Low base-collector conduction band offset (for *n-p-n* devices)

High quality interfaces to suppress base recombination

Small dimensions

New material systems (SiGe, InGaAs)

Large bandgap systems for high power

Table 4.5: Driving forces for improved bipolar devices.

In Fig. 4.18 we show a typical n–p–n BJT. The current flowing from the emitter to the collector can be controlled by raising or lowering the barrier created by the base potential. Thus a small base current can cause a large change in the collector current providing a large current gain.

The performance of bipolar transistors improves tremendously when the base is made from a material with a bandgap smaller than that of the emitter material. The use of a heterojunction device (HBT) greatly reduces the current that flows from the base into the emitter, and thus improves device efficiency. In Fig. 4.19 we show a typical HBT based on the AlGaAs/GaAs system. As can be seen from this figure, the use of the lower gap base increases the barrier for hole injection into the emitter.

Advances in bipolar technology are driven by forces similar to those for FET technology. In Fig. 4.5 we show some of the motivations.

4.7 TRANSPORT IN NON-CRYSTALLINE MATERIALS

As we have noted in the previous section, most of the modern high-performance electronic devices are based on high-quality crystalline materials. In these structures there are very few trap states in the bandgap and transport occurs in the conduction and/or the valence band.

4.7. Transport in non-crystalline materials

As applications diversify away from computing and communication into image display, energy conversion, etc., devices based on non-crystalline materials become useful. In these devices speed is not of primary concern. Cost issues and flexibility in substrate choice may be of greater concern. For example, drivers for a display for a large TV screen may only need to respond in milliseconds, but should be fabricated on substrates which may be as large as a 1 m². Similarly, solar cells fabricated on plastic substrates may not be as efficient as crystalline solar cells, but the lower cost may be the overriding factor.

As discussed in Section 4.2, non-crystalline materials have a high density of trap or localized states. Amorphous and polycrystalline silicon are the most important non-crystalline materials with applications in thin film transistors for display controls and solar cells. Other non-crystalline materials are ferroelectrics and ceramics, which have a wide range of applications as memories and sensors. In all of these materials transport does not occur by "free flight" and scattering, as discussed in the previous section, but through processes where an electron is trapped in a bandgap state and then somehow escapes the trap.

4.7.1 Electron and hole transport in disordered systems

A model for transport in non-crystalline materials is shown in Fig. 4.20. Transport involves combinations of "free flight," "scattering," "capture into a localized or trap state," and "escape from the localized state." A number of models have been developed to address such transport. In general transport is quite complex, since it depends upon the degree of defects, disorder, and temperature. However general qualitative aspects can be described.

Three important conduction mechanisms are shown in Fig. 4.21. We will discuss the transport models for the three processes. In the first mode electrons essentially move in the extended states, i.e., at or above the mobility edge. The electrons are excited to the mobility edge by phonons, and the conduction behavior is described by the thermally activated behavior

$$\sigma = \sigma_0 \exp\left[\frac{-(E_c - E_F)}{k_B T}\right] \quad (4.53)$$

where σ_0 is the conductivity at the mobility edge and has been shown to have the form

$$\sigma_0 = \frac{Ce^2}{\hbar a} \quad (4.54)$$

where $C \sim 0.03$ and a is the minimum distance over which phase coherence could occur. This quantity is also called the minimum metallic conductivity. For $a = 3$ Å, the value is 2×10^2 Ω^{-1} cm^{-1}.

The second process indicated in Fig. 4.21 involves thermal activation from one localized state to the nearest state in space above the Fermi level. This process has been used to explain the impurity conduction in doped semiconductors. The electron is always assumed to move to the nearest empty localized state. To estimate this conductivity, we assume that the wavefunctions are described by (assuming a center at the origin)

$$\psi = e^{-\alpha r} \quad (4.55)$$

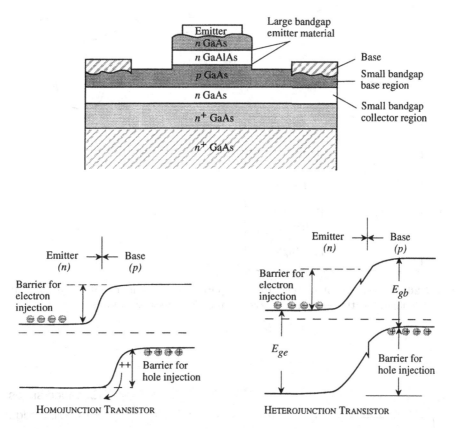

Figure 4.19: A schematic of the n–AlGaAsp–GaAs/n–GaAs heterojunction bipolar transistor under bias. In contrast to a BJT the HBT has a lower base current, due to a higher barrier for hole injection into the emitter, as shown in the bottom panels.

4.7. Transport in non-crystalline materials

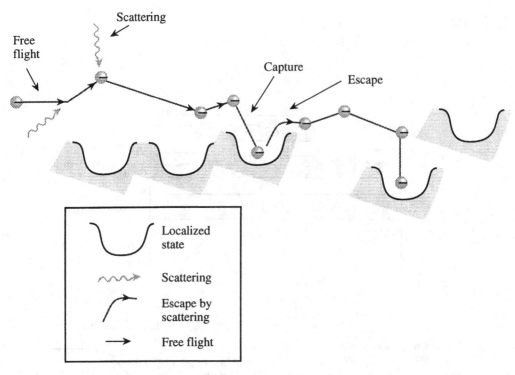

Figure 4.20: A schematic of how electron transport occurs in non-crystalline solids. Transport involves free flight, scattering, capture, and escape (from localized states).

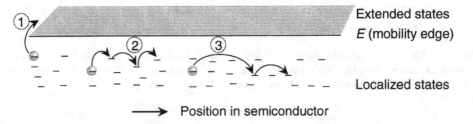

Figure 4.21: Mechanisms for transport in a disordered system. In case 1, the electron is thermal activated to the states above the mobility edge. In case 2, the electron hops to the nearest localized state, while, in the case 3, the electron hops to the "optimum" site, as explained in the text.

where α^{-1} is the localization length. Note that the nature of the localized state is fundamentally different from a Bloch state, which has a plane wave form and propagates in the entire crystal. The current density is proportional to the overlap between the wavefunctions, the density of states at the Fermi level $N(E_F)$, the width of the Fermi distribution $k_B T$, the effective velocity of transport, which is chosen as ν_{ph}, the attempt frequency ($\approx$ phonon frequency) times the average spacing between the states R_0. If ΔE is the average separation between the energies of the two states, and E the applied field, the current density is

$$\begin{aligned} J &\sim ek_B T\, N(E_F)\, R_0\, \nu_{ph}\, \exp(-2\alpha R_0) \\ &\quad \times \left[\exp\left(\frac{-\Delta E + eER_0}{k_B T}\right) - \exp\left(\frac{-\Delta E - eER_0}{k_B T}\right)\right] \\ &= 2ek_B T\, N(E_F)\, R_0\, \nu_{ph}\, \exp\left(-2\alpha R_0 - \frac{\Delta E}{k_B T}\right) \sinh\left(\frac{eER_0}{k_B T}\right) \end{aligned} \qquad (4.56)$$

For small electric fields, the sinh function can be expanded and the conductivity becomes

$$\begin{aligned} \sigma &= \frac{J}{E} \\ &= 2e^2 R_0^2\, \nu_{ph}\, N(E_F)\, \exp\left[-2\alpha R_0 - \frac{\Delta E}{k_B T}\right] \end{aligned} \qquad (4.57)$$

An estimate of the energy spacing of the levels is simply obtained from the definition of the density of states, i.e.

$$\Delta E \approx \frac{1}{N(E_F) R_0^3} \qquad (4.58)$$

where R_0 is the average separation between nearest neighbor states. This kind of nearest neighbor hopping is dominant if nearly all states are strongly localized, e.g. as in impurity states due to dopants.

In many disordered semiconductors, if the disorder is not too strong, we have another important transport process indicated by the third process in Fig. 4.21. This process, known as variable range hopping, was introduced by Mott and is a dominant transport mode at low temperatures. At low temperatures, the hop would not occur to the nearest spatial state, but the electron may prefer to go a potentially farther stat, but one which is closer in energy so that a lower phonon energy is needed.

The density of states per unit energy range near the Fermi level in a sphere of radius R is

$$\left(\frac{4\pi}{3}\right) R^3 N(E_F)$$

Thus, for the hopping process involving a distance within R, the average separation of level energies will be

$$\Delta E = \frac{3}{4\pi R^3\, N(E_F)}$$

As can be seen, the farther the electron hops, the smaller the activation barrier that it needs to overcome. However, a hop of a distance R will involve an overlap function,

4.7. Transport in non-crystalline materials

which falls as $\exp(-2\alpha R)$. Thus, there will be an optimum distance for which the term

$$\exp(-2\alpha R)\exp\left(\frac{-\Delta E}{k_B T}\right)$$

is a maximum. This can easily be seen to occur when

$$\frac{d}{dR}\left[2\alpha R + \frac{3}{4\pi R^3 \, N(E) \, k_B T}\right] = 0$$

or

$$R_m = \left[\frac{1}{8\pi N(E)\,\alpha k_B T}\right]^{1/4}$$

Using this value of R_m, we get for the conductivity behavior

$$\sigma = A\exp\left(-\frac{B}{T^{1/4}}\right) \qquad (4.59)$$

where

$$B = 2\left(\frac{3}{2\pi}\right)^{1/4}\left(\frac{\alpha^3}{k_B N(E_F)}\right)^{1/4}$$

This variable range hopping temperature behavior has been observed in numerous disordered systems.

The formalisms given above for transport in disordered semiconductors provide a mere glimpse into the complexities of transport in disordered materials. We have not discussed high field transport in amorphous materials. It is expected that at high fields, where electron energies are large, carriers will primarily reside in extended states above the mobility edge. Here the transport will be similar to transport in crystalline materials, except the scattering rates would be higher. If we were to express the transport in disordered materials in terms of a mobility, the mobility values are very small. Unlike crystalline materials where mobilities are in the range of $10^3 - 10^4$ cm^2/V· s (for electrons), in disordered systems mobilities are only 0.1 to 1 cm^2/V· s. Nevertheless, as noted earlier, the values are adequate for some applications.

4.7.2 Ionic conduction

So far in this chapter we have discussed transport that occurs due to the motion of electrons or holes. In most semiconductor devices this is the dominant (or only) transport mechanism. There are, however, insulators where the electron (hole) density is negligible due to the large bandgap or due to the very high defect density present. Some such materials include various oxides ($LiNbO_3$, $BaTiO_3$, etc.), which have interesting polarization properties. These materials are used as sensors as well as microwave system elements and optical switches.

In ionic conduction, ions move under an applied electric field through vacancy sites. As shown in Fig. 4.22, there is an energy barrier for an ion to move into a vacant spot. The ion uses a thermal assisted hopping process to overcome the barrier E_i and,

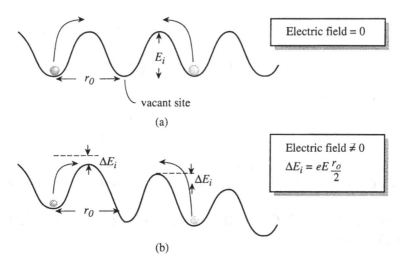

Figure 4.22: A schematic of the configuration energy diagram. (a) In the absence of an electric field hopping, rates to the left and right are equal, since the barriers are E_i. (b) In the presence of a field hopping to the right is more dominant than hopping to the left, due to the changes in the barrier.

therefore, the mobility or diffusion coefficient has very strong temperature dependence (exponential dependence). For ionic conduction to occur, the site into which the ion goes has to be vacant. In the absence of an applied electric field, the diffusion process, which involves hopping from one site to the vacant neighboring site, is random, as shown in Fig. 4.22a. However, when a field is applied, there is a preferential hopping as shown in Fig. 4.22b. The electric field lowers the barrier from left to right hopping by ΔE_i, as shown, while the barrier from right to left is increased. The net particle flow is the difference between the left to right current and the right to left current

$$J = J_{L \rightarrow R} - J_{R \rightarrow L}$$
$$\propto \exp\left[-\frac{(E_i - \Delta E_i)}{k_B T}\right] - \exp\left[-\frac{(E_i + \Delta E_i)}{k_B T}\right]$$

Linearising the $\exp(\Delta E / k_B T)$ term we get ($\Delta E_i = E r_o/2$) as shown in Fig. 4.22b)

$$J \propto \frac{E}{T} \exp\left(\frac{E_i}{k_B T}\right) \qquad (4.60)$$

The current density can be written as

$$J = C \frac{n_v}{N} \frac{E}{T} \exp\left(\frac{-E_i}{k_B T}\right) \qquad (4.61)$$

where C is a constant, n_v/N is the fraction of vacancy sites and is given by

$$\frac{n_v}{N} = \exp\left(\frac{-E_{\text{vac}}}{2 k_B T}\right) \qquad (4.62)$$

where E_{vac} is the energy needed to create a cation and an anion vacancy. The current density is now

$$J = \frac{C}{T} E \exp\left(-\frac{E_{vac}}{2k_B T}\right) \exp\left(\frac{-E_i}{k_B T}\right) \tag{4.63}$$

This gives a conductivity

$$\sigma = \frac{J}{E} = \frac{C}{T} \exp\left\{-\frac{1}{k_B T}\left(E_i + \frac{E_{vac}}{2}\right)\right\} \tag{4.64}$$

Ionic conductivities are very small compared to electronic conductivities, but, as noted earlier in insulators with high defect densities, ionic transport can be important.

4.8 IMPORTANT NON-CRYSTALLINE ELECTRONIC DEVICES

A number of non-crystalline (including crystals with high defect densities) structures are used for device applications. These devices fall into two categories: (i) Devices where conductivity changes occur by the application of an electrical signal (e.g., a voltage to a gate of a transistor), as shown in Fig. 4.23. These devices are similar to the field effect transistors considered in Section 4.6, except their performance is poor because the transport properties (e.g., mobility) is poor in non-crystalline solids. (ii) Devices in which the external perturbation alters the nature of electronic states by changing, say, the density of localized states in the bandgap as shown in Fig. 2.23b. In this case the external perturbation may be a gas or water vapor, depending upon applications. In this category the devices are used as gas sensors and the device conductivity alters due to the changes in the bandgap states.

Devices that fall in the first category mentioned above are made from materials like amorphous or polycrystalline silicon. Also there is a growing interest in using organic semiconductors to make these devices. Organic-semiconductor based devices have a potential advantage that they could be made on flexible (e.g., plastic) substrates. However, many technology issues still need to be resolved to make organic semiconductors a viable alternative to amorphous Si thin film devices.

Gas sensors are usually made from materials traditionally called ceramics. A very large number of materials can be used for gas sensors. Some materials have a greater affinity to detect certain gases and can thus be highly selective in detection.

4.8.1 Thin Film Transistor

Thin film transistors (TFTs) are FETs based on amorphous or polycrystalline materials. They are used in a variety of applications, including the control of display devices, such as laptop screens and flat TVs. More than a million TFTs are required for a full color display. Either polysilicon or amorphous silicon (a-Si) is used to fabricate the device, which is basically an insulated gate field effect transistor. Polycrystalline Si devices have better performance because of superior mobilities of the electrons. Devices used to control display pixels need to switch in timescales of milliseconds instead of nanoseconds

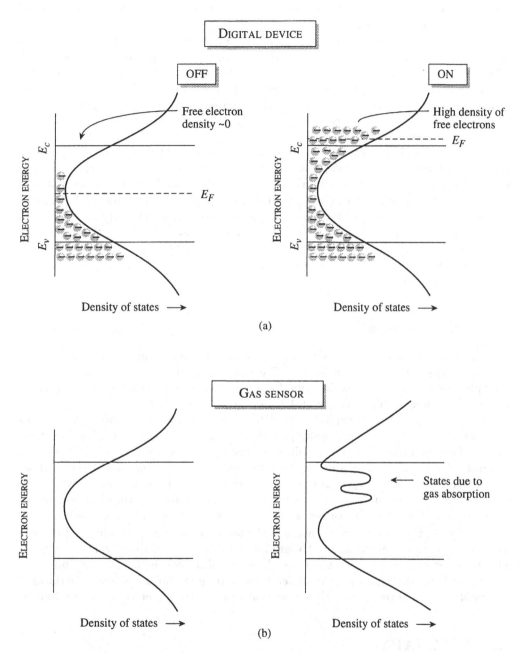

Figure 4.23: (a) A voltage signal shifts the position of the Fermi level and changes the free carrier density. Such a configuration can be used as a switch. (b) In a gas sensor, absorption of a gas molecule alters the density of states and thus alters the device conductivity.

needed for computers and communication devices. Thus even devices with mobilities of 0.1 cm^2/V·s can be useful.

Thin film transistors are essentially similar to MOSFETs, except special designs need to be used due to the substrate, which may be glass or a polymer. The devices do offer considerable challenge because of the difficulty in controlling the material quality. As noted earlier, the material is represented by mobility edges instead of the usual bandedges in crystalline semiconductors. The position of the mobility edges, as well as the density of "localized" states where electrons are essentially trapped, depend upon fabrication processes. At present, the relation between the fabrication process and the exact values of the bandstructure are only understood qualitatively. Since the threshold voltage level in a MOSFET depends critically on the defect and localized interface states, it becomes a challenge to fabricate devices with controllable threshold voltage.

Another important challenge for the TFT is the low mobility of electrons in the transistor. In crystalline MOSFETs, the room temperature mobility of electrons in the channel is ~ 600 cm^2 V^{-1} s^{-1} (in pure silicon it is ~ 1100 cm^2 V^{-1} s^{-1}), but in a-Si TFTs it is only ~ 1 cm^2 V^{-1} s^{-1}. Moreover, the mobility has a strong dependence upon the carrier density in the channel. Considerable research is currently focussing on improving the TFT performance.

4.8.2 Gas sensors

In Fig. 4.23b we have shown a schematic of how a change in surface conditions of a material changes the electronic states. The changes in allowed states can occur, for example, due to the additional bonding from a molecule that has attached itself to the surface. The changes in electronic states can alter the conductivity of the material—an effect that can then be exploited to detect the presence of the molecule. Devices that can detect certain gas molecules provide an extremely important, often lifesaving service. They find use in detecting poisonous gases as well as water vapor for humidity detectors. Ceramic humidity and gas detectors are thin discs of porous material whose conductivity (or resistance) is altered when molecules of a gas are absorbed on the surface. The absorbed molecule interacts with the surface to alter the electronic states as mentioned earlier and this causes a change in the conductivity of the device.

Due to the very complex nature of the transport and gas/solid interactions the models used for gas sensors are mostly empirical. In Fig. 4.24 we show typical results for a humidity sensor and a gas sensor. Through empirical studies (including trial and error) and increasingly through a greater understanding of quantum mechanics of surfaces it is possible to coat surfaces with thin films that can selectively detect certain molecules.

4.9 SUMMARY

In this chapter we have examined transport of charged carriers in materials. We have also examined some of the important devices based on charge flow. In Tables 4.6–4.7 we summarize our findings.

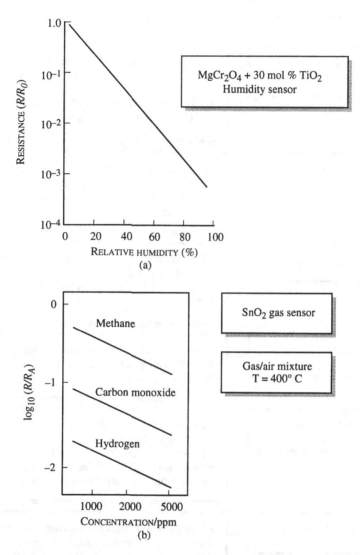

Figure 4.24: Typical responses of (a) humidity and (b) gas sensors.

4.9. Summary

Topics studied	Key findings
Electronic states in crystalline and non-crystalline systems	In perfect crystals there are no allowed states in the bandgap and electrons can propagate without scattering. In non-crystalline systems there are localized states where electrons can be trapped.
Scattering of electrons from fixed potentials	In the presence of potentials, which act as perturbations, electrons can scatter from one state to another. The momentum direction of the electron can change as a result of this scattering.
Scattering time and mobility	• Due to imperfections, electrons scatter as they move through the crystal. • The mobility of electrons (holes) is proportional to the scattering time.
Low field transport	Carrier mobility is constant and velocity increases linearly with applied electric field (for fields ≤ 1 kV/cm).
High field transport	• Carrier velocity tends to saturate and mobility = v/E starts to decrease. • GaAs and other direct-gap semiconductors show regions where velocity decreases as field increases; i.e., they have negative resistance regions.
Very high field transport	• Electrons and holes have so much energy that they can cause impact ionization or carrier multiplication. At a critical field E_{crit} the coefficients approach 10^4 cm^{-1}.

Table 4.6: Summary table.

Topics Studied	Key Findings
Carrier transport by diffusion	Carrier transport by diffusion is described by the diffusion constant that is related to mobility.
Important electronic semiconductor devices	Field effect transistors and bipolar junction transistors are crucial to modern information technology. Device improvements are driven by faster materials, large bandgap materials, and heterostructures.
Transport in non-crystalline materials	There are three important transport mechanisms in non-crystalline materials: (i) hopping to extended states and transport in extended states; (ii) hopping from a localized state to the nearest localized state; and (iii) variable range hopping.
Ionic conduction	In very large bandgap insulators with high defect densities, electron (hole) conduction is very poor and ionic conduction can become dominant. Ionic conduction has a strong dependence on defect density and temperature.
Non-crystalline material based devices	• Amorphous and polycrystalline semiconductors have very poor mobilities (~0.1–1 cm^2/V•s), but can be used as FETs for applications where speed is not an issue. • Absorpton of molecules on the surface of devices can cause changes in conductivity, leading to gas sensor applications.

Table 4.7: Summary table.

4.10 PROBLEMS

Sections 4.3–4.4

4.1 Consider a sample of GaAs with electron effective mass of $0.067m_o$. If an electric field of 1 kV/cm is applied, what is the drift velocity produced if
a) $\tau_{sc} = 10^{-13}$ s
b) $\tau_{sc} = 10^{-12}$ s
c) $\tau_{sc} = 10^{-11}$ s?
How does the drift velocity compare wirh the average thermal speed of the electrons at room temperature?

4.2 Assume that at room temperature the electron mobility in Si is 1300 cm^2/V·s. If an electric field of 100 V/cm is applied, what is the excess energy of the electrons? How does it compare with the thermal energy? If you assume that the mobility is unchanged, how does the same comparison work out at a field of 5 kV/cm? (Excess energy = $\frac{1}{2}m^*v_d^2$ where v_d is the drift velocity.)

4.3 The electron mobility of Si at 300 K is 1400 cm^2/V·s. Calculate the mean free path and the energy gained in a mean free path at an electric field of 1 kV/cm. Assume that the mean free path = $v_{th} \cdot \tau_{sc}$, where v_{th} is the thermal velocity of the electron ($v_{th} \sim 2.0 \times 10^7$ cm/s).

4.4 The mobility of electrons in the material InAs is $\sim$ 35,000 cm^2/V·s at 300K compared to a mobility of 1400 cm^2/V·s for silicon. Calculate the scattering times in the two semiconductors. The electron masses are 0.02 m_0 and 0.26 m_0 for InAs and Si, respectively.

4.5 Calculate the ionized impurity limited mobility ($N_D = 10^{16}$ cm^{-3}; 10^{17} cm^{-3}) in GaAs from 77 K to 300 K.

4.6 If the measured room temperature mobility of electrons in GaAs doped n-type at 5×10^{17} cm^{-3} is 3500 cm^{2}V^{-1}s^{-1} calculate the relaxation time for phonon scattering.

4.7 Calculate the alloy scattering limited mobility in In$_{0.53}$Ga$_{0.47}$As as a function of temperature from 77 K to 400 K. Assume an alloy scattering potential of 1.0 eV.

4.8 Use the velocity–field relations for Si and GaAs to calculate the transit time of electrons in a 1.0 μm region for a field of 1 kVcm^{-1} and 50 kVcm^{-1}.

4.9 The velocity of electrons in silicon remains $\sim$1 $\times$ 10^7 cm s^{-1} between 50 kVcm^{-1} and 200 kVcm^{-1}. Estimate the scattering times at these two electric fields.

4.10 The power output of a device depends upon the maximum voltage that the device can tolerate before impact-ionization-generated carriers become significant (say 10% excess carriers). Consider a device of length L, over which a potential V drops uniformly. What is the maximum voltage that can be tolerated by an Si and a diamond device for $L = 2$ μm and $L = 0.5$ μm? Use the values of the critical fields given in this chapter.

4.11 An electron in a silicon device is injected into a region where the field is 500 kV cm^{-1} The length of this region is 1.0 μm. Calculate the average number of impact ionization events that occur for the incident electron.

4.12 In Table 4.8 we show the resistivity of several metals. Based on these data calculate the mobilities of electrons in (i) Cu; (ii) Ag; (iii) Au; and (iv) Al. Assume that electron effective mass is m_0 and electron density is the density of electrons contributed by the valence electrons.

Section 4.6

4.13 In a silicon sample at 300 K, the electron concentration drops linearly from 10^{18} cm^{-3} to 10^{16} cm^{-3} over a length of 2.0 μm. Calculate the current density due to the electron diffusion current. Use the diffusion constant values given in this chapter.

4.14 In a GaAs sample, it is known that the electron concentration varies linearly. The diffusion current density at 300 K is found to be 100 A/cm^2. Calculate the slope of the electron concentration.

4.15 The electron concentration in a Si sample is given by

$$n(x) = n(0)\exp(-x/L_n); \quad x > 0$$

with $n(0) = 10^{18}$ cm^{-3} and $L_n = 3.0$ μm. Calculate the diffusion current density as a function of position if $D_n = 35$ cm^2/s.

4.16 A silicon sample has the following electron density

$$n(x) = n(0)\exp(-x/L_n) \quad x > 0$$

with $n(0) = 10^{18}$ cm^{-3} and $L_n = 2.0$ μm. There is also a uniform electric field of 2 kV/cm in the material. Calculate the drift and diffusion current densities as a function of position. The diffusion coefficient is 30 cm^2/s and the mobility is 1000 cm^2/V·s.

4.17 In a GaAs sample the electrons are moving under an electric field of 5 kV cm^{-1} and the carrier concentration is uniform at 10^{16} cm^{-3}. The electron velocity is the saturated velocity of 10^7 cm/s. Calculate the drift current density. If a diffusion current has to have the same magnitude, calculate the concentration gradient needed. Assume a diffusion coefficient of 100 cm^2/s.

4.18 Consider a 50 μm long silicon sample with an area of 1.0 μm×20 μm. The sample is doped n-type at 10^{17} cm^{-3}. A bias of 1.0 V is applied across the length and a current of 0.5 mA is observed. (Field is V/L).
- Calculate the Fermi level position of the sample.
- Calculate mobility of the electrons and the average scattering time between collisions.
Assume that the electron effective mass is 0.25 m_0.

4.19 Consider a GaAs sample doped n-type at 10^{16} cm^{-3} on which an experiment is done. At time $t = 0$ an external stimulus introduces excess electrons at a point $x = 0$. The excess charge is detected at $x = 10.0$ μm in the absence of any applied field after 2.5×10^{-9} s.
Use this information to answer the following:
- What is the diffusion coefficient of electrons?
- How much time will electrons travel (by drift) 1.0 μm under an applied field of 1.0 kV/cm? Assume that the velocity–field relation is linear.
- What is the conductivity of this sample? Assume that the electron effective mass is 0.067 m_0.

Element	ρ(μΩ-em)
Na	4.2
Cu	1.56
Ag	1.51
Au	2.04
Fe	8.9
Zn	5.5
Cd	6.8
Al	2.45
In	8.0
Sn	10.6

Table 4.8: Resistivities of some materials at 273 K in micro ohm centimeters.

4.11 FURTHER READING

- **Transport in crystalline materials**
  - M. Lundstrom, *Fundamentals of Carrier Transport* (Modular Series on Solid State Devices), eds. by G.W, Neudeck and R.F. Pierret, Addison-Wesley, Reading, MA, vol. X (1990).
  - J. Singh, *Modern Physics for Engineers*, Wiley-Interscience, New York (1999).
  - K. Seeger, *Semiconductor Physics*, Springer Verlag, New York (1985).
  - J. Singh, *Electronic and Optoelectronic Properties of Semiconductors*, Cambridge University Press (2003).

- **Transport in disordered materials**
  - N.F. Mott and E.A. Davis, *Electronic Processes in Non-Crystalline Materials*, Clarenden Press, Oxford (1971).
  - A.J. Moulson and J.M. Herbert, *Electroceramics: Materials, Properties, and Applications*, Chapman & Hall (1992).

Chapter 5

LIGHT ABSORPTION AND EMISSION

5.1 INTRODUCTION

Light is an integral part of the human experience. Not only do we use light to find our way around this complex world, it also provides us a medium for art and fantasy. And, of course, it has been exploited for the important technology of communication, storage, and display. In contrast to other parts of the electromagnetic spectrum, the human eye directly detects light, giving the spectral range of 0.3 μm to 1.0 μm a special status. In Fig. 5.1 we show the electromagnetic spectrum and the region over which the human eye responds. Of course, while the human eye response window is important for display technology, there is great interest in infrared and far infrared (communication, night visions, etc.) and ultraviolet regions (high-density memory, lithography printing, etc.).

In Fig. 5.2 we show an overview of physical phenomenon governing devices that can detect and emit light. Three general categories are shown:
(i) In conventional inorganic semiconductor devices, light particles or photons create electrons and holes through absorption. The electrons and holes create an electrical signal that can be used to detect the radiation. Conversely, electrons and holes can recombine to emit photons. In semiconductors it is also possible to use bandgap trap levels to alter the photon energy. In this case the semiconductor is intentionally doped with an impurity to create trap levels. In semiconductors, electron–hole pairs can also form excitons that can participate in optical processes.
(ii) While inorganic semiconductors have been the building blocks of most high-performance light detectors and emitters, there is considerable progress in the use of organic semiconductors for such devices. The exact physical process by which light absorption and emission occurs in organic semiconductors is not fully understood. As discussed in Chapter 3 organic semiconductors are described by bonding and antibonding molecular orbitals.

5.1. Introduction

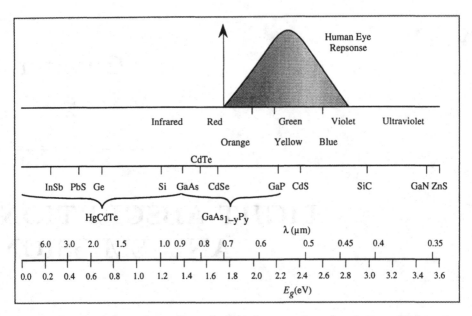

Figure 5.1: The bandgap and cutoff wavelengths for several semiconductors. The semiconductor bandgaps range from 0 (for $Hg_{0.4}Cd_{0.15}Te$) to well above 3 eV, providing versatile detection systems.

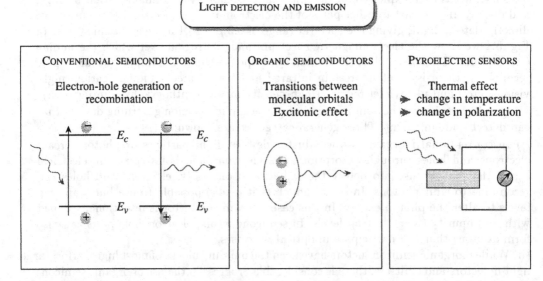

Figure 5.2: Light absorption and emission can occur by various processes, depending upon the material system used.

In these states the electron–hole Coulombic attraction is very strong and plays an important role in optical processes.

iii) Finally, there is the simple technique whereby thermal energy produced by radiation is exploited to detect (or create) light. In the well-known light bulb, a filament is heated to create light, but we will not discuss this aspect. There are materials which are pyroelectric (a property we will discuss in the next chapter) in which a voltage develops when temperature changes. This allows us to use the temperature change as a measure of the optical signal. These devices do not have the high performance (speed, detectivity) that semiconductor devices have, but they are useful for thermal or night vision imaging.

In this chapter we will discuss optical properties of semiconductors. Our focus will be on inorganic semiconductors, although due to the increasing importance of organic semiconductors we will discuss them as well. In the next chapter we will discuss polarization-related effects, some of which can be used for optical detection.

5.2 IMPORTANT MATERIAL SYSTEMS

A large number of semiconductor systems are exploited for optoelectronic devices for detection and emission. The choice of the material depends upon a number of issues; some dependent on physics, some on technology, and some on market forces. We will discuss some of the important factors that play a role in the choice of a material for light detector or emitter. We will discuss inorganic semiconductors in this section. Organic semiconductors will be discussed later in this chapter.

Bandstructure

Optical processes are strongest for direct gap semiconductors because of the momentum conservation law. Materials like Si and Ge have weak optical properties. Indirect gap semiconductors can be used for light detection, but they are not suitable for light emission.

During the emission process the energy of the photon is very close to the bandgap energy, E_g

$$\hbar\omega = E_g \quad (5.1)$$

and the wavelength is

$$\lambda_c = \frac{\hbar c}{E_g} = \frac{1.24}{E_g(\text{eV})\mu m} \quad (5.2)$$

The wavelength λ_c is also the cutoff wavelength if the material is used as a detector, i.e., the material is transparent and, therefore, unresponsive to longer wavelengths. The need for a particular optical wavelength thus determines the bandgap needed and the materials used.

In Tables 5.1 and 5.2 we show some of the important materials and their bandgaps. Factors that are crucial in determining the desired wavelength include:
(i) As shown in Fig. 5.3, demands from optical communication, where 1.55 μm and 1.3 μm light has the lowest optical fiber loss and lowest dispersion, respectively.
(ii) Blue light emission is needed for display applications.

5.2. Important material systems

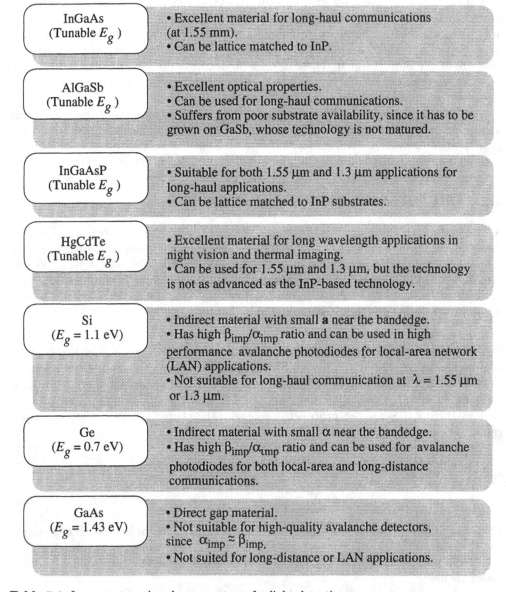

InGaAs (Tunable E_g)	• Excellent material for long-haul communications (at 1.55 mm). • Can be lattice matched to InP.
AlGaSb (Tunable E_g)	• Excellent optical properties. • Can be used for long-haul communications. • Suffers from poor substrate availability, since it has to be grown on GaSb, whose technology is not matured.
InGaAsP (Tunable E_g)	• Suitable for both 1.55 µm and 1.3 µm applications for long-haul applications. • Can be lattice matched to InP substrates.
HgCdTe (Tunable E_g)	• Excellent material for long wavelength applications in night vision and thermal imaging. • Can be used for 1.55 µm and 1.3 µm, but the technology is not as advanced as the InP-based technology.
Si (E_g = 1.1 eV)	• Indirect material with small a near the bandedge. • Has high β_{imp}/α_{imp} ratio and can be used in high performance avalanche photodiodes for local-area network (LAN) applications. • Not suitable for long-haul communication at λ = 1.55 µm or 1.3 µm.
Ge (E_g = 0.7 eV)	• Indirect material with small α near the bandedge. • Has high β_{imp}/α_{imp} ratio and can be used for avalanche photodiodes for both local-area and long-distance communications.
GaAs (E_g = 1.43 eV)	• Direct gap material. • Not suitable for high-quality avalanche detectors, since $\alpha_{imp} \approx \beta_{imp}$. • Not suited for long-distance or LAN applications.

Table 5.1: Important semiconductor systems for light detection.

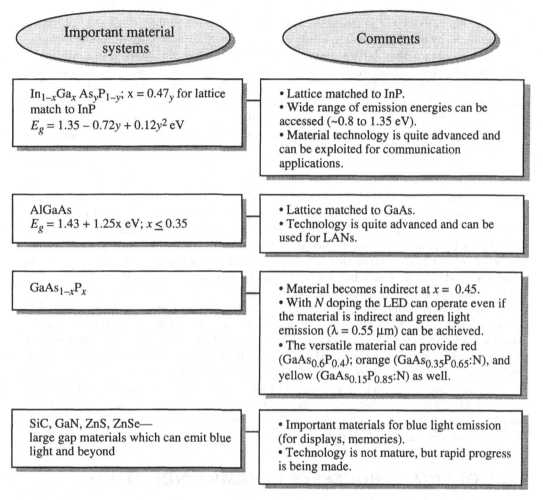

Important material systems	Comments
$In_{1-x}Ga_x As_y P_{1-y}$; $x = 0.47y$ for lattice match to InP $E_g = 1.35 - 0.72y + 0.12y^2$ eV	• Lattice matched to InP. • Wide range of emission energies can be accessed (~0.8 to 1.35 eV). • Material technology is quite advanced and can be exploited for communication applications.
AlGaAs $E_g = 1.43 + 1.25x$ eV; $x \leq 0.35$	• Lattice matched to GaAs. • Technology is quite advanced and can be used for LANs.
$GaAs_{1-x}P_x$	• Material becomes indirect at $x = 0.45$. • With N doping the LED can operate even if the material is indirect and green light emission ($\lambda = 0.55$ μm) can be achieved. • The versatile material can provide red ($GaAs_{0.6}P_{0.4}$); orange ($GaAs_{0.35}P_{0.65}$:N), and yellow ($GaAs_{0.15}P_{0.85}$:N) as well.
SiC, GaN, ZnS, ZnSe— large gap materials which can emit blue light and beyond	• Important materials for blue light emission (for displays, memories). • Technology is not mature, but rapid progress is being made.

Table 5.2: Important material systems for light emission. LEDs can be made from indirect materials with an appropriate impurity, but the emission efficiency is low.

(iii) Ultraviolet light is needed for reducing resolution for optical storage and laser printer applications.
iv) Long wavelength light ($\lambda \sim 5-14$ μm) is needed for night vision or thermal imaging. As can be seen from this brief overview, a wide range of material systems is needed.

Substrate availability
Almost all optoelectronic light sources depend upon epitaxial crystal growth techniques, where a thin active layer (a few microns) is grown on a substrate (which is ~ 200 μm). The availability of a high-quality substrate is extremely important in epitaxial technology. If a substrate that lattice matches with the active device layer is not available, the device layer may have dislocations and other defects in it. These can seriously hurt the device performance. The important substrates that are available for light emitting technology are GaAs and InP. A few semiconductors and their alloys can match with these substrates. The lattice constant of an alloy is the weighted mean of the lattice constants of the individual components; i.e., the lattice constant of the alloy $A_x B_{1-x}$ is

$$a_{\text{all}} = x a_A + (1-x) a_B \tag{5.3}$$

where a_A and a_B are the lattice constants of A and B.

Important semiconductor materials exploited in optoelectronics are the alloy $Ga_x Al_{1-x}As$, which is lattice matched very well to GaAs substrates; $In_{0.53}Ga_{0.47}As$ and $In_{0.52}Al_{0.48}As$, which are lattice matched with InP; InGaAsP which, is a quaternary material whose composition can be tailored to match with InP and can emit at 1.55 μm; and GaAsP, which has a wide range of bandgaps available. Recently there has been a considerable interest in large bandgap materials such GaN (with InN and AlN) to produce devices that emit blue or green light. The motivation is for superior display technology and for high-density optical memory applications (a shorter wavelength allows reading of smaller features). Reliable nitride-based LEDs are now available in the commercial market, although the substrate technology is still not fully developed. Silicon carbide or sapphire serve as substrates for the nitride devices.

It is important to keep in mind that alloys such as GaAlAs and GaAsP become indirect at certain compositions, as shown in Fig. 5.4. For efficient light emission, we need to work in the direct gap region. However, with a suitable impurity, we can obtain light emission in an indirect bandgap material.

5.3 OPTICAL PROCESSES IN SEMICONDUCTORS

In this section we will consider the basic photon absorption and emission processes in semiconductors. To understand the optical processes we need not only to use the quantum description of electrons (i.e., band theory), but also the quantum description for light. In quantum mechanics, classical waves, such as electromagnetic waves, sound waves, etc., behave as particles, and this particle nature becomes critical under some conditions. Light waves, which are electromagnetic waves, behave as photons of energy $\hbar\omega$. In simple processes electron–photon interaction changes the electron energy by $+\hbar\omega$ (absorption) or $-\hbar\omega$ (emission), as shown in Fig. 5.5.

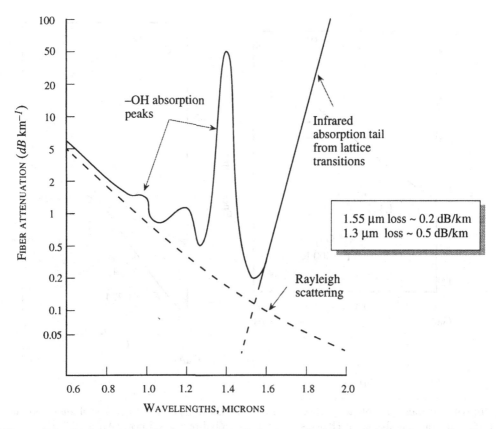

Figure 5.3: Optical attenuation vs. wavelength for an optical fiber. Primary loss mechanisms are identified as absorption and scattering.

5.3. Optical processes in semiconductors

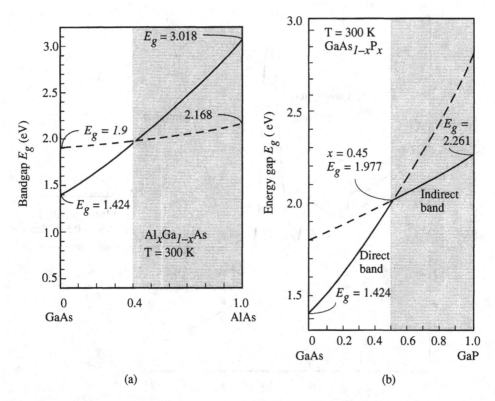

Figure 5.4: Bandgaps of (a) $Al_xGa_{1-x}As$ and (b) $GaAs_{1-x}P_x$ as a function of alloy composition. Note that the bandgap changes from direct to indirect as shown. (After H.C. Casey and M. Panish, *Heterostructure Lasers*, Academic Press, New York 1978.)

The full understanding of electron–photon interaction requires details of quantum mechanics, which are beyond the scope of this book. Here we will provide some physical motivation for the processes and then discuss the results.

5.3.1 Optical absorption and emission

We will now discuss how electrons in a semiconductor respond to electromagnetic fields or photons. The interaction of electrons and photons is the basis of all semiconductor optoelectronic devices. There are two kinds of events that occur when electron–photon interactions occur: (i) absorption of photons, where the electron gains energy by absorbing a photon; and (ii) emission, where the electron emits a photon and loses energy. These processes are schematically shown in Fig. 5.5. The emission process itself is characterized as spontaneous emission and stimulated emission. Spontaneous emission occurs even if there are no photons present, while stimulated emission occurs because of the presence of photons. In this section we will give quantitative expressions for the various processes characterizing electron–photon interactions. These expressions will not be derived in detail, but will be simply introduced to the reader. The operation of the various optoelectronic devices is governed by the expressions presented in this and the next few sections.

Light is represented by electromagnetic waves, which travel through a medium such as a semiconductor. This is described by Maxwell's equations, which show that the waves have a form given by the electric field vector dependence

$$\mathbf{E} = \mathbf{E}_o \exp\left\{i\omega\left(\frac{n_r}{c} - t\right)\right\} \exp\left(-\frac{\alpha z}{2}\right) \tag{5.4}$$

Here z is the propagation direction, ω the frequency, n_r the refractive index, and α the absorption coefficient of the medium. If α is zero, the wave propagates without attenuation with a velocity $\frac{c}{n_r}$. However, for non-zero α, the photon flux I ($\sim E^*E$) falls as

$$I(z) = I(0) \exp\{-\alpha z\} \tag{5.5}$$

The absorption of light can occur for a variety of reasons, including absorption by impurities in the material, or absorption where electrons in the conduction band or holes in the valence band absorb the radiation (i.e., intraband absorption). However, the most important optoelectronic interaction in semiconductors, as far as devices are concerned, is the band-to-band transition shown in Fig. 5.6. In the photon absorption process, a photon interacts with an electron in the valence band, causing the electron to go into the conduction band. In the reverse process, the electron in the conduction band recombines with a hole in the valence band to generate a photon. These two processes are of obvious importance for light-detection and light–emission devices. As has been noted above, the detailed scattering theory is beyond the scope of this text. However, the important expressions that control the performance of detectors and lasers will be examined in this section. The rates for the light emission and absorption processes are determined by quantum mechanics. The scattering involves the following issues:

5.3. Optical processes in semiconductors

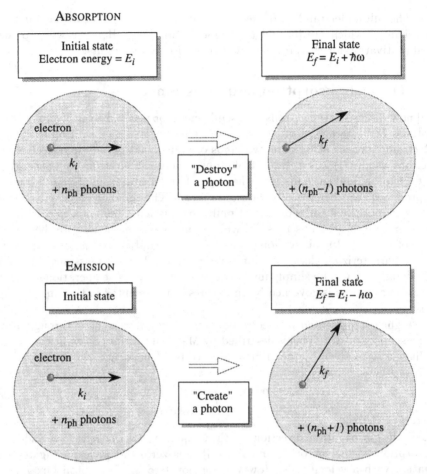

Figure 5.5: (a) A schematic of an absorption process, where a photon is absorbed (destroyed) and the energy and momentum of the electron is altered. (b) The emission of a photon, where a photon is created.

(i) **Conservation of energy**: In the absorption and emission process we have for the initial and final energies of the electrons E_i and E_f

$$\text{Absorption}: \quad E_f = E_i + \hbar\omega \quad (5.6)$$
$$\text{Emission}: \quad E_f = E_i - \hbar\omega \quad (5.7)$$

where $\hbar\omega$ is the photon energy. Since the minimum energy difference between the conduction and valence band states is the bandgap E_g, *the photon energy must be larger than the bandgap in order for absorption to occur.*
(ii) **Conservation of momentum**: In addition to energy conservation, we also need to conserve the momentum $\hbar\mathbf{k}$ for the electrons and the photon system. The photon k_{ph}

value is given by

$$k_{\text{ph}} = \frac{2\pi}{\lambda} \tag{5.8}$$

Since 1 eV photons correspond to a wavelength of 1.24 μm, the k-values of relevance are $\sim 10^{-4}$ Å^{-1}, which is essentially zero compared to the k-values for electrons. Thus k-conservation ensures that the initial and final electrons should have the same k-value. Another way to say this is that *only "vertical" k transitions are allowed* in the band-structure picture, as shown in Fig. 5.6.

Because of this constraint of k-conservation, in semiconductors where the valence band and conduction bandedges are at the same $k = 0$ value (the direct semiconductors), the optical transitions are quite strong. In indirect materials like Si, Ge, etc., the optical transitions are very weak. This makes a tremendous difference in the optical properties of these two kinds of materials. The transitions are very weak near the bandedges of indirect semiconductors, and such materials cannot be used for high-performance optoelectronic devices, such as lasers.

Because the k-values for the electron and hole are the same in vertical transitions, we have, as shown in Fig. 5.7

$$\begin{aligned}\hbar\omega &= E_g + \frac{\hbar^2 k^2}{2}\left(\frac{1}{m_e^*} + \frac{1}{m_h^*}\right) \\ &= E_g + \frac{\hbar^2 k^2}{2m_r^*}\end{aligned} \tag{5.9}$$

where m_r^* is the reduced mass of the electron–hole system. Thus the *relevant density of states function in the scattering process is that where the mass used is the reduced mass*. This density of states is called the *joint density of states*.

Before providing details on absorption and emission processes let us briefly describe some important relations. The absorption coefficient α described above tells us how photons are absorbed as a function of distance traveled. We can relate this to the absorption rate in time W_{abs} via the relation

$$W_{\text{abs}} = \alpha v n_{ph} \tag{5.10}$$

where v is the speed of light in the material given by c/n_r with n_r being the refractive index. The parameter n_{ph} is the photon number in the photon mode.

In quantum mechanics the electromagnetic radiation is described by photon number rather than intensity or electric (magnetic) field values. The classical and quantum description can be related however. If P_{op} denotes the power density (W/cm^2 of light we have the relation

$$P_{op} = \frac{n_{ph}}{V}\hbar\omega c \tag{5.11}$$

where n_{ph}/V is the photon number density. The photon number can also be related to the electric field in the radiation since the electric field amplitude E_0 and power density

5.3. Optical processes in semiconductors

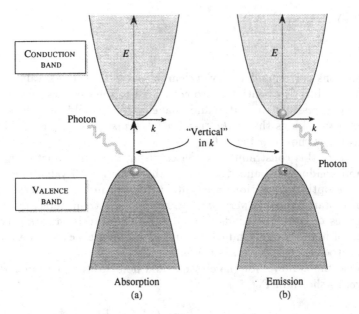

Figure 5.6: Band-to-band absorption in semiconductors. (a) An electron in the valence band absorbs a photon and moves into the conduction band. (b) In the reverse process, the electron recombines with a hole to emit a photon. Momentum conservation ensures that only vertical transitions are allowed.

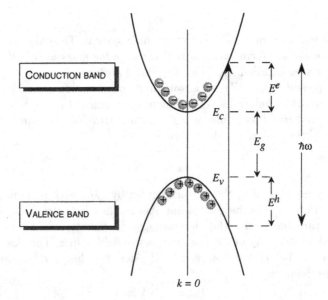

Figure 5.7: A schematic showing the relationship between photon energy and electron and hole energies in momentum conserving transitions.

are related. The relation is

$$E_0^2 = \frac{2\hbar\omega}{\epsilon_0} \frac{n_{ph}}{V} \quad (5.12)$$

Each photon that is absorbed by inter-band absorption creates an electron–hole pair. The generation rate of electron–hole pairs, G_L, is

$$G_L = \frac{\alpha P_{op}}{\hbar\omega} \quad (5.13)$$

Keeping all these issues in mind, we can calculate the absorption coefficient for a semiconductor using Fermi golden rule. For light polarized along direction **a**, the absorption coefficient can be shown to be

$$\alpha = \frac{\pi e^2 \hbar}{m_0^2 c n_r \epsilon_0} \frac{1}{\hbar\omega} |(\mathbf{a}\cdot\mathbf{p})_{if}|^2 N_{cv}(\hbar\omega) \quad (5.14)$$

With N_{cv}, the joint density of states is given by

$$N_{cv}(\hbar\omega) = \frac{\sqrt{2}(m_r^*)^{3/2}(\hbar\omega - E_g)^{1/2}}{\pi^2 \hbar^3} \quad (5.15)$$

The quantity $|(\mathbf{a}\cdot\mathbf{p})_{if}|^2$ is called the momentum or dipole matrix element between the conduction and the valence band. The polarization averaged matrix element turns out to be $(2/3)p_{cv}^2$, where it is found that, for most semiconductors

$$\frac{2p_{cv}^2}{m_0} \cong 20 \text{ to } 24 \text{ eV} \quad (5.16)$$

If we plug in some numbers for the absorption coefficient of unpolarized light, we get for GaAs ($E_g \cong 1.5$ eV) (see Example 5.4)

$$\alpha(\hbar\omega) \cong 4.7 \times 10^4 \frac{(\hbar\omega - E_g)^{1/2}}{\hbar\omega} \text{ cm}^{-1} \quad (5.17)$$

where $\hbar\omega$ and E_g are in units of electron volts. The prefactor for any other direct semiconductor can be obtained by scaling the reduced mass and the photon energy (see Eqs. 5.10 and 5.11) according to the reduced density of states.

In Fig. 5.8 we show the absorption coefficient for several direct and indirect bandgap semiconductors. In the case of indirect bandgap materials, the absorption coefficient is considerably smaller at the bandedges.

If we have an electron in a conduction band state and a hole in the valence band state with the same k-value, the two can recombine and emit a photon. There are two important classes of emission processes. *In spontaneous emission an electron recombines with a hole, even though no photons are present in the region, and emits a photon.* The rate for this emission process is given by

$$W_{em}(\hbar\omega) = \frac{e^2 n_r \hbar\omega}{3\pi\epsilon_0 m_o^2 c^3 \hbar^2} |p_{cv}|^2 \quad (5.18)$$

5.3. Optical processes in semiconductors

If photons with frequency ω are already present in the cavity with the semiconductor, the recombination rate is enhanced by the presence of these photons. If $n_{ph}(\hbar\omega)$ is the photon occupation (i.e., the number of photons in a particular mode), the emission (called stimulated emission) rate is given by

$$W_{em}^{st}(\hbar\omega) = \frac{e^2 n_r \hbar\omega}{3\pi\epsilon_0 m_0^2 c^3 \hbar^2} |p_{cv}|^2 \cdot n_{ph}(\hbar\omega) \qquad (5.19)$$

Thus the rate is increased in proportion to the photon density already present in the cavity.

In spontaneous emission, the photons that are emitted have no particular phase relationship with each other and are thus incoherent. Light emission in light-emitting diodes (LEDs) is due to spontaneous emission. In stimulated emission, however, the photons that are emitted are in phase with the original photons that are present. The radiation is thus coherent. Laser diodes depend upon stimulated emission. The acronym *laser* stands for *light amplification by stimulated emission radiation*.

The radiative recombination time of an electron having a momentum $\hbar k$ to recombine with a hole (in the absence of photons) having the same momentum is

$$\tau_o = \frac{1}{W_{em}} \qquad (5.20)$$

In the definition of τ_o it is assumed that the electron can find a hole with which to recombine. If the probability of finding the hole is small, the radiative time can be much longer, as discussed in Section 5.3.2. For materials like GaAs, the value of τ_o is about 1 ns (see Example 5.1). However, for indirect materials, the recombination time can be as large as 1 μs. The recombination process is not only important for optical emission devices, but the rate also plays a key role in the speed of many electronic devices, e.g., bipolar devices, diodes, etc.

The electron–hole recombination during stimulated emission can be quite a bit smaller than τ_o, depending upon the photon intensity present.

EXAMPLE 5.1 A 1.7 eV photon is absorbed by a valence band electron in GaAs. If the bandgap of GaAs is 1.41 eV, calculate the energy of the electron and heavy hole produced by the photon absorption.

The electron, heavy-hole, and reduced mass of GaAs are 0.067 m_0, 0.45 m_0, and 0.058 m_0, respectively. The electron and the hole generated by photon absorption have the same momentum. The energy of the electron is

$$E^e = E_c + \frac{m_r^*}{m_e^*}(\hbar\omega - E_g)$$

$$E^e - E_c = \frac{0.058}{0.067}(1.7 - 1.41) = 0.25 \text{ eV}$$

The hole energy is

$$E^h - E_v = -\frac{m_r^*}{m_h^*}(\hbar\omega - E_g) = -\frac{0.058}{0.45}(1.7 - 1.41)$$
$$= -0.04 \text{ eV}$$

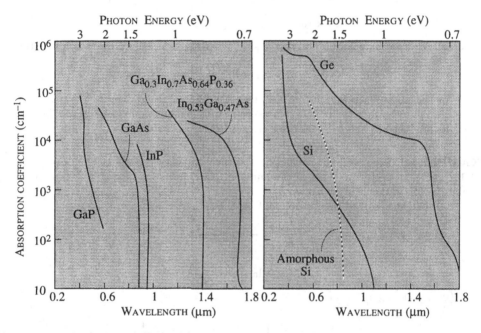

Figure 5.8: Absorption coefficient of several direct and indirect semiconductors. For the direct gap material, the absorption coefficient is very strong once the photon energy exceeds the bandgap. For indirect materials, the absorption coefficient rises much more gradually. Once the photon energy is more than the direct gap, the absorption coefficient increases rapidly.

5.3. Optical processes in semiconductors

The electron, by virtue of its lower mass, is created with a much greater energy than the hole.

EXAMPLE 5.2 In silicon, an electron from the top of the valence band is taken to the bottom of the conduction band by photon absorption. Calculate the change in the electron momentum. Can this momentum difference be provided by a photon?

The conduction band minima for silicon are at a k-value of $\frac{2\pi}{a}$ (0.85, 0, 0). There are five other similar bandedges. The top of the valence band has a k-value of 0. The change in the momentum is thus

$$\hbar \Delta k = \hbar \frac{2\pi}{a}(0.85) = (1.05 \times 10^{-34})\left(\frac{2\pi}{5.43 \times 10^{-10}}\right)(0.85)$$

$$= 1.03 \times 10^{-24} \text{ kg.m.s}^{-1}$$

A photon which has an energy equal to the silicon bandgap can only provide a momentum of

$$\hbar k_{ph} = h \cdot \frac{2\pi}{\lambda}$$

The λ for silicon bandgap is 1.06 μm and thus the photon momentum is about a factor of 1800 too small to balance the momentum needed for the momentum conservation. The lattice vibrations produced by thermal vibration are needed for the process.

EXAMPLE 5.3 The absorption coefficient near the bandedges of GaAs and Si are $\sim 10^4$ cm^{-1} and 10^3 cm^{-1}, respectively. What is the minimum thickness of a sample in each case which can absorb 90% of the incident light?

The light absorbed in a sample of length L is

$$\frac{I_{abs}}{I_{inc}} = -\exp(-\alpha L)$$

or

$$L = \frac{1}{\alpha} \ln\left(1 - \frac{I_{abs}}{I_{inc}}\right)$$

Using $\frac{I_{abs}}{I_{inc}}$ equal to 0.9, we get

$$L(\text{GaAs}) = -\frac{1}{10^4} \ln(0.1) = 2.3 \times 10^{-4} \text{ cm}$$

$$= 2.3 \ \mu\text{m}$$

$$L(\text{Si}) = -\frac{1}{10^3} \ln(0.1) = 23 \ \mu\text{m}$$

Thus a Si detector requires a very thick active absorption layer to function.

EXAMPLE 5.4 Calculate the absorption coefficient of GaAs as a function of photon frequency.

The joint density of states for GaAs is (using a reduced mass of $0.058m_0$)

$$N_{cv}(E) = \frac{\sqrt{2}(m_r^*)^{3/2}(E - E_g)^{1/2}}{\pi^2 \hbar^3}$$

$$= \frac{1.414 \times (0.058 \times 0.91 \times 10^{-30} \text{ kg})^{3/2}(E - E_g)^{1/2}}{9.87 \times (1.05 \times 10^{-34})^3}$$

$$= 1.5 \times 10^{54}(E - \hbar\omega)^{1/2} \text{ J}^{-1}\text{m}^{-3}$$

The absorption coefficient for unpolarized light is

$$\alpha(\hbar\omega) = \frac{\pi e^2 \hbar}{2n_r c \epsilon_0 m_0} \left(\frac{2p_{cv}^2}{m_0}\right) \frac{N_{cv}(\hbar\omega)}{\hbar\omega} \cdot \frac{2}{3}$$

The term $\frac{2p_{cv}^2}{m_0}$ is ~23.0 eV for GaAs. This gives

$$\alpha(\hbar\omega) = \frac{3.1416 \times (1.6 \times 10^{-19}\ \text{C})^2 (1.05 \times 10^{-34}\ \text{J.s})}{2 \times 3.4 \times (3 \times 10^8\ \text{m/s})(8.84 \times 10^{-12}\ (\text{F/m})^2)}$$

$$\cdot \frac{(23.0 \times 1.6 \times 10^{-19}\ \text{J})}{(0.91 \times 10^{-30}\ \text{kg})} \frac{(\hbar\omega - E_g)^{1/2}}{\hbar\omega} \times 1.5 \times 10^{54} \times \frac{2}{3}$$

$$\alpha(\hbar\omega) = 1.9 \times 10^{-3} \frac{(\hbar\omega - E_g)^{1/2}}{\hbar\omega}\ \text{m}^{-1}$$

Here the energy and $\hbar\omega$ are in units of Joules. It is usual to express the energy in eV, and the absorption coefficient in cm^{-1}. This is obtained by multiplying the result by

$$\left[\frac{1}{(1.6 \times 10^{-19})^{1/2}} \times \frac{1}{100}\right]$$

$$\alpha(\hbar\omega) = 4.7 \times 10^4 \frac{(\hbar\omega - E_g)^{1/2}}{\hbar\omega}\ \text{cm}^{-1}$$

For GaAs the bandgap is 1.5 eV at low temperatures and 1.43 eV at room temperatures. From the value of α, we can see that a few microns of GaAs are adequate to absorb a significant fraction of light above the bandgap.

EXAMPLE 5.5 Calculate the electron–hole recombination time in GaAs.

The recombination rate is given by

$$W_{em} = \frac{e^2 n_r}{6\pi\epsilon_0 m_0 c^3 \hbar^2} \left(\frac{2p_{cv}^2}{m_0}\right) \hbar\omega$$

with $\frac{2p_{cv}^2}{m_0}$ being 23 eV for GaAs.

$$W_{em} = \frac{(1.6 \times 10^{-19}\ \text{C})^2 \times 3.4 \times (23 \times 1.6 \times 10^{-19}\ \text{J})\hbar\omega}{6 \times 3.1416 \times (8.84 \times 10^{-12}\ \text{F/m}) \times (0.91 \times 10^{-30}\ \text{kg})}$$

$$\cdot \frac{1}{(3 \times 10^8\ \text{m/s})^3 \times (1.05 \times 10^{-34}\ \text{J.s})^2}$$

$$= 7.1 \times 10^{27} \hbar\omega\ \text{s}^{-1}$$

If we require the value of $\hbar\omega$ in eV instead of J we get

$$W_{em} = 7.1 \times 10^{27} \times (1.6 \times 10^{-19})\hbar\omega\ \text{s}^{-1}$$
$$= 1.14 \times 10^9\ \hbar\omega\ \text{s}^{-1}$$

For GaAs, $\hbar\omega \sim 1.5$ eV so that

$$W_{em} = 1.71 \times 10^9\ \text{s}^{-1}$$

5.3. Optical processes in semiconductors

The corresponding recombination time is

$$\tau_o = \frac{1}{W_{em}} = 0.58 \text{ ns}$$

Remember that this is the recombination time when an electron can find a hole to recombine with. This happens when there is a high concentration of electrons and holes; i.e., at high injections of electrons and holes or when a minority carrier is injected into a heavily doped majority carrier region, as discussed in the next subsection.

5.3.2 Charge injection, quasi-Fermi levels, and recombination

In the previous subsection we have discussed how light absorption and emission can occur. In semiconductors we can alter these processes by changing the electron–hole density. When electrons or holes are injected into a semiconductor a question arises: What kind of distribution function describes the electron and hole occupation? We know that in equilibrium the electron and hole occupation is represented by the Fermi function. A new function is needed to describe the system when the electrons and holes are not in equilibrium.

Quasi-Fermi levels

As discussed in Section 3.2, under equilibrium conditions the distribution of electrons and holes is given by the Fermi function, which is defined once one knows the Fermi level. Also the product of electrons and holes, np, is approximately constant. If excess electrons and holes are injected into the semiconductor, clearly the same function will not describe the occupation of states. Under certain assumptions the electron and hole occupation can be described by the use of *quasi-Fermi levels*. These assumptions are:

(i) The electrons are in thermal equilibrium in the conduction band and the holes are in equilibrium in the valence band. This means that the carriers in each band are neither gaining nor losing energy from the crystal lattice atoms.

(ii) The electron–hole recombination time is much larger than the time for the electrons and holes to reach equilibrium within the conduction and valence band, respectively.

In most problems of interest, the time to reach equilibrium in the same band is approximately a few picoseconds, while the e–h recombination time is anywhere from a nanosecond to a microsecond. Thus the above assumptions are usually met. In this case, the quasi-equilibrium electron and holes can be represented by an electron Fermi function f^e (with electron Fermi level) and a hole Fermi function f^h (with a *different* hole Fermi level). We now have

$$n = \int_{E_c}^{\infty} N_e(E) f^e(E) dE$$

$$p = \int_{-\infty}^{E_v} N_h(E) f^h(E) dE \qquad (5.21)$$

where

$$f^e(E) = \frac{1}{\exp\left(\frac{E-E_{Fn}}{k_BT}\right)+1} \tag{5.22}$$

If $f^v(E)$ is the electron occupation in the valence band, the hole occupation is

$$f^h(E) = 1 - f^v(E) = 1 - \frac{1}{\exp\left(\frac{E-E_{Fp}}{k_BT}\right)+1}$$

$$= \frac{1}{\exp\left(\frac{E_{Fp}-E}{k_BT}\right)+1} \tag{5.23}$$

At equilibrium $E_{Fn} = E_{Fp}$. If excess electrons and holes are injected into the semiconductor, the electron quasi-Fermi level E_{Fn} moves toward the conduction band, while the hole quasi-Fermi level E_{Fp} moves toward the valence band. The ability to define quasi-Fermi levels E_{Fn} and E_{Fp} provides us with a very useful approach to solve non-equilibrium problems which are, of course, of greatest interest in devices.

By defining separate Fermi levels for the electrons and holes, we can study the properties of excess carriers using the same relationship between Fermi level and carrier density as we developed for the equilibrium problem. Thus, in the approximation where the Fermi distribution is replaced by an exponential, we have

$$n = N_c \exp\left[\frac{(E_{Fn}-E_c)}{k_BT}\right]$$

$$p = N_v \exp\left[\frac{(E_v-E_{Fp})}{k_BT}\right] \tag{5.24}$$

In the more accurate Joyce–Dixon approximation we have

$$(E_{Fn}-E_c) = k_BT\left[\ln\frac{n}{N_c} + \frac{n}{\sqrt{8}N_c}\right]$$

$$(E_v-E_{Fp}) = k_BT\left[\ln\frac{p}{N_v} + \frac{p}{\sqrt{8}N_v}\right] \tag{5.25}$$

The dependence of the quasi-Fermi levels on the electron and hole densities in GaAs at 300 K are shown in Fig. 5.9. Note that for the same carrier injection ($n = p$), the electron quasi-Fermi level moves a greater amount than the hole quasi-Fermi level. This is because of the smaller electron effective mass.

EXAMPLE 5.6 Using Boltzmann statistics calculate the position of the electron and hole quasi-Fermi levels when an e–h density of 10^{17} cm^{-3} is injected into pure silicon at 300 K.
At room temperature for Si we have

$$N_c = 2.8 \times 10^{19} \text{ cm}^{-3}$$
$$N_v = 1.04 \times 10^{19} \text{ cm}^{-3}$$

5.3. Optical processes in semiconductors

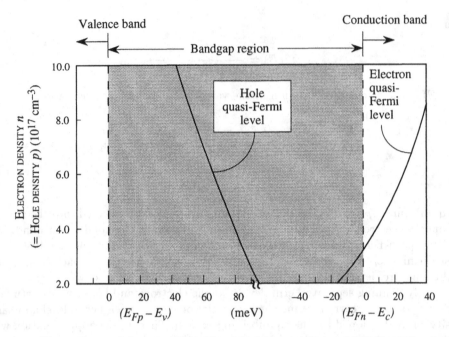

Figure 5.9: Dependence of the quasi-Fermi level positions in GaAs at 300 K on the electron and hole density.

If $n_c = p_v = 10^{17}$ cm^{-3}, we obtain ($k_B T = 0.026$ eV)

$$E_{Fn} = k_B T \, \ell n \left[\frac{n}{N_c}\right] + E_c$$
$$= E_c - 0.146 \text{ eV}$$
$$E_{Fp} = E_v - k_B T \left[\ell n \, \frac{p}{N_v}\right]$$
$$= E_v + 0.121 \text{ eV}$$

Since in Si, the bandgap is $E_c - E_v = 1.1$eV, we have

$$E_{Fn} - E_{Fp} = (E_c - E_v) - (0.146 + 0.121)$$
$$= 1.1 - 0.267 = 0.833 \text{ eV}$$

Electron–hole recombination

We have discussed the absorption and emission processes above. What happens if we have some electrons in the conduction band and some holes in the valence band? What is the recombination rate?

If an electron is available in a state k and a hole is also available in the state k (i.e., if the Fermi functions for the electrons and holes satisfy $f^e(k) = f^h(k) = 1$), the radiative recombination rate is found to be approximately (see Example 5.5)

$$W_{em} \sim 1.5 \times 10^9 \hbar\omega(\text{eV}) \text{ s}^{-1} \tag{5.26}$$

and the recombination time becomes ($\hbar\omega$ is expressed in electron volts)

$$\tau_o = \frac{0.67}{\hbar\omega(\text{eV})} \text{ ns} \tag{5.27}$$

The recombination time discussed above is the shortest possible spontaneous emission time since we have assumed that the electron has a unit probability of finding a hole with the same k-value.

When carriers are injected into the semiconductors, the occupation probabilities for the appropriate quasi-Fermi levels give the electron and hole states. The emitted photons leave the device volume so that the photon density never becomes high in the e–h recombination region. In a laser diode the situation is different, as we shall see later. The photon emission rate is given by integrating the emission rate W_{em} over all the electron–hole pairs, after introducing the appropriate Fermi functions.

There are several important limits of the spontaneous rate:

(i) Low injection: In the case where the electron and hole densities n and p are small, the Fermi functions have a Boltzmann form $(\exp(-E/k_BT))$. The recombination rate is found to be

$$R_{\text{spon}} = \frac{1}{2\tau_o} \left(\frac{2\pi\hbar^2 m_r^*}{k_B T m_e^* m_h^*} \right)^{3/2} np \tag{5.28}$$

The rate of photon emission depends upon the product of the electron and hole densities. If we define the lifetime of a single electron injected into a lightly doped ($p = N_a \leq 10^{17} \text{cm}^{-3}$) p-type region with hole density p, it would be given from Eq. 5.28 by

$$\frac{R_{\text{spon}}}{n} = \frac{1}{\tau_r} = \frac{1}{2\tau_o} \left(\frac{2\pi\hbar^2 m_r^*}{k_B T m_e^* m_h^*} \right)^{3/2} p \tag{5.29}$$

The time τ_r in this regime is very long (hundreds of nanoseconds), as shown in Fig. 5.10, and becomes smaller as p increases.

(ii) Injection into heavily doped materials: In the case where electrons are injected into a heavily doped p–region (or holes are injected into a heavily doped n–region), the function $f^h(f^e)$ can be assumed to be unity. The spontaneous emission rate is

$$R_{\text{spon}} \sim \frac{1}{\tau_o} \left(\frac{m_r^*}{m_h^*} \right)^{3/2} n \tag{5.30}$$

for electron concentration n injected into a heavily doped p-type region and

$$R_{\text{spon}} \sim \frac{1}{\tau_o} \left(\frac{m_r^*}{m_e^*} \right)^{3/2} p \tag{5.31}$$

for hole injection into a heavily doped n-type region.

The minority carrier lifetimes (i.e., n/R_{spon}) play a very important role not only in LEDs, but also in diodes and bipolar devices. In this regime the lifetime of a

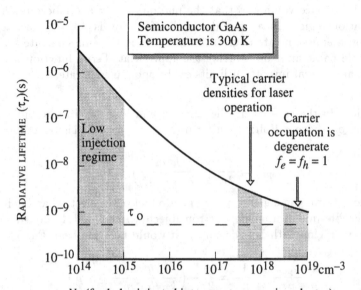

N_d (for holes injected into an n-type semiconductor)

$n = p$ (for excess electron-hole pairs injected into a region)

Figure 5.10: Radiative lifetimes of electrons or holes in a direct gap semiconductor as a function of doping or excess charge. The figure gives the lifetimes of a minority charge (a hole) injected into an n-type material. The figure also gives the lifetime behavior of electron–hole recombination when excess electrons and holes are injected into a material as a function of excess carrier concentration.

single electron (hole) is independent of the holes (electrons) present since there is always a unity probability that the electron (hole) will find a hole (electron). The lifetime is now essentially τ_o, as shown in Fig. 5.10.

(iii) High injection: Another important regime is that of high injection, where $n = p$ is so high that we can assume $f^e = f^h = 1$ in the integral for the spontaneous emission rate. The spontaneous emission rate is

$$R_{\text{spon}} \sim \frac{n}{\tau_o} \sim \frac{p}{\tau_o} \qquad (5.32)$$

and the radiative lifetime ($n/R_{\text{spon}} = p/R_{\text{spon}}$) is τ_o.

(iv) Inversion conditions: A regime that is quite important for laser operation is one where sufficient electrons and holes are injected into the semiconductor to cause "inversion." As will be discussed later, this occurs if $f^e + f^h \geq 1$. If we make the approximation $f^e \sim f^h = 1/2$ for all the electrons and holes at inversion, we get the relation

$$R_{\text{spon}} \sim \frac{n}{4\tau_o} \qquad (5.33)$$

or the radiative lifetime at inversion is

$$\tau \sim \frac{\tau_o}{4} \qquad (5.34)$$

This value is a reasonable estimate for the spontaneous emission rate in lasers near the threshold.

The radiative recombination depends upon the radiative lifetime τ_r and the non-radiative lifetime τ_{nr}. To improve the efficiency of photon emission we need a value of τ_r as small as possible and τ_{nr} as large as possible. To increase τ_{nr} we must reduce the material defect density. This includes improving surface and interface qualities.

EXAMPLE 5.7 Calculate the e–h recombination time when an excess electron and hole density of 10^{15} cm^{-3} is injected into a GaAs sample at room temperature.

Since 10^{15} cm^{-3} or 10^{21} m^{-3} is a very low level of injection, the recombination time is given by Eq. 5.26 as

$$\frac{1}{\tau_r} = \frac{1}{2\tau_o}\left(\frac{2\pi\hbar^2 m_r^*}{k_B T m_e^* m_h^*}\right)^{3/2} p$$

$$= \frac{1}{2\tau_o}\left(\frac{2\pi\hbar^2}{k_B T m_e^* + m_h^*}\right)^{3/2} p$$

Using $\tau_o = 0.6$ ns and $k_B T = 0.026$ eV, we get for $m_e^* = 0.067\, m_o$, $m_h^* = 0.45\, m_o$,

$$\frac{1}{\tau_r} = \frac{10^{21}\text{ m}^{-3}}{2 \times (0.6 \times 10^{-9}\text{ s})}\left[\frac{2 \times 3.1416 \times (1.05 \times 10^{-34}\text{ Js})^2}{(0.026 \times 1.6 \times 10^{-19}\text{ J}) \times (0.517 \times 9.1 \times 10^{-31}\text{ kg})}\right]^{3/2}$$

$$\tau_r = 5.7 \times 10^{-6}\text{ s} \cong 9.5 \times 10^3\, \tau_o$$

We see from this example that at low injection levels, the carrier lifetime can be very long. Physically, this occurs because at such a low injection level, the electron has a very small probability of finding a hole to recombine with.

5.3.3 Optical absorption, loss, and gain

The intensity associated with an electromagnetic wave traveling through a semiconductor is described by

$$I_{\text{ph}} = I_{\text{ph}}^0 \exp(-\alpha x) \tag{5.35}$$

where α (the absorption coefficient) is usually positive and I_{ph}^0 is the incident light intensity at $x = 0$. The optical intensity, which is the photon current multiplied by the photon energy $\hbar\omega$, falls as the wave travels if α is positive. However, if electrons are pumped in the conduction band and holes in the valence band, the electron–hole recombination process (photon emission) can be stronger than the reverse process of electron–hole generation (photon absorption). In general, the gain coefficient is defined by gain = emission coefficient−absorption coefficient. If $f^e(E^e)$ and $f^h(E^h)$ denote the electron and hole occupation, the emission coefficient depends upon the product of $f^e(E^e)$ and $f^h(E^h)$, while the absorption coefficient depends upon the product of $(1-f^e(E^e))$ and $(1-f^h(E^h))$. Here the energies E^e and E^h are related to the photon energy by the condition of vertical k-transitions. For these transitions we have

$$E^e = E_c + \frac{m_r^*}{m_e^*}(\hbar\omega - E_g)$$

$$E^h = E_v - \frac{m_r^*}{m_h^*}(\hbar\omega - E_g) \tag{5.36}$$

The occupation probabilities f^e and f^h are determined by the quasi-Fermi levels for electrons and holes, as discussed in Section 5.3.2.

The gain, which is the difference of the emission and absorption coefficient, is now proportional to

$$g(\hbar\omega) \sim f^e(E^e) \cdot f^h(E^h) - \{1 - f^e(E^e)\}\{1 - f^h(E^h)\} = \{f^e(E^e) + f^h(E^h)\} - 1 \tag{5.37}$$

The optical wave has a general spatial intensity dependence

$$I_{\text{ph}} = I_{\text{ph}}^0 \exp(g(\hbar\omega)x) \tag{5.38}$$

and, *if g is positive, the intensity grows because additional photons are added by emission to the intensity*. The condition for positive gain requires "inversion" of the semiconductor system; i.e., from Eq. 5.37,

$$f^e(E^e) + f^h(E^h) > 1 \tag{5.39}$$

The quasi-Fermi levels must penetrate their respective bands for this condition to be satisfied. It is found that gain is approximately given by (compare with Eq. 5.17)

$$g(\hbar\omega) \cong 4.7 \times 10^4 \frac{(\hbar\omega - E_g)^{1/2}}{\hbar\omega} \left[f^e(E^e) + f^h(E^h) - 1\right] \text{ cm}^{-1} \tag{5.40}$$

If there are no electrons in the conduction band and no holes in the valence band the gain becomes the inverse of the absorption coefficient. To evaluate the actual gain in a material as a function of carrier injection n ($= p$), we have to find the electron and hole quasi-Fermi levels and the occupation probabilities $f^e(E^e)$ and $f^h(E^h)$, where E^e and E^h are related to $\hbar\omega$ by Eq. 5.9.

It must be noted that the laser operates under conditions where f^e and f^h are larger than 0.5. In this high-injection limit, the occupation probabilities are not given accurately by the Boltzmann statistics. A useful approach is to use the Joyce–Dixon approximation for the position of the Fermi levels. For a given injection density n $(= p)$, the position of the quasi-Fermi levels is given by

$$E_{Fn} = E_c + k_B T \left[\ln \frac{n}{N_c} + \frac{1}{\sqrt{8}} \frac{n}{N_c} \right]$$

$$E_{Fp} = E_v - k_B T \left[\ln \frac{p}{N_v} + \frac{1}{\sqrt{8}} \frac{p}{N_v} \right] \tag{5.41}$$

where N_c and N_v are the effective density of states at the conduction and valence bands. Note that these results are for three-dimensional systems. For quantum wells the relation between the Fermi level and carrier density is different.

With these expressions the gain can be calculated as a function of photon energy for various levels of injection densities n $(= p)$. At low injections, f^e and f^h are quite small and the gain is negative. However, as injection is increased, for electrons and holes near the bandedges, f^e and f^h increase and gain can be positive. However even at high injections, for $\hbar\omega \gg E_g$ the gain is negative. The general form of the gain-energy curves for different injection levels are shown in Fig. 5.11.

The gain discussed above is called the material gain and comes only from the active semiconductor region where the recombination is occurring. As we will discuss later when we describe the laser usually has a very small dimension. In this case, we need to define the cavity gain, which is given by

$$\text{Cavity gain} = g(\hbar\omega)\Gamma \tag{5.42}$$

where Γ is the fraction of the optical intensity overlapping with the active gain medium. The value of Γ is almost unity for double heterostructure lasers based on bulk (3D) materials and ~ 0.01 for quantum-well lasers. The issue of lasers will be discussed in the next section.

5.4 OPTICAL PROCESSES IN QUANTUM WELLS

In the discussions of optical properties in semiconductors we have seen these are closely related to the density of states. In the gain term we have

$$g(\hbar\omega) \propto N_{cv}(\hbar\omega)(f^e + f^h - 1) \tag{5.43}$$

In three-dimensional systems the joint density of states is zero at the bandgap energy and increases monotonically. As a result the carrier distribution in a three-dimensional system (i.e. the product of density of states and occupation) has a form shown in Fig. 5.12a. The carrier distribution is given by

$$\begin{aligned} n(E) &= f^e(E) N_c(E) \\ p(E) &= f^h(E) N_v(E) \end{aligned} \tag{5.44}$$

5.4. Optical processes in quantum wells

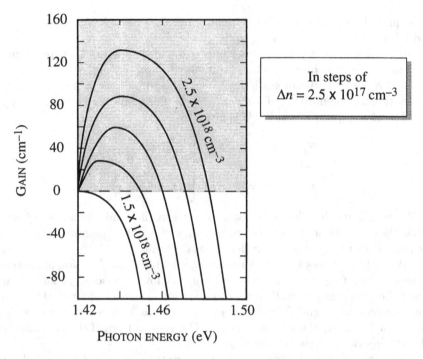

Figure 5.11: Gain vs. photon energy curves for a variety of carrier injections for GaAs at 300 K. The electron and hole injections are the same.

The Fermi function is independent of dimensionality, but density of states can be modified by altering dimensionality. The gain in the three-dimensional system starts at zero at bandgap energy and peaks away from the bandgap as shown in Fig. 5.12b. For most optoelectronic applications we would prefer that the carrier distribution and gain peaks occur at the bandedge. This requires a modification of the density of states.

In Chapter 3, Section 3.7 we have seen that quantum wells made from semiconductors can alter electronic properties and density of states. Such modifications improve device performance as seen later and many high performance devices use quantum well systems. The reader should review Section 3.7.2 on quantum wells.

In Fig. 5.12 we show a schematic of carrier distribution and a gain curve in three- and two-dimensional systems. It is important to note that in the two-dimensional system the gain curve peaks at the bandedge and falls off at higher energies. Since the photon energy in light emitters is controlled by e–h energy, in a quantum well system the photon spectra is peaked towards the bandege. This has the effect of higher peak gain at lower injection. This in turn allows lasers to have lower threshold currents, as discussed later.

As discussed in Section 3.7.2 the two-dimensional quantum well structure creates electron energies that can be described by *subbands* ($n = 1, 2, 3 \cdots$). The subbands for conduction band and valence band are shown schematically in Fig. 5.13. The solution of the quantum well problem gives energy levels for the electron, heavy hole and

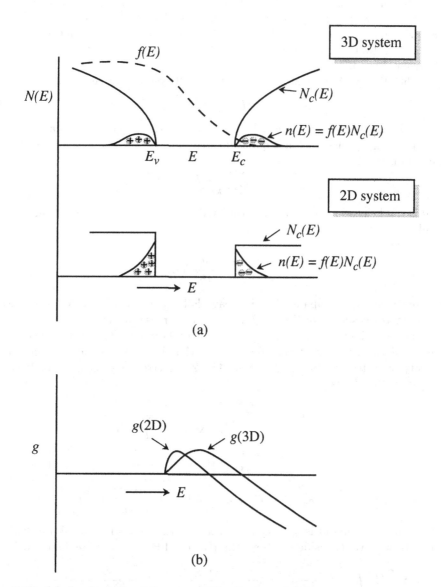

Figure 5.12: (a) A schematic of the energy distribution of carriers in a 3D and 2D system. (b) A schematic of the gain curve in a 3D and 2D system.

5.4. Optical processes in quantum wells

light hole. Each level E_1, E_2, etc., is actually a subband due to the electron energy in the x-y plane. As shown in Fig. 5.13 we have a series of subbands in the conduction and valence band. In the valence band we have a subband series originating from heavy holes and another one originating from light holes. The subband structure has important consequences for optical and transport properties of heterostructures. As noted above an important manifestation of this subband structure is the density of states of the electronic bands.

The effective bandgap of the quantum well system is given by

$$E_g(eff) = E_g(well) + E_1^e + E_1^{hh} \tag{5.45}$$

The well size and barrier height can be used to alter the effective bandgap.

The density of states in a quantum well is
- Conduction band

$$N(E) = \sum_i \frac{m^*}{\pi\hbar^2} \theta(E - E_i) \tag{5.46}$$

where θ is the heavyside step function (unity if $E > E_i$; zero otherwise) and E_i are the subband energy levels.
- Valence band

$$N(E) = \sum_i \sum_{j=1}^{2} \frac{m_j^*}{\pi\hbar^2} \theta(E_{ij} - E) \tag{5.47}$$

where i represents the subbands for the heavy hole ($j = 1$) and light holes ($j = 2$). The density of states is shown in Fig. 5.13 and has a staircase-like shape.

The relationship between the electron or hole density (areal density for 2D systems) and the Fermi level is different from that in 3-dimensional systems because the density of states function is different. The 2D electron density in a single subband starting at energy E_1^e is

$$n = \frac{m_e^*}{\pi\hbar^2} \int_{E_1^e}^{\infty} \frac{dE}{\exp\left(\frac{E - E_{Fn}}{k_B T}\right) + 1}$$

$$= \frac{m_e^* k_B T}{\pi\hbar^2} \left[\ell n \left\{1 + \exp\left(\frac{E_{Fn} - E_1^e}{k_B T}\right)\right\}\right]$$

or $\quad E_{Fn} = E_1^e + k_B T \ell n \left[\exp\left(\frac{n\pi\hbar^2}{m_e^* k_B T}\right) - 1\right] \tag{5.48}$

If more than one subband is occupied one can add their contribution similarly. For the hole density we have (considering both the HH and LH ground state subbands)

$$p = \frac{m_{hh}^*}{\pi\hbar^2} \int_{E_1^{hh}}^{-\infty} \frac{dE}{\exp\left(\frac{E_{Fp} - E}{k_B T}\right) + 1} + \frac{m_{\ell h}^*}{\pi\hbar^2} \int_{E_1^{\ell h}}^{-\infty} \frac{dE}{\exp\left(\frac{E_{Fp} - E}{k_B T}\right) + 1} \tag{5.49}$$

where m_{hh}^* and $m_{\ell h}^*$ are the in-plane density of states masses of the HH and LH subbands. We have then

$$p = \frac{m_{hh}^* k_B T}{\pi\hbar^2} \left[\ell n \left\{1 + \exp\frac{(E_1^{hh} - E_{Fp})}{k_B T}\right\}\right]$$

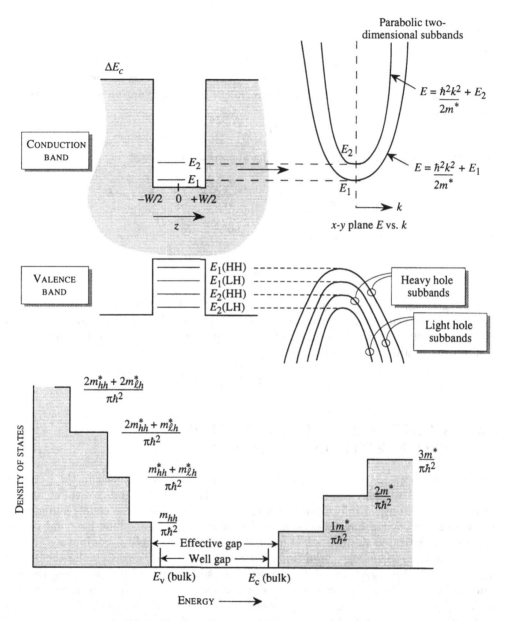

Figure 5.13: Schematic of density of states in a three-, two- and one-dimensional system with parabolic energy momentum relations.

$$+ \frac{m^*_{\ell h} k_B T}{\pi \hbar^2} \left[\ell n \left\{ 1 + \exp \frac{(E_1^{\ell h} - E_{Fp})}{k_B T} \right\} \right] \quad (5.50)$$

If $\quad E_1^{hh} - E_1^{\ell h} > k_B T$

The occupation of the light hole subband can be ignored.

The changes in density of states discussed above directly effect the optical properties in quantum wells. The main change that is needed is to replace the three-dimensional density of states N_{cv} by an equivalent term for quantum wells. This transformation is

$$N_{cv}(3D) \rightarrow \frac{N_{cv}(2D)}{W} \quad (5.51)$$

where W is the well size. This gives for the absorption coefficient

$$\alpha(\hbar\omega) = \frac{\pi e^2 \hbar}{m_0^2 c n_r \epsilon_0} \frac{1}{\hbar\omega} |\mathbf{a} \cdot \mathbf{p}_{if}|^2 \frac{N_{2D}}{W} \sum_{n,m} f_{nm} \theta(E_{nm} - \hbar\omega) \quad (5.52)$$

where the overlap integral f_{nm} is

$$f_{nm} = \int g_n^e(z) g_m^h(z) dz \quad (5.53)$$

The integral f_{nm} is essentially unity if $n = m$ (i.e., the symmetry of the envelope functions is the same) and is close to zero otherwise. In Fig. 5.14 we show a typical absorption coefficient in a quantum well. The absorption coefficient, just like the density of states has a staircase-like form, starting at

$$\hbar\omega = E_g(\text{well}) + E_1^e + E_1^{hh}. \quad (5.54)$$

5.5 IMPORTANT SEMICONDUCTOR OPTOELECTRONIC DEVICES

Semiconductor optoelectronic devices based on the concepts discussed in the previous section have led to many important technologies. These include image display, optical communication, optical information storage, desktop printing, bar code reading, etc. Semiconductor devices have also been used to demonstrate logic functions and to implement sophisticated mathematical algorithms. However, "optical computing" is not yet a commercial technology. Semiconductor energy conversion devices (solar cells) have also found niche markets, but due to mostly market forces and politics, they are not a major force in global energy markets. But advances in solar cell fabrication (including the potential use of organic solar cells), along with environmental concerns may gradually make semiconductors a dominant force in addressing the energy needs of the world.

An area where semiconductor light-emitting diodes (LED) can have a major impact is in the lighting market. The LED, a device with a far greater lifetime and efficiency, can in principal, replace the light bulb that currently dominates this industry.

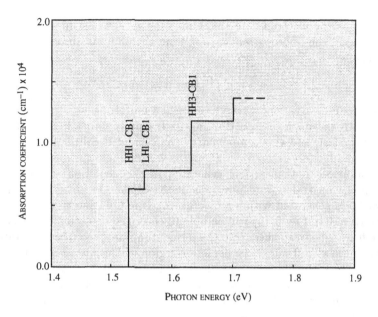

Figure 5.14: Calculated absorption coefficient in a 100 Å GaAs/Al$_{0.3}$Ga$_{0.7}$As quantum well structure.

5.5.1 Light detectors and solar cells

We have seen in Section 5.3 that when light impinges on a semiconductor with photon energy larger than the bandgap, electron–hole pairs are generated. If the electron–holes can be swept away to create an electrical signal (a current or a voltage) we can have a detection or energy conversion device. A device structure in which the photogenerated electrons and holes can be collected is the p–n diode. A metal-semiconductor junction can also provide a structure for a detector, but it is not as versatile as the p–n diode structure. Due to the importance of the p–n diode in optoelectronics we have provided a summary of the underlying theory in Appendix B. The reader may want to review this Appendix. Typically when the diode is used as a detector it is reverse biased, while when it is used as a solar cell it is unbiased (the solar energy creates a voltage across the diode as discussed later). When a diode is used as a light emitter it is forward biased.

As discussed in Appendix B, the p–n diode has a neutral region where the electric field is negligible, and a depletion region, where there is a strong electric field. When light impinges upon the diode to create electron–hole pairs, some of the carriers are collected at the contact, and lead to the photocurrent. Let us consider a long p–n diode in which excess carriers are generated at a rate G_L. The generation rate for an optical power density P_{op} is

$$G_L = \frac{\alpha P_{op}}{\hbar \omega} \tag{5.55}$$

Fig. 5.15 shows a p–n diode with a depletion region of width W. The electron–hole pairs generated in the depletion region are swept out rapidly by the electric field existing in

5.5. Important semiconductor optoelectronic devices

the region. Thus the electrons are swept into the n-region, while the holes are swept into the p-region. The photocurrent arising from the photons absorbed in the depletion region is thus

$$I_{L1} = A \cdot e \int_0^{x'} G_L \cdot dx = A \cdot e G_L W \qquad (5.56)$$

where A is the diode area and we have assumed a uniform generation rate in the diode. Since the electrons and holes contributing to I_{L1} move under high electric fields, the response is very fast, and this component of the current is called the prompt photocurrent.

In addition to the carriers generated in the depletion region, e–h pairs are generated in the neutral n- and p-regions of the diode. On physical grounds, we may expect that holes generated within a distance L_p (the diffusion length) of the depletion region edge ($x = 0$ of Fig. 5.15) will be able to enter the depletion region, from whence the electric field will sweep them into the p-side. Similarly, electrons generated within a distance L_n of the $x' = 0$ side of the depletion region will also be collected and contribute to the current. Thus the photocurrent should come from all carriers generated in a region $(W + L_n + L_p)$. A quantitative analysis reaches the same conclusion.

The total current due to carriers in the neutral region and the depletion region is

$$I_L = I_{nL} + I_{pL} + I_{L1} = eG_L(L_p + L_n + W)A \qquad (5.57)$$

It must be noted that the e–h pair generation is not uniform with penetration depth, but decreases with it. Thus G_L has to be replaced by an average generation rate for an accurate description. It is also important to note that the photocurrent flows in the direction of the reverse current of the diode.

The total current in the diode connected to the external load, as shown in Fig. 5.16, is given by the light-generated current and the diode current in the absence of light. In general, if the voltage across the diode is V, the total current is (note that the photocurrent flows in the opposite direction to the forward-bias diode current)

$$I = I_L + I_0 \left[1 - \exp\left\{\frac{e(V + R_s I)}{nk_B T}\right\}\right] \qquad (5.58)$$

where R_s is the diode series resistance, n the ideality factor, and V the voltage across the diode. As shown in Fig. 5.16, the photodiode can be used in one of two configurations. In the photovoltaic mode, used for solar cells, there is no external bias applied. The photocurrent passes through an external load to generate power. In the photoconductive mode, used for detectors, the diode is reverse biased and the photocurrent is collected.

Application to a solar cell

An important use of the p–n diode is to convert optical energy to electrical energy as in a solar cell. The solar cell operates without an external power supply and relies on the optical power to generate current and voltage. To calculate the important parameters of a solar cell, consider the case where the diode is used in the open circuit mode so that the current I is zero. This gives, for Eq. 5.58,

$$I = 0 = I_L - I_0 \left[\exp\left(\frac{eV_{oc}}{nk_B T}\right) - 1\right] \qquad (5.59)$$

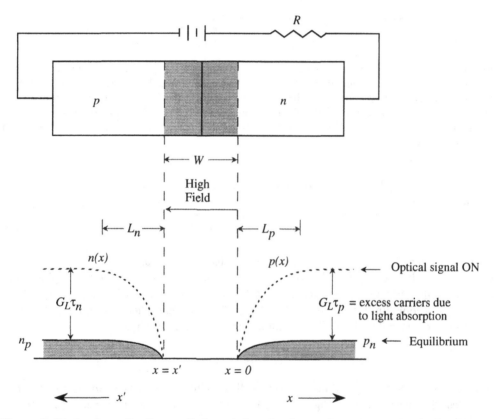

Figure 5.15: A schematic of a p–n diode and the minority carrier concentration in the absence and presence of light. The minority charge goes to zero at the depletion region edge due to the high field, which sweeps the charge away.

5.5. Important semiconductor optoelectronic devices

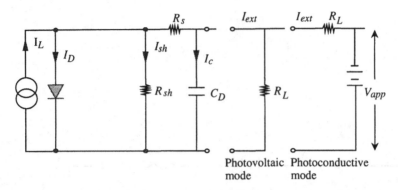

Figure 5.16: The equivalent circuit of a photodiode. The device can be represented by a photocurrent source I_L feeding into a diode. The device's internal characteristics are represented by a shunt resistor R_{sh} and a capacitor C_D. R_s is the series resistance of the diode. In the photovoltaic mode (used for solar cells and other devices) the diode is connected to a high-resistance R_L, while in the photoconductive mode (used for detectors) the diode is connected to a load R_L and a power supply.

where V_{oc} is the voltage across the diode, known as the open-circuit voltage. We get for this voltage

$$V_{oc} = \frac{nk_BT}{e}\ln\left(1 + \frac{I_L}{I_0}\right) \tag{5.60}$$

At high optical intensities the open-circuit voltage can approach the semiconductor bandgap. In the case of Si solar cells for solar illumination (without atmospheric absorption), the value of V_{oc} is roughly 0.7 eV.

A second limiting case in the solar cell is the one where the output is short-circuited; i.e., $R = 0$ and $V = 0$. The short-circuit current is then

$$I = I_{sc} = I_L \tag{5.61}$$

A plot of the diode current in the solar cell as a function of the diode voltage then provides the curve shown in Fig. 5.17a. In general, the electrical power delivered to the load is given by

$$P = I \times V = I_L V - I_0 \left[\exp\left(\frac{eV}{k_BT}\right) - 1\right] V \tag{5.62}$$

The maximum power is delivered at voltage and current values of V_m and I_m, as shown in Fig. 5.17a.

The conversion efficiency of a solar cell is defined as the rate of the output electrical power to the input optical power. When the solar cell is operating under maximum power conditions, the conversion efficiency is

$$\eta_{conv} = \frac{P_m}{P_{in}} \times 100 \text{ (percent)} = \frac{I_m V_m}{P_{in}} \times 100 \text{ (percent)} \tag{5.63}$$

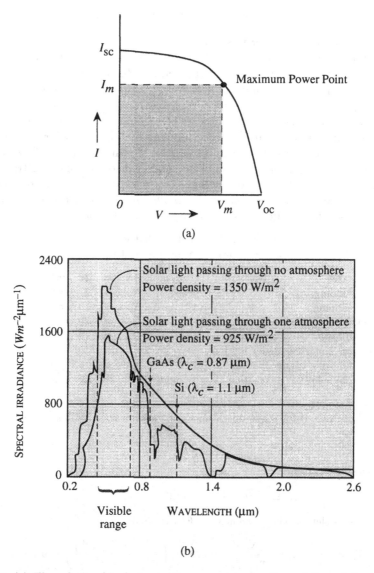

Figure 5.17: (a) The relationship between the current and voltage delivered by a solar cell. The open-circuit voltage is V_{oc} and the short-circuit current is I_{sc}. The maximum power is delivered at the point shown. (b) The spectral irradiance of the solar energy. The spectra are shown for no absorption in the atmosphere and for the sea-level spectra. Also shown are the cutoff wavelengths for GaAs and Si.

5.5. Important semiconductor optoelectronic devices

Another useful parameter in defining solar cell parameters is the fill factor F_f, defined as

$$F_f = \frac{I_m V_m}{I_{sc} V_{oc}} \qquad (5.64)$$

In most solar cells the fill factor is ~ 0.7.

In the solar cell conversion efficiency, it is important to note that photons that have an energy $\hbar\omega$ smaller than the semiconductor bandgap will not produce any electron–hole pairs. *Also, photons with energy greater than the bandgap will produce electrons and holes with the same energy (E_g), regardless of how large ($\hbar\omega - E_g$) is.* The excess energy $\hbar\omega - E_g$ is simply dissipated as heat. Thus the solar cell efficiency depends quite critically on how the semiconductor bandgap matches the solar energy spectra. In Fig. 5.17b we show the solar energy spectra. Also shown are the cutoff wavelengths for silicon and GaAs. GaAs solar cells are better matched to the solar spectra and provide greater efficiencies. However, the technology is more expensive than Si technology. Thus GaAs solar cells are used for space applications, while silicon (or amorphous silicon) solar cells are used for applications where cost is a key factor.

EXAMPLE 5.8 Consider a long Si p–n junction that is reverse biased with a reverse bias voltage of 2 V. The diode has the following parameters (all at 300 K):

Diode area,	A	$= 10^4\,\mu m^2$
p-side doping,	N_a	$= 2 \times 10^{16}\,cm^{-3}$
n-side doping,	N_d	$= 10^{16}\,cm^{-3}$
Electron diffusion coefficient	D_n	$= 20\,cm^2/s$
Hole diffusion coefficient,	D_p	$= 12\,cm^2/s$
Electron minority carrier lifetime,	τ_n	$= 10^{-8}\,s$
Hole minority carrier lifetime,	τ_p	$= 10^{-8}\,s$
Electron–hole pair generation rate by light,	G_L	$= 10^{22}\,cm^{-3}\,s^{-1}$

Calculate the photocurrent.

See Appendix B for diode operation. The electron length diffusion length is

$$L_n = \sqrt{D_n \tau_n} = \left[(20)(10^{-8})\right]^{1/2} = 4.5\,\mu m$$

The hole diffusion length is

$$L_p = \sqrt{D_p \tau_p} = \left[(12)(10^{-8})\right]^{1/2} = 3.46\,\mu m$$

To calculate the depletion width, we need to find the built-in voltage

$$V_{bi} = \frac{k_B T}{e} \ln\left(\frac{N_a N_d}{n_i^2}\right) = 0.026 \ln\left(\frac{(2 \times 10^{16})(10^{16})}{(1.5 \times 10^{10})^2}\right) = 0.715\,V$$

The depletion width is now

$$W = \left\{\frac{2\epsilon_s}{e}\left(\frac{N_a + N_d}{N_a N_d}\right)(V_{bi} + V_R)\right\}^{1/2}$$

$$= \left\{\frac{2(11.9)(8.85 \times 10^{-14})}{(1.6 \times 10^{-19})}\left(\frac{(2 \times 10^{16} + 10^{16})}{(2 \times 10^{16})(10^{16})}\right)(2.715)\right\}^{1/2}$$

$$= 0.73\,\mu m$$

We see in this case that L_n and L_p are larger than W. The prompt photocurrent is thus a small part of the total photocurrent. The photocurrent is now

$$\begin{aligned} I_L &= eAG_L(W + L_n + L_p) \\ &= (1.6 \times 10^{-19} \text{ C})(10^4 \times 10^{-8} \text{ cm}^2)(10^{22} \text{ cm}^{-3} \text{ s}^{-1}) \\ &\quad (0.73 \times 10^{-4} \text{ cm} + 4.5 \times 10^{-4} \text{ cm} + 3.46 \times 10^{-4} \text{ cm}) \\ &= 0.137 \text{ mA} \end{aligned}$$

The photocurrent is much larger than the reverse saturation current I_0 and its direction is the same as the reverse current.

EXAMPLE 5.9 Consider an Si solar cell at 300 K with the following parameters:

Area,	A	=	1.0 cm^2
Acceptor doping,	N_a	=	$5 \times 10^{17} \text{ cm}^{-3}$
Donor doping,	N_d	=	10^{16} cm^{-3}
Electron diffusion coefficient,	D_n	=	$20 \text{ cm}^2/\text{s}$
Hole diffusion coefficient,	D_p	=	$10 \text{ cm}^2/\text{s}$
Electron recombination time,	τ_n	=	$3 \times 10^{-7} \text{ s}$
Hole recombination time,	τ_p	=	10^{-7} s
Photocurrent,	I_L	=	25 mA

Calculate the open-circuit voltage of the solar cell.

To find the open-circuit voltage, we need to calculate the saturation current I_0, which is given by

$$I_0 = A\left[\frac{eD_n n_p}{L_n} + \frac{eD_p p_n}{L_p}\right] = Aen_i^2\left[\frac{D_n}{L_n N_a} + \frac{D_p}{L_p N_d}\right]$$

Also

$$L_n = \sqrt{D_n \tau_n} = [(20)(3 \times 10^{-7})]^{1/2} = 24.5 \ \mu\text{m}$$
$$L_p = \sqrt{D_p \tau_p} = [(10)(10^{-7})]^{1/2} = 10.0 \ \mu\text{m}$$

Thus,

$$\begin{aligned} I_0 &= (1)(1.6 \times 10^{-19})(1.5 \times 10^{10})^2 \\ &\quad \left[\frac{20}{(24.5 \times 10^{-4})(5 \times 10^{17})} + \frac{10}{(10 \times 10^{-4})(10^{16})}\right] \\ &= 3.66 \times 10^{-11} \text{ A} \end{aligned}$$

The open-circuit voltage is now

$$V_{oc} = \frac{k_B T}{e}\ln\left(1 + \frac{I_L}{I_0}\right) = (0.026)\ln\left(1 + \frac{25 \times 10^{-3}}{3.66 \times 10^{-11}}\right) = 0.53 \text{ V}$$

5.5.2 Light-emitting diode

The simplicity of the light-emitting diode (LED) makes it a very attractive device for display and communication applications. The basic LED is a p–n junction that is forward biased to inject electrons and holes into the p- and n-sides respectively. The

5.5. Important semiconductor optoelectronic devices

injected minority charge recombines with the majority charge in the depletion region or the neutral region. In direct band semiconductors, this recombination leads to light emission, since radiative recombination dominates in high-quality materials. In indirect gap materials, the light emission efficiency is quite poor and most of the recombination paths are non-radiative, which generates heat rather than light.

In general, the electron–hole recombination process can occur by radiative and non-radiative channels. If τ_r and τ_{nr} are the radiative and non-radiative lifetimes, the total recombination time is (for, say, an electron)

$$\frac{1}{\tau_n} = \frac{1}{\tau_r} + \frac{1}{\tau_{nr}} \qquad (5.65)$$

The internal quantum efficiency for the radiative processes is then defined as

$$\eta_{Qr} = \frac{\frac{1}{\tau_r}}{\frac{1}{\tau_r} + \frac{1}{\tau_{nr}}} = \frac{1}{1 + \frac{\tau_r}{\tau_{nr}}} \qquad (5.66)$$

In high-quality direct gap semiconductors, the internal efficiency is usually close to unity. In indirect materials the efficiency is of the order of 10^{-2} to 10^{-3}.

Before starting the discussion of light emission, let us remind ourselves of some important definitions and symbols used in this chapter:

I_{ph} : photon current = number of photons passing a cross-section/sec.
J_{ph} : photon current density = number of photons passing a unit area/sec.
P_{op} : optical power intensity = energy carried by photons per sec/area.

Carrier injection and spontaneous emission

The LED is essentially a forward-biased p–n diode, as shown in Fig. 5.18. Electrons and holes are injected as minority carriers across the diode junction and they recombine by either radiative recombination or non-radiative recombination. The diode must be designed so that the radiative recombination can be made as strong as possible.

The theory of the p–n diode is discussed in detail in Appendix B. In the forward-bias conditions the electrons are injected from the n-side to the p-side while holes are injected from the p-side to the n-side. The forward-bias current is dominated by the minority charge diffusion current. The diffusion current, in general, has three components: (i) minority carrier electron diffusion current; (ii) minority carrier hole diffusion current; and (iii) trap-assisted recombination current in the depletion region of width W. These current densities have the following forms, respectively

$$J_n = \frac{eD_n n_p}{L_n}\left[\exp\left(\frac{eV}{k_B T}\right) - 1\right] \qquad (5.67)$$

$$J_p = \frac{eD_p p_n}{L_p}\left[\exp\left(\frac{eV}{k_B T}\right) - 1\right] \qquad (5.68)$$

$$J_{GR} = \frac{en_i W}{2\tau}\left[\exp\left(\frac{eV}{2k_B T}\right) - 1\right] \qquad (5.69)$$

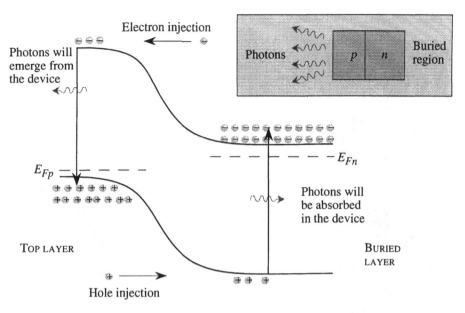

Figure 5.18: In a forward-biased p–n junction, electrons and holes are injected as shown. In the figure, the holes injected into the buried n region will generate photons that will not emerge from the surface of the LED. The electrons injected will generate photons that are near the surface and have a high probability to emerge.

where τ is the recombination time in the depletion region and depends upon the trap density. The LED is designed so that the photons are emitted close to the top layer and not in the buried layer, as shown in Fig. 5.18. The reason for this choice is that photons emitted deep in the device have a high probability of being reabsorbed. Thus we prefer to have only one kind of carrier injection for the diode current. Usually the top layer of the LED is p-type, and for photons to be emitted in this layer one must require the diode current to be dominated by the electron current (i.e., $J_n \gg J_p$). The ratio of the electron current density to the total diode current density is called the injection efficiency γ_{inj}. Thus we have

$$\gamma_{\text{inj}} = \frac{J_n}{J_n + J_p + J_{GR}} \qquad (5.70)$$

If the diode is pn^+, $n_p \gg p_n$ and, as can be seen from Eqs. 5.67 and 5.68, J_n becomes much larger than J_p. If, in addition, the material is high-quality, so that the recombination current is small, the injection efficiency approaches unity.

Once the minority charge (electrons) is injected into the doped neutral region (p-type), the electrons and holes will recombine to produce photons. They may also recombine non-radiatively via defects or via phonons.

The LED is a very simple device to produce optical signals. Due to its low cost it is quite important in display applications. However, it is not optimum for high-speed applications, such as optical communications. This is due to two reasons. The spectral

5.5. Important semiconductor optoelectronic devices

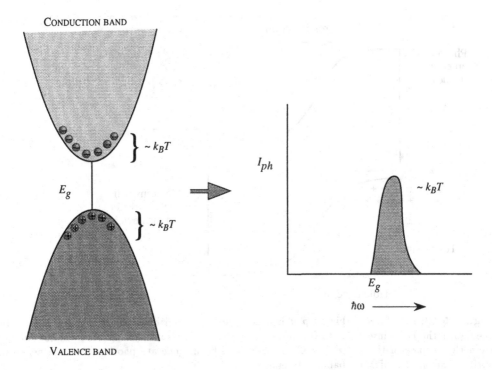

Figure 5.19: In a LED the electrons and holes are distributed over an energy width of $\sim k_B T$. Since all e–h pairs contribute to the optical output, the LED output is quite broad with a width roughly equal to $k_B T$. The shape of the output depends upon the carrier occupation function and the density of states function.

width of the LED is essentially controlled by the carrier distribution, as shown in Fig. 5.19, and is given by

$$\hbar \Delta \omega \sim k_B T \tag{5.71}$$

or

$$\Delta \lambda \sim \frac{k_B T}{h} \frac{\lambda^2}{c} \tag{5.72}$$

At room temperature this leads to $\Delta \lambda \sim 100$ Å, a width too large for high-speed long-distance communication. Additionally, LED speeds are controlled by spontaneous emission time τ_o. Thus the upper frequencies of modulation are a few GHz.

EXAMPLE 5.10 In two $n^+ p$ GaAs LEDs, $n^+ \gg p$ so that the electron injection efficiency is 100% for both diodes. If the non-radiative recombination time is 10^{-7}s, calculate the 300 K internal radiative efficiency for the diodes when the doping in the p-region for the two diodes is 10^{16} cm^{-3} and 5×10^{17} cm^{-3}.

When the p-type doping is 10^{16} cm^{-3}, the hole density is low and the e–h recombination time for the injected electrons is given by

$$\frac{1}{\tau_r} = \frac{1}{2\tau_o} \left(\frac{2\pi \hbar^2 m_r^*}{k_B T m_e^* m_h^*} \right)^{3/2} p$$

This gives
$$\tau_r = 5.7 \times 10^{-7} \text{ s}$$
In the case where the p-doping is high, the recombination time is given by the high-density limit

$$\frac{1}{\tau_r} = \frac{R_{spon}}{n} = \frac{1}{\tau_o}\left(\frac{m_r^*}{m_h^*}\right)^{3/2}$$

$$\tau_r = \frac{\tau_o}{0.05} \sim 20\tau_o \sim 12 \text{ ns}$$

For the low-doping case, the internal quantum efficiency for the diode is

$$\eta_{Qr} = \frac{1}{1 + \frac{\tau_r}{t_{nr}}} = \frac{1}{1+(5.7)} = 0.15$$

For the more heavily doped p-region diode, we have

$$\eta_{Qr} = \frac{1}{1 + \frac{10^{-7}}{20 \times 10^{-9}}} = 0.83$$

Thus there is an increase in the internal efficiency as the p doping is increased.

EXAMPLE 5.11 Consider a GaAs $p-n$ diode with the following parameters at 300 K:

Electron diffusion coefficient, $D_n = 30 \text{ cm}^2/\text{V}\cdot\text{s}$
Hole diffusion coefficient, $D_p = 15 \text{ cm}^2/\text{V}\cdot\text{s}$
p-side doping, $N_a = 5 \times 10^{16} \text{ cm}^{-3}$
n-side doping, $N_d = 5 \times 10^{17} \text{ cm}^{-3}$
Electron minority carrier lifetime, $\tau_n = 10^{-8}$ s
Hole minority carrier lifetime, $\tau_p = 10^{-7}$ s

Calculate the injection efficiency of the LED assuming no recombination due to traps.
The intrinsic carrier concentration in GaAs at 300 K is $1.84 \times 10^6 \text{ cm}^{-3}$. This gives

$$n_p = \frac{n_i^2}{N_a} = \frac{(1.84 \times 10^6)^2}{5 \times 10^{16}} = 6.8 \times 10^{-5} \text{ cm}^{-3}$$

$$p_n = \frac{n_i^2}{N_d} = \frac{(1.84 \times 10^6)^2}{5 \times 10^{17}} = 6.8 \times 10^{-6} \text{ cm}^{-3}$$

The diffusion lengths are

$$L_n = \sqrt{D_n \tau_n} = \left[(30)(10^{-8})\right]^{1/2} = 5.47 \text{ μm}$$

$$L_p = \sqrt{D_p \tau_p} = \left[(15)(10^{-7})\right]^{1/2} = 12.25 \text{ μm}$$

The injection efficiency is now (assuming no recombination via traps)

$$\gamma_{inj} = \frac{\frac{eD_n n_{po}}{L_n}}{\frac{eD_n n_{po}}{L_n} + \frac{eD_p p_{no}}{L_p}} = 0.98$$

EXAMPLE 5.12 Consider the $p-n^+$ diode of the previous example. The diode is forward biased with a forward-bias potential of 1 V. If the radiative recombination efficiency $\eta_{Qr} = 0.5$, calculate the photon flux and optical power generated by the LED. The diode area is 1 mm^2.

5.5. Important semiconductor optoelectronic devices

The electron current injected into the p-region will be responsible for the photon generation. This current is

$$\begin{aligned}
I_n &= \frac{AeD_n n_{po}}{L_n}\left[\exp\left(\frac{eV}{k_BT}\right) - 1\right] \\
&= \frac{(10^{-2}\text{ cm}^2)(1.6 \times 10^{-19}\text{ C})(30\text{ cm}^2/\text{s})(6.8 \times 10^{-5}\text{ cm}^{-3})}{5.47 \times 10^{-4}\text{ cm}}\left[\exp\left(\frac{1}{0.026}\right) - 1\right] \\
&= 0.30\text{ mA}
\end{aligned}$$

The photons generated per second are

$$\begin{aligned}
I_{ph} = \frac{I_n}{e} \cdot \eta_{Qr} &= \frac{(0.30 \times 10^{-3}\text{ A})(0.5)}{1.6 \times 10^{-19}\text{ C}} \\
&= 9.38 \times 10^{14}\text{ s}^{-1}
\end{aligned}$$

Each photon has an energy of 1.41 eV (= bandgap of GaAs). The optical power is thus

$$\begin{aligned}
\text{Power} &= (9.38 \times 10^{14}\text{ s}^{-1})(1.41)(1.6 \times 10^{-19}\text{ J}) \\
&= 0.21\text{ mW}
\end{aligned}$$

5.5.3 Laser diode

The laser diode overcomes the problem faced by LEDs of spectral width and low speed (discussed in the previous subsection). This is done by exploiting the basic physics of stimulated emission discussed earlier. The laser diode differs from the LED in two respects: (i) Mirrors are used to create a feedback effect so that photons with a desired frequency are selectively built-up in the laser cavity. (ii) Use of stimulated emission to generate photons allow not only coherent photons, but also an optical output that can be modulated at high speeds.

As in the case of the LED, electrons and holes are injected into an active region by forward biasing the laser diode. At low injection, these electrons and holes recombine radiatively via the spontaneous emission process to emit photons. However, the laser structure is so designed that at higher injection emission occurs by stimulated emission. As we will discuss below, the stimulated emission process provides spectral purity to the photon output, provides coherent photons, and offers high-speed performance. *Thus the key difference between the LED and the laser diode arises from the difference between spontaneous and stimulated emission.*

Spontaneous and stimulated emission: need for optical cavity

The key to understanding the semiconductor laser diode is the physics behind spontaneous and stimulated emission. Let us develop this understanding using Fig. 5.20. Consider an electron with wave vector k and a hole with a wave vector k in the conduction and valence bands, respectively, of a semiconductor. In the case shown in Fig. 5.20a, initially there are no photons in the semiconductor. The electron and hole recombine to emit a photon as shown, and this process is the spontaneous emission. The spontaneous emission rate was discussed in the context of the LED.

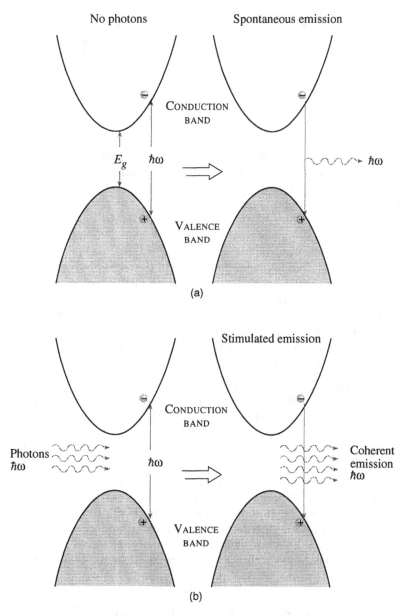

Figure 5.20: (a) In spontaneous emission, the e–h pair recombines in the absence of any photons present to emit a photon. (b) In simulated emission, an e–h pair recombines in the presence of photons of the correct energy $\hbar\omega$ to emit coherent photons. In coherent emission the phase of the photons emitted is the same as the phase of the photons causing the emission.

5.5. Important semiconductor optoelectronic devices

In Fig. 5.20b, we show the electron–hole pair, along with photons of energy $\hbar\omega$ equal to the electron–hole energy difference. In this case, in addition to the spontaneous emission rate, we have an additional emission rate, called the stimulated emission process. The stimulated emission process is proportional to the photon density (of photons with the correct photon energy to cause the e–h transition). *The photons that are emitted are in phase (i.e., same energy and wave vector) with the incident photons.* Quantum-mechanical calculations show that the rate for the stimulated emission is

$$W_{em}^{st}(\hbar\omega) = W_{em}(\hbar\omega) \cdot n_{ph}(\hbar\omega) \tag{5.73}$$

where $n_{ph}(\hbar\omega)$ is the photon density and W_{em} is the spontaneous emission rate we have already discussed. Thus if $n_{ph}(\hbar\omega) \sim 0$, there is no stimulated emission process. In the LED, when photons are emitted by spontaneous emission, they are either lost by reabsorption or simply leave the structure. Thus $n_{ph}(\hbar\omega)$ remains extremely small and stimulated emission cannot get started.

Consider now the possibility that, when photons are emitted via spontaneous emission, an optical cavity is designed so that photons with a well-defined energy are selectively confined in the semiconductor structure. If this is possible, two important effects occur: (i) the photon emission for photons with the chosen energy becomes stronger due to stimulated emission; (ii) the e-h recombination rate increases, as can be seen from Eq. 5.14. These two effects are highly desirable since they produce an optical spectrum with very narrow emission lines and the light output can be modulated at high speeds.

The challenge for the design of the laser is, therefore, to incorporate an optical cavity that ensures that the photons emitted are allowed to build up in the semiconductor device so that stimulated emission can occur.

The laser structure: optical cavity

While both the LED and the laser diode use a forward-biased p–n junction to inject electrons and holes to generate light, the laser structure is designed to create an "optical cavity" that can "guide" the photons generated. The optical cavity is essentially a resonant cavity in which the photons have multiple reflections. Thus, when photons are emitted, only a small fraction is allowed to leave the cavity. As a result, the photon density in the cavity starts to build up. A number of important cavities is used for solid-state lasers. These are the Fabry–Perot cavity, cylindrical cavity, rectangular cavity, etc. For semiconductor lasers, the most widely used cavity is the Fabry–Perot cavity shown in Fig. 5.21a. The important ingredient of the cavity is a polished mirror surface that assures that resonant modes are produced in the cavity, as shown in Fig. 5.21b. These resonant modes are those for which the wave vectors of the photon satisfy the relation

$$L = q\lambda/2 \tag{5.74}$$

where q is an integer, L is the cavity length, and λ is the light wavelength in the material and is related to the free-space wavelength by

$$\lambda = \frac{\lambda_o}{n_r} \tag{5.75}$$

where n_r is the refractive index of the cavity.

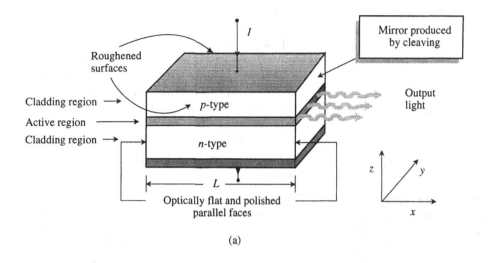
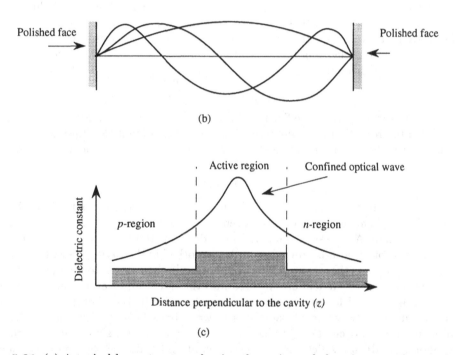

Figure 5.21: (a) A typical laser structure showing the cavity and the mirrors used to confine photons. The active region has a smaller bandgap than the cladding layers. (b) The stationary states of the cavity. The mirrors are responsible for these resonant states. (c) The variation in the dielectric constant is responsible for the optical confinement. The structure for the optical cavity shown in this figure is called the Fabry–Perot cavity.

5.5. Important semiconductor optoelectronic devices

As can be seen from Fig. 5.21a, the Fabry–Perot cavity has mirrored surfaces on two sides. The other sides are roughened so that photons emitted through these sides are not reflected back and are not allowed to build up. Thus only the resonant modes are allowed to build up and participate in the stimulated emission process.

If a planar heterostructure of the form shown in Fig. 5.21c is used, the optical wave is confined in the z-direction, as shown. This requires the light-confining or cladding layers to be made from a large bandgap material. An important parameter of the laser cavity is the optical confinement factor Γ, which gives the fraction of the optical wave in the active region. This confinement factor is almost unity for "bulk" double heterostructure lasers, where the active region is $\gtrsim 1.0$ μm, while it is as small as 1% for advanced quantum-well lasers.

Laser below and above threshold

When the p–n diode of a laser is forward biased, the injected electrons and holes recombine to emit photons. The photons emerging in the laser mode show an interesting transition at a threshold current. In Fig. 5.22 we show the light output in the laser mode as a function of injected current density in a laser diode (LD). If we compare this with the output from an LED we notice an important difference. The light output from a laser diode displays a rather abrupt change in behavior below the "threshold" condition and above this condition. The threshold condition is defined as the condition where the cavity gain overcomes the cavity loss arising from photon absorption in the cavity α_{loss} usually in the cladding region, and photon loss through transmission through the mirrors. The condition is

$$\Gamma g(\hbar\omega) = \alpha_{\text{loss}} - \frac{\ln R}{L} \tag{5.76}$$

where R is the reflection coefficient of the mirrors. In high-quality lasers $\alpha_{loss} \sim 10$ cm^{-1} and the reflection loss may contribute a similar amount. Another useful definition in the laser is the condition of transparency when the light suffers no absorption or gain, i.e.

$$\Gamma g(\hbar\omega) = 0 \tag{5.77}$$

It is important to identify two distinct regions of operation of the laser. Referring to Fig. 5.23, when the forward-bias current is small, the number of electrons and holes injected is small. As a result, the gain in the device is too small to overcome the cavity loss. The photons that are emitted are either absorbed in the cavity or lost to the outside. Thus, in this regime there is no buildup of photons in the cavity. However, as the forward bias increases, more carriers are injected into the device until eventually the threshold condition is satisfied for some photon energy. As a result, the photon number starts to build up in the cavity. As the device is forward biased beyond threshold, stimulated emission starts to occur and dominates the spontaneous emission. The light output in the photon mode, for which the threshold condition is satisfied, becomes very strong.

Below the threshold, the device essentially operates as an LED, except that there is a higher cavity loss in the laser diode, since photons cannot escape from the device due to the mirrors. Figure 5.23a shows this regime of operation.

Once the carrier density of the electrons and holes is high enough so that the threshold condition given by Eq. 5.77 is met, the photons generated in the laser cavity

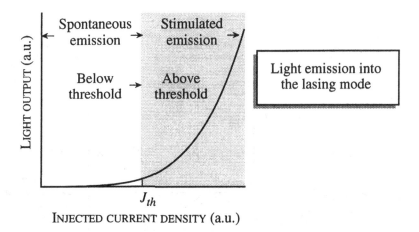

Figure 5.22: The light output as a function of current injection in a semiconductor laser. Above threshold, the presence of a high photon density causes stimulated emission to dominate.

grow in intensity after emission. Of course, out of all the optical modes that are allowed in the cavity, one or two will have the highest gain, since the gain curves have a peak at some energy, as seen from Fig. 5.23b. Since the gain is positive, the photon density in the laser cavity starts to increase rapidly. As a result, the stimulated emission process starts to grow. As noted in Section 5.3, the stimulated emission rate is related to the spontaneous emission rate by

$$W_{\text{em}}^{st}(\hbar\omega) = W_{\text{em}}(\hbar\omega) \cdot n_{\text{ph}}(\hbar\omega) \tag{5.78}$$

where $n_{\text{ph}}(\hbar\omega)$ is the photon density in the mode.

In order to study the laser characteristics around and above threshold, let us establish a simple relation between the injected current density, radiative lifetime, dimensions of the active region where recombination occurs, and the carrier density ($n = p$) in the active region. The rate of arrival of electrons (holes) into the active region is

$$\frac{JA}{e}$$

The rate at which the injected e–h pairs recombine is

$$\frac{nAd_{\text{las}}}{\tau_r(J)}$$

where $\tau_r(J)$ is the current-density-dependent radiative lifetime. Assuming a radiative efficiency of unity, we equate the two results given above to get

$$n = \frac{J\tau_r(J)}{ed_{\text{las}}} \tag{5.79}$$

5.5. Important semiconductor optoelectronic devices

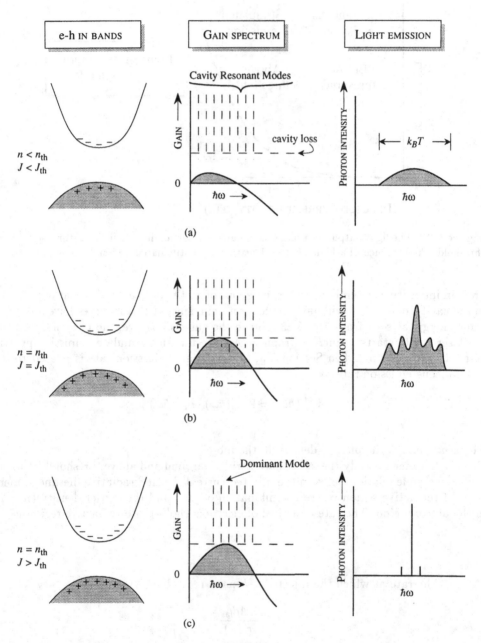

Figure 5.23: (a) The laser below threshold. The gain is less than the cavity loss and the light emission is broad as in an LED. (b) The laser at threshold. A few modes start to dominate the emission spectrum. (c) The laser above threshold. The gain spectrum does not change but, due to the stimulated emission, a dominant mode takes over the light emission.

At threshold we have

$$n_{th} = \frac{J_{th}\tau_r(J_{th})}{ed_{las}} \quad (5.80)$$

As discussed in Section 5.3, at threshold $\tau_r(J_{th}) \sim 4\tau_o$ (~2 ns for a GaAs laser).

As the current density exceeds J_{th}, the photon density in the dominant mode builds up, as discussed above, and the value of τ_r starts to become smaller. As a result, even though the injected charge density increases, the carrier density in the active region saturates close to the threshold density n_{th}.

The light output is given by ($n = n_{th}$; use Eq. 5.79 for the current density)

$$I_{ph} = \frac{I}{e} = \frac{n_{th} A d_{las}}{\tau_r} \quad (5.81)$$

Upon comparing this equation with the results for light output from an LED, it may be seen that the photon current is similar for the same injected current for an LED and LD (biased above threshold). However, in the case of the laser diode, the entire photon output emerges only in one or two photon modes, rather than in a broad spectrum of width $k_B T$. This spectral purity that arises, because of the importance of stimulated emission, distinguishes the LD from an LED. Also, the light output is highly collimated and coherent for similar reasons.

EXAMPLE 5.13 According to the Joyce–Dixon approximation, the relation between the Fermi level and carrier concentration is given by

$$E_F - E_c = k_B T \left[\ln \frac{n}{N_c} + \frac{1}{\sqrt{8}} \frac{n}{N_c} \right]$$

where N_c is the effective density of states for the band. Calculate the carrier density needed for the transparency condition in GaAs at 300 K and 77 K. The transparency condition is defined at the situation where the maximum gain is zero (i.e., the optical beam propagates without loss or gain).

At room temperature the valence and conduction band effective density of states are

$$N_v = 7 \times 10^{18} \text{ cm}^{-3}$$
$$N_c = 4.7 \times 10^{17} \text{ cm}^{-3}$$

The values at 77 K are

$$N_v = 0.91 \times 10^{18} \text{ cm}^{-3}$$
$$N_c = 0.61 \times 10^{17} \text{ cm}^{-3}$$

In the semiconductor laser, an equal number of electrons and holes are injected into the active region. We will look for the transparency conditions for photons with energy equal to the bandgap. The approach is very simple: (i) choose a value of n or p; (ii) calculate μ from the Joyce–Dixon approximation; (iii) calculate $f^e + f^h - 1$ and check if it is positive at the bandedge. The same approach can be used to find the gain as a function of $\hbar\omega$.

For 300 K we find that the material is transparent when $n \sim 1.1 \times 10^{18}$ cm^{-3} at 300 K and $n \sim 2.5 \times 10^{17}$ cm^{-3} at 77 K. Thus a significant decrease in the injected charge occurs as temperature is decreased.

EXAMPLE 5.14 Consider a GaAs laser at 300 K. The optical confinement factor is unity. Calculate the threshold carrier density assuming that it is 20% larger than the density for transparency. If the active layer thickness is 2 μm, calculate the threshold current density.

From the previous example we see that at transparency

$$n = 1.1 \times 10^{18} \text{ cm}^{-3}$$

The threshold density is then

$$n_{th} = 1.32 \times 10^{18} \text{ cm}^{-3}$$

The radiative recombination time is approximately four times τ_o, i.e., $\sim$2.4 ns. The current density then becomes

$$J_{th} = \frac{e \cdot n_{th} \cdot d}{\tau_r} = \frac{(1.6 \times 10^{-19} \text{ C})(1.32 \times 10^{18} \text{ cm}^{-3})(2 \times 10^{-4} \text{ cm})}{2.4 \times 10^{-9} \text{ s}}$$
$$= 1.76 \times 10^4 \text{ A/cm}^2$$

5.6 ORGANIC SEMICONDUCTORS: OPTICAL PROCESSES AND DEVICES

Over the last 50 years, insulating polymers or plastics have transformed society, replacing wood, metals, and ceramics, because of their structural strength, lightweight, and fabrication ease. Until recently, while it was recognized that the electronic properties of polymers could also be interesting, they were not seriously considered for information technology applications. Recently however, as synthesis techniques have improved, some polymers have shown semiconductor-like properties; i.e., they can be doped, their conductance can be controlled, and light can be detected and emitted. This is not to imply that these materials have electronic properties approaching those of inorganic semiconductors. Mobility for instance is still $\sim 10^{-2}$ cm^2/Vs in organic semiconductors, i.e., nearly a millionth of what may be found in materials like GaAs. However, for many applications these materials have good enough optoelectronic properties and with continuous improvement it is expected that they will play an important role in future technologies.

Rapid progress in doped polymers, especially doped polyacetylene was made in the 1970s, but the material remained a curiosity due to the difficulties in processing. In the late 1980s the interest in organic semiconductors surged because Eastman Kodak and Cambridge University demonstrated electroluminescent devices, and FETs made from polythiophene were demonstrated. Polymer-based devices are now used for backlights of liquid crystal displays, displays of devices, such as cell phones or watches. It is expected that commercial technologies, such as televisions, solar cells, etc., will benefit from new advances.

Polymer LEDs were first demonstrated in 1990 and are very attractive because of potential large area applications and mechanical flexibility. The diode is usually not fabricated by doping the polymer itself. Instead an undoped film is placed between an anode (indium tin oxide) and a cathode (e.g., calcium).

We remind ourselves that in organic semiconductors, the nature of atomic bonding is such that instead of energy bands we have very narrow range energy levels. The ones that are relevant for us and are equivalent to the valence and conduction bands are the HOMO and LUMO states. Figure 5.24 shows a schematic of a metal-organic-semiconductor junction. As shown in Fig. 5.24a, if the metal Fermi energy is close to the HOMO level, electrons from the HOMO can move into the metal (a hole is injected into the semiconductor). Thus in this case the metal acts as a p-type contact. If however, the metal workfunctions are close to the LUMO state, as shown in Fig. 5.24b, an n-type contact results. In Fig. 5.24c we show how current flows in a p–n diode using an organic semiconductor film.

We see that the n-contact injects an electron, which diffuses into the semiconductor, eventually recombining with a hole in the HOMO level. An electron then leaves the HOMO state (i.e., a hole is injected).

The junction described above is one approach to creating electrons and holes. It is possible also to have heterojunctions between different organic semiconductors as well as junctions between n-type and p-type organic semiconductors.

The process of photon emission by electron–hole recombination in organic semiconductors is not fully understood, but it is clear that it is significantly different from that in inorganic semiconductors. The difference has to do with the importance of the exciton state which is an e–h system interacting via Coulombic interaction. We will briefly review the exciton state before examining the optical properties of organic semiconductors.

5.6.1 Excitonic state

In a semiconductor (or an insulator) the electronic spectra are such that the valence band is completely filled with electrons at 0 K and the conduction band is empty. The two bands are separated by the bandgap. As noted above, for organic materials these are the HOMO and LUMO bands. This gives a "single-electron energy-momentum" relation of the form shown in Fig. 5.25a. Now consider an electron being removed from the valence band (the ground state of the problem is the filled valence band state) and excited to a higher energy state. If we ignore the interaction of the electron that is removed from the valence band with the hole in the valence band produced by the absence of the electrons, the lowest energy needed to excite the system is the bandgap energy E_g. However, the electron and hole interact with each other, since the hole responds as if it is a positively charged particle, as shown in Fig. 5.25b. The electron–hole system or exciton can be represented by the hydrogen-like model. The exciton problem can be written as

$$\left[-\frac{\hbar^2}{2m_e^*}\nabla_e^2 - \frac{\hbar^2}{2m_h^*}\nabla_h^2 - \frac{e^2}{4\pi\epsilon|\mathbf{r}_e - \mathbf{r}_h|} \right]\psi_{ex} = E\psi_{ex} \tag{5.82}$$

Here m_e^* and m_h^* are the electron and hole effective masses and $|\mathbf{r}_e - \mathbf{r}_h|$ is the difference in coordinates defining the Coulombic interaction between the electron and the hole. The problem is now the standard two-body problem like the electron–proton problem of the H atom. The relative coordinate problem has the form (m_r^* is the reduced mass of the electron–hole pair)

$$\left(\frac{\hbar^2 k^2}{2m_r^*} - \frac{e^2}{4\pi\epsilon|\mathbf{r}|} \right) F(\mathbf{r}) = E F(\mathbf{r}) \tag{5.83}$$

5.6. Organic semiconductors: optical processes and devices

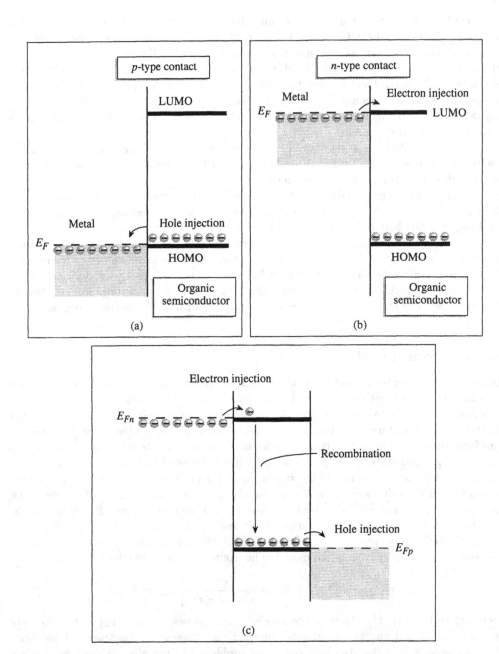

Figure 5.24: A schematic of a metal–organic–semiconductor junction. The metal workfunction (the relative position of the metal) Fermi level determines if the contact is p-type (a) or n-type (b). (c) In this figure we show how current flows in a junction diode.

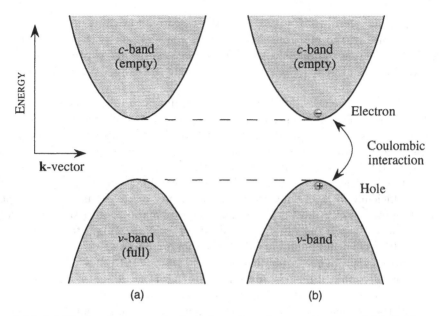

Figure 5.25: (a) The bandstructure in the independent electron picture and (b) the Coulombic interaction between the electron and hole, which would modify the band picture. For organic semiconductors the valence band is the HOMO and the conduction band is the LUMO.

This is the usual hydrogen atom problem and $F(\mathbf{r})$ can be obtained from the mathematics of that problem. The general exciton solution is now (writing $\mathbf{K}_{ex} = \mathbf{K}$)

$$\psi_{n\mathbf{K}_{ex}} = e^{i\mathbf{K}_{ex}\cdot\mathbf{R}}F_n(\mathbf{r})\phi_c(\mathbf{r})\phi_v(\mathbf{r}_h) \quad (5.84)$$

where ϕ_c and ϕ_v represent the electron and hole bandedge states. The excitonic energy levels are then

$$E_{n\mathbf{K}_{ex}} = E_n + \frac{\hbar^2}{2(m_e^* + m_h^*)}K_{ex}^2 \quad (5.85)$$

with E_n being the eigenvalues of the hydrogen atom-like problem (the energy is referenced to the bandgap energy E_g)

$$E_n = -\frac{m_r^* e^4}{2(4\pi\epsilon)^2\hbar^2}\frac{1}{n^2} \quad (5.86)$$

and the second term in Eq. 5.85 represents the kinetic energy of the center of mass of the electron–hole pair.

The exciton energy is thus slightly lower than the bandgap energy of the semiconductor. Excitonic states can be observed in optical absorption spectra and are exploited for many optoelectronic devices.

In inorganic semiconductors like GaAs, the relative dielectric constant is ~ 10 and the reduced mass is ~ 0.1. The exciton binding energy is then ~ 4 meV; i.e., quite small compared to the single electron bandgap. The excitonic state has an envelope function that has a radius of ~ 100 Å. In the case of the organic semiconductors, the molecular nature of the electronic state causes the electron–hole state to be tightly localized. The exciton binding energy is quite large as a result (in the range of $\sim$ eV). The exciton creates photons, which are quite different in energy from the LUMA-HOMO energy difference.

In Fig. 5.26 we show a schematic of how photons are generated in the organic LED. The process of light emission involves (i) injection of electrons (holes) from the contacts, (ii) diffusion of the carrier in the LUMO or HOMO states (this involves hopping as discussed in the previous chapter's discussion of transport in disordered systems), (iii) exciton formation, and (iv) exciton recombination to emit a photon. Experiments have shown that, while the e–h pair formation and recombination efficiency is essentially 100%, the efficiency of light emission (i.e., photon per e–h pair) is only a few percent ($< 5\%$). This implies that most of the excitonic recombination occurs through generation of heat or energy transfer to other non-radiative processes. This is to be contrasted to the nearly 100% efficiency found in direct gap inorganic LEDs.

A serious problem so far in organic LEDs or OLEDs is the lifetime of the devices. A problem that limits the lifetime of OLEDs is the loss of efficiency with time. This appears to be linked to the presence of defects like oxygen, which cause chemical reactions in the presence of electrons. The chemical reactions lead to new molecules degrading the polymer. As the fabrication processes improve defect-related lifetime, problems will become less critical and the use of organic semiconductors for optical applications will grow. The tremendous flexibility in being able to tailor the optical response by exploiting chemistry makes this class of semiconductors very versatile for future applications.

5.7 SUMMARY

In this chapter we have discussed important optical properties of semiconductors and how these are exploited for devices. The topics covered are summarized in Tables 5.1 to 5.2.

5.8 PROBLEMS

5.1 The bandgap of the $Hg_{1-x}Cd_x Te$ alloy is given by the expression

$$E_g(x) = -0.3 + 1.9x \text{ (eV)}$$

Calculate the composition of an alloy which gives a cutoff wavelength of (a) 10 μm; (b) 5.0μm.

5.2 Calculate the cutoff wavelength for a GaAs detector. If the cutoff wavelength is to be decreased to 0.7 μm, how much AlAs must be added to a GaAs? Assume that the bandgap of $Ga_{1-x}Al_x$ is given by

$$E_g(x) = 1.43 + 1.25x \text{ (eV)}$$

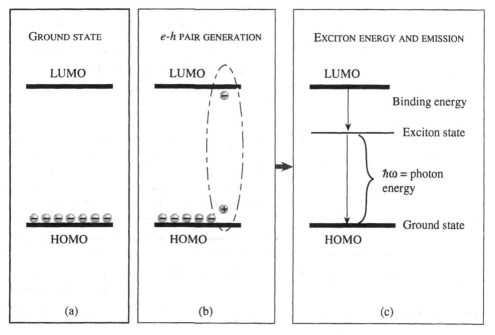

Figure 5.26: A schematic of the electronic states in (a) an unexcited organic semiconductor system where the electrons are in the HOMO state. (b) A system where an e–h pair is created. (c) The e–h pair forms an exciton with energy lower than the separation of the LUMO and HOMO levels. A photon emitted by the exciton, therefore has a different energy.

5.8. Problems

Topics studied	Key findings
Important semiconductor materials	• Choice of semiconductors used for applications depends upon: (i) bandgap, (ii) substrate availability, and (iii) special application needs. • Direct gap materials are needed for light emission devices with high efficiency.
Optical absorption	• A light particle (photon) can cause an electron in the valence band to go into the conduction band. The photon is absorbed and the process is defined by an absorption coefficient α ($\hbar\omega \geq E_g$). • The absorption process is strong in direct gap semiconductors due to momentum conservation. Thus the optical transitions are vertical in k-space.
Optical emission	An electron with a certain momentum can recombine with a hole with the same momentum to emit a photon. The emission rate is the radiative lifetime τ_0 and has a value of ~0.6 ns for GaAs.
Quasi-Fermi levels	For non-equilibrium electron and hole concentrations, the occupations of the carriers are given by electron and hole Fermi functions independent of each other. Note that in equilibrium the same Fermi function describes both electrons and holes.

Table 5.3: Summary table.

Topics studied	Key findings
Electron–hole recombination of excess carriers	• Electrons and holes can recombine by emission of photons. The process depends upon the electron finding a hole and has a strong dependence on charge density. • Carriers can also recombine via traps. Such recombination is non-radiative and depends upon the trap density.
Solar cell	• A p–n diode operated with no external bias. • Presence of optical intensity creates an output voltage and current which can be used to convert photons to electrical power.
p–i–n detector	• A reverse biased p–i–n detector. • The device has high speed, although it has no gain. • Carriers generated in the i-region are collected very efficiently to provide a high-performance detector.
Light-emitting diode, LED	An LED emits light as a forward biased p–n diode. Electrons and holes recombine to emit bandgap photons with an energy spread of $\sim k_B T$. Electron–hole recombination time, determined by spontaneous emission, limits the switching time for LEDs.

Table 5.4: Summary table (cont.)

Topics Studied	Key Findings
The laser diode	A forward biased p–n junction in which photons emitted are confined in an optical cavity. Certain resonant modes of the cavity are allowed to get strong feedback. This allows the preferred mode to have light emission by stimulated emission.
Stimulated versus spontaneous emission	e–h recombination by the presence of photons produces a coherent beam of photons in contrast to spontaneous emission. Stimulated emission is critical for laser performance.
The optical cavity	The role of the optical cavity is to provide resonant states in which photon density can grow preferentially. This provides selectivity in the stimulated emission.
The laser below and above threshold	Below threshold, the emission of photons is by spontaneous emission–incoherent photons are emitted over a broad spectral range. Above threshold, some modes start to dominate the photon output. The spectrum becomes sharp and coherent.
Organic semiconductor based optical processes	Excitonic effects play an important role in organic semiconductors. Exciton formation and recombination creates photons with energies smaller than the electronic bandgap between the LUMO and HOMO states.

Table 5.5: Summary table (cont.)

5.3 Calculate the absorption coefficient for GaAs for photons with energy 1.8 eV. Calculate the fraction of this light absorbed in a GaAs sample of thickness of 0.5 μm.

5.4 An optical power density of 1 W/cm^2 is incident on a GaAs sample. The photon energy is 2.0 eV and there is no reflection from the surface. Calculate the excess electron–hole carrier densities at the surface and 0.5 μm from the surface. The e–h recombination time is 10^{-8} s.

5.5 The performance of a silicon detector is poor for photons of wavelength 1.0 μm, but are quite good for photons of wavelength shorter than 0.4 μm. Explain this observation by examining the bandstructure of silicon.

5.6 In long-distance fiber optics communication, it is important that photons with energy with low absorption in the fiber be used. The lowest absorption for silica fibers occurs at 1.55 μm. Find a semiconductor alloy using InAs, GaAs, and InP that has a bandgap corresponding to this wavelength (semiconductor lasers emit at close to bandgap energy). Also find an alloy with a bandgap corresponding to 1.3 μm, the wavelength at which fiber dispersion is minimum.

5.7 Calculate and plot the absorption coefficients for In$_{0.53}$Ga$_{0.47}$As and InP. Assume that the momentum matrix element values are the same as those for GaAs.

5.8 Calculate the rate (per second) at which photons with energy 1.6 eV are absorbed in GaAs ($E_g = 1.43$ eV).

5.9 Calculate the electron densities at which the quasi-Fermi level for electrons just enters the conduction band in Si and GaAs at 77 K and 300 K.

5.10 Calculate the hole densities at which the hole quasi-Fermi level enters the valence band in Si and GaAs at 77 K and 300 K.

5.11 Assume that equal densities of electrons and holes are injected into GaAs. Calculate the electron or hole density at which $f^e(E^e = E_c) + f^h(E^h = E_v) = 1$. Calculate the densities at 77 K and 300 K.

5.12 In a p-type GaAs doped at $N_a = 10^{18}$ cm^{-3}, electrons are injected to produce a constant electron density of 10^{15} cm^{-3}. Calculate the rate of photon emission, assuming that all electron–hole recombination results in photon emission. What is the optical power output if the device volume is 10^{-7} cm^3?

5.13 Electrons and holes are injected into a GaAs device where they recombine to produce photons. The volume of the active region of the device where recombination occurs is 10^{-6} cm^2 and the temperature is 300 K. Calculate the output power for the cases: i) $n = p = 10^{15}$ cm^{-3}; and ii) $n = p = 10^{18}$ cm^{-3}.

5.14 The photon number in a GaAs laser diode is found to be 100. Calculate the electron–hole recombination time for carriers corresponding to the lasing energy.

5.15 Calculate the gain versus energy curves for GaAs at injection densities of:

(1) $n = p = 0$
(2) $n = p = 5 \times 10^{16}$ cm^{-3}
(3) $n = p = 1.0 \times 10^{17}$ cm^{-3}
(4) $n = p = 1.0 \times 10^{18}$ cm^{-3}
(5) $n = p = 2.0 \times 10^{18}$ cm^{-3}

5.8. Problems

Plot the results for $\hbar\omega$ going from E_g to $E_g + 0.3$ eV. Calculate the results for 77 K and 300 K.

5.16 In wavelength division multiplexing (WDM) schemes for high data transmission on optical networks a number of laser beams emitting at slightly different wavelengths ($\lambda_1, \lambda_2, \cdots$) are used to send data through the same fiber.

Design a series of 6 GaAs/Al$_{0.3}$Ga$_{0.7}$As wells (i.e. find the well sizes) which emit light at a wavelength spacing of 20 Å, with one of them emitting at 8260 Å.

The bandgap of GaAs is 1.43 eV. The bandgap of Al$_x$Al$_{1-x}$ is given by

$$E_g(x) = 1.43 + 1.2x \text{ eV}; \ x < 0.4$$

$\Delta E_c : \Delta E_v = 0.65 : 0.35$ for this system. What can you say about the fabrication problems of such lasers?

5.17 For an application in an optical reader, a laser is needed with a wavelength of $\lambda = 0.82 \ \mu$m at 300 K. Design a GaAs/AlGaAs quantum well in which the effective bandgap corresponds to the desired wavelength. You can adjust the Al composition as long as the composition is below 0.4 and the well size as long as it is between 50 Å and 150 Å.

- What are the parameters (well size, composition of the barrier) for the design?

- Plot the density of states in the quantum well you have designed for the conduction and valence band.

$\Delta E_c : \Delta E_v = 0.65 : 0.35$ for this system. Do not go above $x = 0.4$, since the alloy becomes indirect.

5.18 GaAs is used as a well region to form a quantum well and an electron density of 10^{12} cm^{-2} is placed in the conduction band well. Estimate the maximum widths of the quantum wells that will display two-dimensional effects for electrons at (i) 77 K and at (ii) 300 K. for 2D effects to occur, 90 % of the electrons have to be in the ground state. Use the infinite barrier model for your results. Assume that the mass of electrons in GaAs is 0.067 m_0.

5.19 Consider a GaAs p-n^+ junction LED with the following parameters at 300 K:

Electron diffusion coefficient,	D_n =	25 cm^2/s
Hole diffusion coefficient,	D_p =	12 cm^2/s
n-side doping,	N_d =	5×10^{17} cm^{-3}
p-side doping,	N_a =	10^{16} cm^{-3}
Electron minority carrier lifetime,	τ_n =	10 ns
Hole minority carrier lifetime,	τ_p =	10 ns

Calculate the injection efficiency of the LED assuming no trap-related recombination.

5.20 The diode in Problem 5.19 is to be used to generate an optical power of 1 mW. The diode area is 1 mm^2 and the external radiative efficiency is 20%. Calculate the forward bias voltage required.

5.21 Consider the GaAs LED of Problem 5.19. The LED has to be used in a communication system. The binary data bits 0 and 1 are to be coded so that the optical pulse output is 1 nW and 50 μW. If the external efficiency factor is 10%, calculate the forward bias voltages required to send the 0s and 1s. The LED area is 1 mm^2.

5.22 Consider the semiconductor alloy InGaAsP with a bandgap of 0.8 eV. The electron and hole masses are 0.04 m_0 and 0.35 m_0, respectively. Calculate the injected electron and hole densities needed at 300 K to cause inversion for the electrons and holes at the bandedge energies. How does the injected density change if the temperature is 77 K? Use the Joyce–Dixon approximation.

5.23 Consider a GaAs-based laser at 300 K. Calculate the injection density required at which the inversion condition is satisfied at (i) the bandedges; (ii) at an energy of $\hbar\omega = E_g + k_BT$. Use the Joyce–Dixon approximation.

5.24 Consider a GaAs-based laser at 300 K. A gain of 30 cm^{-1} is needed to overcome cavity losses at an energy of $\hbar\omega = E_g + 0.026$ eV. Calculate the injection density required. Also, calculate the injection density if the laser is to operate at 400 K.

5.25 Consider the laser of the previous problem. If the time for e–h recombination is 2.0 ns at threshold, calculate the threshold current density at 300 K and 400 K. The active layer thickness is 2.0 μm and the optical confinement is unity.

5.26 Two GaAs/AlGaAs double heterostructure lasers are fabricated with active region thicknesses of 2.0 μm and 0.5 μm. The optical confinement factors are 1.0 and 0.8, respectively. The carrier injection density needed to cause lasing is 1.0×10^{18} cm^{-3} in the first laser and 1.1×10^{18} cm^{-3} in the second one. The radiative recombination times are 1.5 ns. Calculate the threshold current densities for the two lasers.

5.27 Consider a laser in which the carrier masses could be tuned. Assume that the hole density of states mass is 0.5 m_0, while the electron density of states mass changes from 0.02 m_0 to 0.2 m_0. Calculate and plot the transparency density needed at an energy of $E_g + k_BT$, where E_g is 1.4 eV. The temperature is 300 K.

5.9 FURTHER READING

- **Optical processes in semiconductors**

    - J. Gowar, *Optical Communication Systems*, Prentice-Hall, Englewood Cliffs, NJ (1989).
    - J.I. Pankove, *Optical Processes in Semiconductors*, Prentice-Hall, Englewood Cliffs, NJ (1971).
    - J.I. Pankove, *Optical Processes in Semiconductors*, Dover Publications, New York (1977).
    - J. Singh, *Electronic and Optoelectronic Properties of Semiconductor Structures*, Cambridge University Press (2003).
    - F. Stern, "Elementary Theory of the Optical Properties of Solids," *Solid State Physics*, Academic Press, New York (1963), vol. 15.
    - V.F. Weisskopf, "How Light Interacts with Matter," *Scientific American*, **219**, 60 (1968).
    - J. Wilson and J.F.B. Hawkes, *Optoelectronics: An Introduction*, Prentice-Hall, Englewood Cliffs, NJ (1983).

5.9. Further reading

- **Devices**
    - Articles in *Semiconductors and Semimetals*, ed. W.T. Tsang, vol. 22, part C, Academic Press, New York (1985).
    - R. Baets, "Heterostructures in III-V Optoelectronic Devices," *Solid State Electronics*, **30**, 1175 (1987).
    - H. Kressel and J.K. Butler, *Semiconductor Lasers and Heterojunction LEDs*, Academic Press, New York (1977).
    - J. Shinar, editor, *Organic Light Emitting Devices*, Springer Verlag (2003).
    - W.T. Tsang, "High Speed Photonic Devices," *High Speed Semiconductor Devices*, ed. S.M. Sze, Wiley-Interscience, New York (1990).

Chapter 6

DIELECTRIC RESPONSE: POLARIZATION EFFECTS

6.1 INTRODUCTION

In the previous chapter we have examined optical processes in semiconductors and discussed optical absorption, gain, recombination, etc. In this chapter we will discuss the origins of dielectric response and how one can modify the response. It is well known that the dielectric response of a solid is different from that of vacuum. This difference arises from the presence of charges in the solid so that the "local field" felt inside a solid is different from the externally applied field. The presence of an external electric field polarizes the solid by creating relative displacement of charges. The atomic electron charge cloud can be disturbed, the ionic charge (for solids with cations and anions) can be shifted, the ions themselves can move physically, the "free" charge in the conduction band can be disturbed, etc.

In this chapter we will develop several models for polarization and dielectric response. Particularly interesting are cases where the response can be controlled by external stimulus and thus be exploited for device applications. The dielectric response can change as a function of applied field, applied strain, temperature, optical intensity, etc. As a result we can use the dielectric response change for strain sensors, tunable capacitors, interference-based optical devices, etc.

6.2 POLARIZATION IN MATERIALS: DIELECTRIC RESPONSE

We have seen that solids can be characterized as insulators, semiconductors, and metals. Insulators, where current flow is negligible in the presence of an electric field (due to the absence of mobile charges) show very interesting dielectric response. When an external electric field is applied the charges in the solid are disturbed, creating a dipole moment and polarization (dipole moment per unit volume). The polarization influences the dielectric response, which, in general, is a function of frequency.

In Fig. 6.1 we show form different origins to polarization in a material. Each source of polarization has its own time constant associated with its response time. There are several sources of polarization as shown.
• The external field can polarize the atoms in the solid. The atomic polarization response times are very short so that this source plays a role at low and high frequencies (e.g., optical frequencies–10^{15} Hz), as shown in Fig. 6.2. The reason atomic polarization can respond at very high frequencies is that the electron cloud (with very low mass) adjusts rapidly to applied fields.
• In ionic polarization the cations and anions forming the crystal are displaced physically as a result of the external field. This polarization has a response time of ~1 ps, due to the large mass of the ions that have to be displaced. As a result, at frequencies above ~ 10^{12} Hz, ionic polarization does not play a role, as shown in Fig. 6.2.
• In dipolar polarization molecules have to rotate to create polarization. This is usually a slow process, and, as a result, is only important at low frequencies.
• Free charge polarization effects are important in metals and doped semiconductors respond in very short times, due to the small electron mass.

The schematic of the frequency dependence of polarizability shown in Fig. 6.2 will be discussed in more detail later in this chapter and the presence of the resonances shown will be explained on the basis of simple models.

6.2.1 Dielectric response: some definitions

In this section we will recall some basic definitions used in understanding the dielectric response of a solid. Consider a solid, subject to an external electric field which has a value $\mathbf{E}_0$, as shown in Fig. 6.3. This field causes charges in the solid to shift, causing a polarization given by

$$\mathbf{P} = \sum_n q_n \mathbf{r}_n \tag{6.1}$$

where the index n represents the various sites on the solid (could be continuous or discrete), and q and r represent the charge and its position. In free space the electric field $\mathbf{E}_0$ would be related to charge density ρ by the Maxwell equation

$$\nabla \cdot \mathbf{E}_0 = \rho/\epsilon_0$$

However, in a medium the presence of the external field $\mathbf{E}_0$ causes a polarization of the internal charge. The polarization causes a field $\mathbf{E}_1$ as shown in Fig. 6.3, which opposes

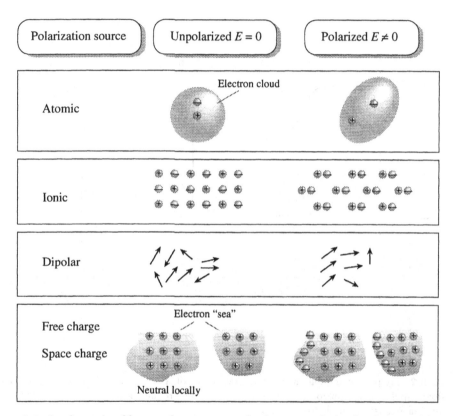

Figure 6.1: A schematic of how various processes lead to net polarization in materials.

6.2. Polarization in materials: dielectric response

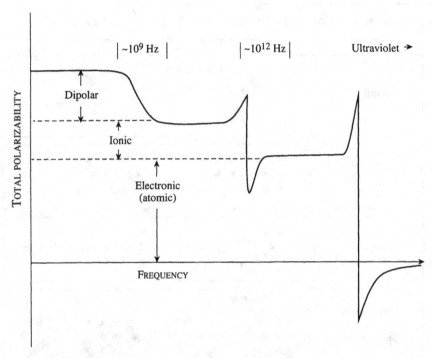

Figure 6.2: Frequency dependence of polarizability of a solid and its various components. The slow dipolar contribution is present only at low frequencies, while the fast electronic contribution persists at high frequencies.

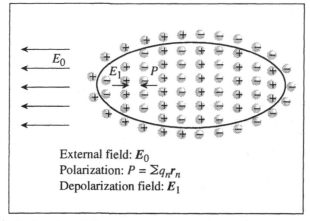

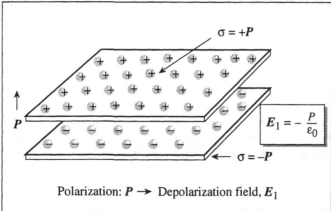

Figure 6.3: (a) A schematic of how an external field polarizes a solid creating a depolarization field, which tends to reduce the net field in the solid. (b) A uniformly polarized dielectric slab creates a depolarizing electric field E_1. The surface charge density is $\sigma = \mathbf{P}$ as shown.

the external field $\mathbf{E}_0$. If we consider a uniformly polarized dielectric slab, as shown in Fig. 6.3b, the depolarization field is given by the field created by sheet charges $\sigma = +P$ and $\sigma = -P$ as shown. The total macroscopic field inside the solid may now be represented by the sum of $\mathbf{E}_0$ and $\mathbf{E}_1$ (assume the field is along the z-direction).

$$\mathbf{E} = \mathbf{E}_0 + \mathbf{E}_1 = \mathbf{E}_0 - \frac{P}{\epsilon_0}\hat{z} \tag{6.2}$$

The dielectric susceptibility χ is defined through the relation between the polarization and the total field

$$\mathbf{P} = \epsilon_0 \chi \mathbf{E} \tag{6.3}$$

In a solid the internal polarization effects are described by the relative dielectric

6.2. Polarization in materials: dielectric response

constant ϵ_r and the equation

$$\nabla \cdot \mathbf{D} = \rho$$
$$\mathbf{D} = \epsilon_r \epsilon_0 \mathbf{E} \tag{6.4}$$

where $\mathbf{D}$ is the displacement vector (which accounts for the internal polarization effects), ρ is the total charge density, and $\mathbf{E}$ is the electric field.

The Maxwell equation in a medium can be described through a relative dielectric constant. The electric field produced by charges are reduced by a factor of the relative dielectric constant, ϵ_r, where

$$\epsilon_r = \frac{\epsilon_0 E + P}{\epsilon_0 E} = 1 + \chi \tag{6.5}$$

In case the material does not have isotropic properties the relevant relation is described by the susceptibility tensor

$$P_\mu = \chi_{\mu\nu} \epsilon_0 E_\nu \tag{6.6}$$

and

$$\epsilon_{\mu\nu} = 1 + \chi_{\mu\nu} \tag{6.7}$$

Dielectric response in alternating electric fields

Insulators are often used in capacitors, which form key components of microwave circuitry. We will examine how dielectric materials respond in the presence of a time varying field. In Fig. 6.4 we show a simple circuit for a capacitor, subject to an ac field given by

$$V = V_0 e^{i\omega t} \tag{6.8}$$

The current through the capacitor is

$$I = C\frac{dV}{dt} = i\omega C V \tag{6.9}$$

To see the effects of the dielectric response explicitly we write the capacitance as

$$C = \frac{\epsilon_r \epsilon_0 A}{h} \tag{6.10}$$

where A and h are the area and width of the capacitor. We also write the dielectric constant as a complex term having a real part ϵ'_r and an imaginary part ϵ''_r

$$\epsilon_r = \epsilon'_r - i\epsilon''_r \tag{6.11}$$

The current in the capacitor is

$$I = i\omega \epsilon'_r C_0 V + \omega \epsilon''_r C_0 V \tag{6.12}$$

where C_0 is the *capacitance if the dielectric was replaced by a vacuum*. The current is made up of two components – one a "lossless" term, which is 90° out of phase with the

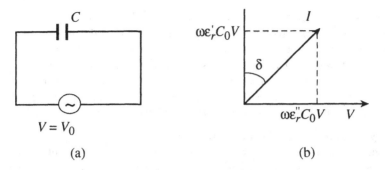

Figure 6.4: (a) A circuit where an alternating field is applied to a capacitor. (b) A complex representation of the current-voltage across the capacitor.

applied field, and one that is a "lossy" term and is in phase with the field. This is shown in Fig. 6.4b.

From Fig. 6.4b we can see that the angle δ is defined by

$$\tan \delta = \frac{\epsilon_r''}{\epsilon_r'} \tag{6.13}$$

The average power dissipated in the circuit is given by

$$\bar{P} = \frac{1}{T} \int_0^T VI dt \tag{6.14}$$

Taking the real part, we get for the average power dissipated

$$\bar{P} = \frac{1}{2}\omega \epsilon_r'' C_0 V_0^2 \tag{6.15}$$

Writing the potential amplitude as $V_0 = E_0 h$ where E_0 is the electric field amplitude we get

$$\begin{aligned}\bar{P} &= \frac{1}{2}\omega \epsilon_r'' \epsilon_0 \frac{A}{h} E_0^2 h^2 \\ &= \frac{1}{2}\omega \epsilon_r'' \epsilon_0 E_0^2 V_{\text{cap}} \end{aligned} \tag{6.16}$$

where V_{cap} is the volume of the capacitor. The power density dissipated is then

$$\begin{aligned}\frac{\bar{P}}{V_{\text{cap}}} &= \frac{1}{2}\omega \epsilon_r'' \epsilon_0 E_0^2 \\ &= \frac{1}{2}\omega E_0^2 \epsilon_0 \epsilon_r' \tan \delta \end{aligned} \tag{6.17}$$

We can see that if $\delta = 0$, i.e. if the imaginary part of the dielectric constant is zero, there is no dissipation. The imaginary part of ϵ comes from scattering processes that occur

6.2. Polarization in materials: dielectric response

as the charges in the system respond to the applied field. For microwave frequencies the scattering processes involve scattering of charges by lattice vibrations or defects. For optical frequencies we can have charges scattering from the valence band into the conduction band, resulting in optical absorption as discussed in Chapter 5.

The dielectric function itself has interesting dependence on frequency. The mobile charge in the solid has several resonant frequencies. When the external frequency of the ac field approaches these frequencies the dielectric constant shows unusual behavior.

Frequency dependence of dielectric response

As discussed earlier atomic and ionic polarization are an important component of the dielectric response of a solid. A simple model can be used to describe these charges as bound to their equilibrium positions by a linear restoring force – in other words, by a simple harmonic oscillator model, as shown in Fig. 6.5. In the absence of any external force, the equation of motion for the ideal charges may be written as

$$m\frac{d^2x}{dt^2} + m\omega_0^2 x = 0 \tag{6.18}$$

where m is the mass of the oscillator and ω_0 is the natural frequency of the oscillation of the oscillator. In a real system we need to introduce a loss term arising from the various scattering processes. The scattering processes can be represented by a damping factor γ, which is proportional to the velocity of the charge. If an external field $E_0 \exp(i\omega t)$ is applied, we have the equation

$$m\frac{d^2x}{dt^2} + m\gamma\frac{dx}{dt} + m\omega_0^2 x = qE_0 e^{i\omega t} \tag{6.19}$$

where q is the charge (see Fig. 6.5).

Assuming a general solution of the form $x = x_0 e^{i\omega t}$, we get

$$x(t) = \frac{qE_0 e^{i\omega t}}{m\{(\omega_0^2 - \omega^2) + i\gamma\omega\}} \tag{6.20}$$

If we assume that there are N charges per volume, the time-dependent polarization is

$$P(t) = Nqx(t) \tag{6.21}$$

The susceptibility is then

$$\chi(\omega) = \frac{Nq^2}{m\epsilon_0}\left\{\frac{1}{(\omega_0^2 - \omega^2) + i\gamma\omega}\right\} \tag{6.22}$$

The dielectric constant is now

$$\epsilon_r(\omega) = 1 + \frac{Nq^2}{m\epsilon_0}\frac{1}{(\omega_0^2 - \omega^2) + i\gamma\omega} \tag{6.23}$$

At optical frequencies the dielectric response is denoted by $\epsilon_{r\infty}$, since the frequencies are very high. The real and imaginary part of the dielectric response is now

$$\epsilon'_{r\infty} - 1 = \frac{Nq^2}{m\epsilon_0}\left\{\frac{\omega_0^2 - \omega^2}{(\omega_0^2 - \omega^2)^2 + \gamma^2\omega^2}\right\}$$

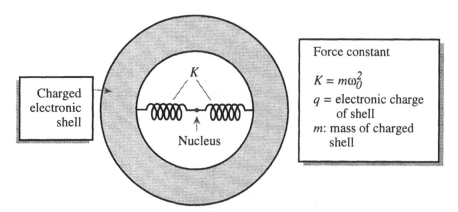

Figure 6.5: A simple classical model to represent the polarization of atoms. A charged electronic shell (total charge q) is bound to a nucleus by a "spring" of force constant K.

$$\epsilon''_{r\infty} = \frac{Nq^2}{m\epsilon_0}\left\{\frac{\gamma\omega}{(\omega_0^2 - \omega^2)^2 + \gamma^2\omega^2}\right\} \qquad (6.24)$$

Typical frequency responses for the dielectric function are shown in Fig. 6.6.

The expression for the dielectric response derived above is modified in solids, since the local field is different from the external field. However the general form of the dielectric response is unchanged.

Relaxation time effects

In the treatment for dielectric response, it has been assumed that the polar charge can respond to the local field instantly. In many materials this is not the case, especially at high frequencies. It is useful to define a time constant τ, called the relaxation time, which represents the time it takes for a polarization disturbance to reach its equilibrium value. The time dependence of the polarization may be represented by

$$\frac{d\mathbf{P}}{dt} = \frac{1}{\tau}\{\mathbf{P}_s - \mathbf{P}(t)\} \qquad (6.25)$$

where $\mathbf{P}_s$ is the steady state value. It is useful to define the static dielectric constant ϵ'_{rs} and the high frequency dielectric constant $\epsilon'_{r\infty}$. The static constant describes the response of the system under dc conditions; i.e., where the system has enough time to respond. It is possible to show that, if the polarization time dependence is described by Eq. 6.25, the dielectric response has the following form (for a field $\mathbf{E} = \mathbf{E}_0 \exp(i\omega t)$)

$$\epsilon'_r = 1 + \frac{\epsilon'_{rs} - \epsilon'_{r\infty}}{1 + \omega^2\tau^2}$$
$$\epsilon''_r = (\epsilon'_{rs} - \epsilon'_{r\infty})\frac{\omega\tau}{1 + \omega^2\tau^2} \qquad (6.26)$$

When the relaxation time effects are important, the dielectric constant has strong temperature dependence. This is because the time constant τ has a strong dependence on temperature. Usually τ has a temperature dependence given by an activation

6.3. Ferroelectric dielectric response

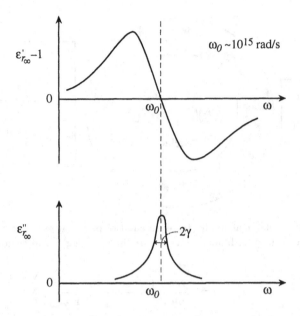

Figure 6.6: Variation in $\epsilon'_{r\infty}$ and $\epsilon''_{r\infty}$ with frequency close to a resonance frequency ω_0.

energy E_A

$$\tau = \tau_B \exp\left(\frac{E_A}{k_B T}\right) \tag{6.27}$$

6.3 FERROELECTRIC DIELECTRIC RESPONSE

In Chapter 1 we have discussed the structures of several classes of materials. In materials with cations and anions it is possible to have crystal structures, where the polarization in the material is non-zero. Ferroelectric materials exhibit polarization even in the absence of an applied electric field. In Fig. 6.7 we show a typical hysterisis loop for ferroelectric materials. The ferroelectric effect disappears above a temperature called the Curie temperature. Ferroelectric materials may be classified into two categories: order–disorder, or displacive. In "order–disorder" case the polarization arises from the ordering of the ions to create net polarization. In a displacive type, the cations are physically displaced with respect to the anions to create net polarization. Crystals containing H bonds are usually order–disorder type and include materials like KH_2PO_4, RbH_2PO_4, etc. Ionic crystals, like $BaTiO_3$, $LiNbO_3$, $GeTe$, etc., are displacive type ferroelectrics.

As can be seen in the hysterisis curve of a typical ferroelectric, the polarization properties of the material are described by the spontaneous polarization P_s and the critical field E_c. In Table 6.1 we show the values of the spontaneous polarization and Curie temperature for several ferroelectrics. If we convert the spontaneous polarization value to ionic displacement, it is seen that displacements as large as 0.1 Å are obtained. Thus in the ferroelectric effect the ions move by a significant fraction of the lattice

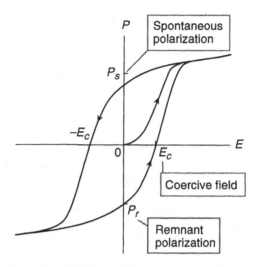

Figure 6.7: A typical schematic of the hysterisis curve seen in ferroelectric materials.

MATERIAL		T_c (K)	P_s (μCcm^{-2})	POLAR CHARGE (cm^{-2})
KDP type	KH$_2$PO$_4$	123	5.33	3.3 x 10^{13}
	KH$_2$AsO$_4$	96	5.0	3.1 x 10^{13}
Perovskites	BaTiO$_3$	393	26.0	1.62 x 10^{14}
	SrTiO$_3$	32(?)	3.0	1.87 x 10^{13}
	PbTiO$_3$	763	>50.0	3.1 x 10^{14}
	KNbO$_3$	712	30.0	1.87 x 10^{14}
	LiNbiO$_3$	1470	300.0	1.87 x 10^{15}
	LiTaO$_3$		23.3	1.45 x 10^{14}

Table 6.1: A list of several ferroelectric materials and their properties (from *Introduction to Solid State Physics*, C. Kittel, John Wiley and Sons, New York, 1971). Spontaneous polarization, Curie temperature, and fields for some ferroelectrics.

constant.

An analytical expression that describes the hysterisis loop can be written. If the field is increasing from a large negative value to a large positive value (the forward cycle) the polarization can be written as

$$P^+(E) = P_s \tanh\left(\frac{E - E_c}{2\delta}\right)$$

$$\delta = E_c \left[\ell_n \left(\frac{1 + P_r/P_s}{1 - P_r/P_s}\right)\right]^{-1} \quad (6.28)$$

where E_c is called the coercive field and is the field at which polarization switches sign. The polarization parameters P_r and P_s are called the remnant and spontaneous polarization, respectively. In the negative cycle we have

$$P^-(E) = -P^+(E) \quad (6.29)$$

Detailed theories behind the ferroelectric effect are quite complex, but a simple model can be developed in terms of the free energy configuration of atoms in the crystal. In Fig. 6.8 we show a schematic of the free energy versus separation between anions and cations at different temperatures. The origin represents the case where the anions and cations are arranged so that the net polarization is zero. We can see that below Curie temperature there are two equilibrium positions corresponding to opposite polarization values. The application of an electric field can pull the material from one state to the other. The spontaneous polarization value drops to zero at $T \geq T_c$ in a manner that depends upon the nature of the free energy diagram. The dielectric response is approximately given by

$$\epsilon \sim \frac{\chi}{T - T_c} \quad (6.30)$$

and extremely large values of ϵ can be observed near T_c. A typical result for BaTiO$_3$ is shown in Fig. 6.9.

In ferroelectric crystalline solids domains of varying polarization orientation are formed, as shown schematically in Fig. 6.10. Such domains are to minimize the electrostatic energy in the crystal. The multidomain state can be transformed into a single domain by applications of a strong electric field along a polar direction. In addition to 180° domains shown in Fig. 6.10, 90°, domains are also formed as shown in Fig. 6.10 to minimize the strain energy of the system.

In polycrystalline ferroelectrics, grown materials are usually non-polar due to the random orientation of domains. However, the material can be rendered polar by application of a field (the poling field).

6.4 TAILORING POLARIZATION: PIEZOELECTRIC EFFECTS

Polarization in a material can be exploited in a number of ways in device design. Polarization influences the dielectric constant and, therefore, the electromagnetic properties

276 Dielectric response: polarization effects

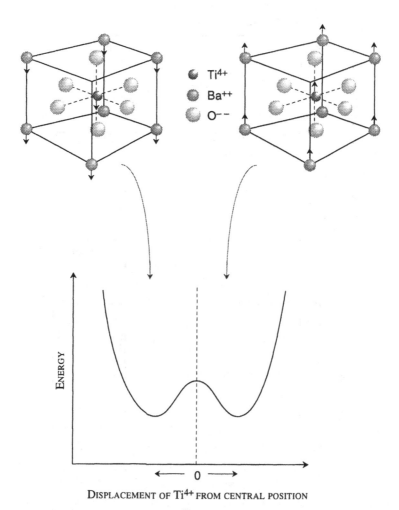

Figure 6.8: A schematic of how the energy of a ferroelectric crystal (such as $BaTiO_3$) changes as a function of the relative displacement of the positive charges.

6.4. Tailoring polarization: piezoelectric effects

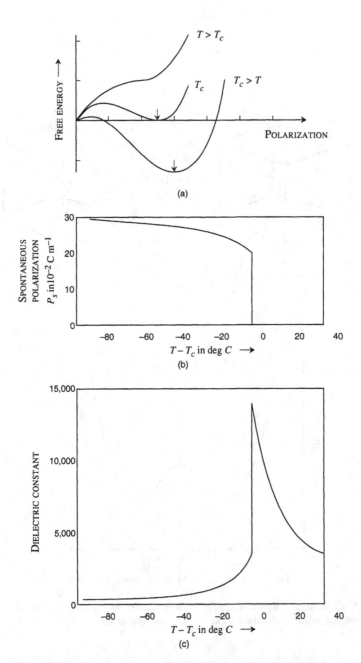

Figure 6.9: (a) A schematic of free energy as a function of polarization for various temperatures. (b) Spontaneous polarization in BaTiO$_3$ as a function of temperature. (c) Dielectric constant as a function of temperature in BaTiO$_3$.

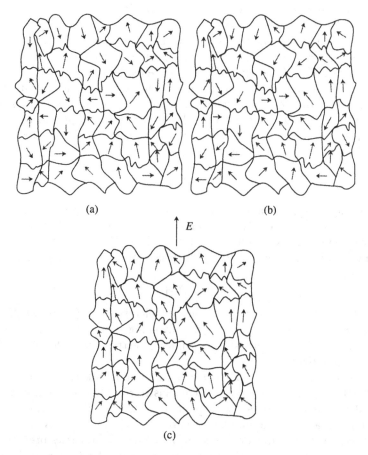

Figure 6.10: A schematic of how poling of a ferroelectric material occurs. (a) Unoriented material. (b) Oriented by 180° domain changes. (c) Oriented by 180° and 90° domain changes.

6.4. Tailoring polarization: piezoelectric effects

can be influenced. It can be used to create voltage changes in devices to be used as sensors and detectors. A number of stimuli can alter polarization, including pressure, temperature, and optical inputs. The piezoelectric effect, exploited for many device technologies, is one way the polarization in a material can be tailored.

Piezoelectric effect describes the polarization change that occurs when a strain is present in a material. It also describes the reverse effect where an electric field applied to the material creates a strain. In Fig. 6.11 we show schematically how stress causes atomic sublattices in a crystal to shift to create net polarization. Strain-related polarization can arise in ferroelectric materials (where spontaneous polarization is present without a field) or in non-ferroelectric materials, as shown in Fig. 6.11a and b. Out of the total of 32 crystal groups, 11 have center of symmetry and do not display the piezoelectric effect. Of the remaining 21, 20 do display piezoelectric effect.

Device applications exploiting the piezoelectric effect depend upon the generation of current or voltage due to strain, as shown in Fig. 6.12, or the reverse process of using electric field to generate strain. The piezoelectric properties of a material are described through several response parameters. The general relations between displacement field D, stress X, and electric field E, is

$$D = dX + \epsilon^x E \tag{6.31}$$

where d is the piezoelectric strain constant (units: pC/N) and ϵ^x is the dielectric constant under constant strain. Conversely, the relation between strain ϵ, electric field, and stress is (note that the symbol ϵ is used here for strain. The reader should be able to distinguish between strain and dielectric response, which uses the same symbol, from the context of the equation)

$$\epsilon = s^E X + dE \tag{6.32}$$

where s^E is the elastic compliance at constant electric field (units: m^2/N). As noted earlier a material may be piezoelectric without being ferroelectric. However, the piezoelectric coefficient is considerably stronger in ferroelectric materials, due to the presence of spontaneous polarization.

The strain in a solid is described by a second rank tensor. The stress tensor is also, similarly, a second rank tensor. The elastic compliance s is thus a fourth rank tensor. The dielectric response is also a second rank tensor. We can see that overall there are several tensors and a large number of parameters in the full description of s, d, ϵ in the equations given above. However, due to the various symmetries present in a general crystal and in a specific crystal, not all the components of a tensor are independent or non-zero. In general we expect to have 21 elastic compliance components, six dielectric response components, and 18 piezoelectric coefficients.

We use the contracted notation, where an index i goes from 1 to 6, the variables 1,2, and 3 represent xx, yy and zz, respectively, and 4,5, and 6 represent yz, zx, and xy, respectively. The 18 piezoelectric coefficients are given by

$$d_{ik} = \left.\frac{\partial \epsilon_k}{\partial E_i}\right|_x \tag{6.33}$$

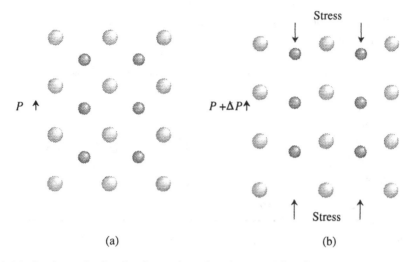

Figure 6.11: A schematic showing how a ferroelectric material with a net polarization P alters its polarization due to stress. Non-ferroelectric materials can also display net polarization in the presence of stress.

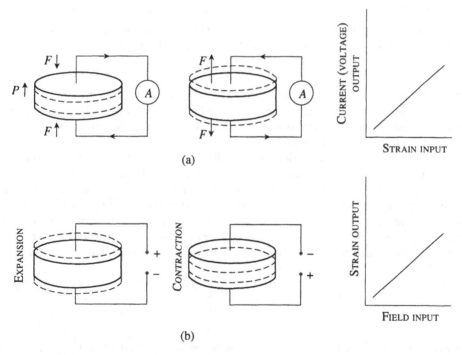

Figure 6.12: (a) Application of stress alters the current or voltage across a piezoelectric material. (b) An electric field applied across a piezoelectric material can cause contraction or expansion.

6.4. Tailoring polarization: piezoelectric effects

PIEZOELECTRIC CONSTANT PCN-1	α-QUARTZ	BaTiO$_3$	PZT	LiNbO$_3$
d_{31}		−79	−119	−0.85
d_{33}		190	268	6
d_{15}		270	335	69
d_{11}	2.3			
d_{14}	0.67			

Table 6.2: Piezoelectric constants for some materials.

In addition to the equation relating E to the strain, the piezoelectric effect is also described by the relation between polarization and stress or between polarization and strain. We can write

$$P = d'X + \epsilon_0 \chi E \qquad (6.34)$$

where d' is the piezoelectric constant (units: m/V), X is the stress and χ is the susceptibility. The polarization can also be written in terms of strain as

$$P = e\epsilon + \epsilon_0 \chi E \qquad (6.35)$$

where e is the piezoelectric coefficient (units: C/m^2). We see that an applied electric field can produce strain and strain can produce an electric field. These effects are used, for example, in ultrasonic generations and strain sensors. In Table 6.2 we show piezoelectric constants for some important materials.

In most applications of piezoelectric effects the strain is applied from an external source (strain sensor applications) or an electric field is applied to distort a crystal. However, it is possible to have built-in strain in thin epilayers grown on a substrate. For layer-by-layer growth, the epitaxial semiconductor layer is biaxially strained in the plane of the substrate, by an amount $\epsilon_\parallel$, and uniaxially strained in the perpendicular direction, by an amount $\epsilon_\perp$. For a thick substrate, the in-plane strain of the layer is determined from the bulk lattice constants of the substrate material, a_S, and the layer material, a_L:

$$\epsilon_\parallel = \frac{a_S}{a_L} - 1$$
$$= \epsilon \qquad (6.36)$$

Since the layer is subjected to no stress in the perpendicular direction, the perpendicular strain, $\epsilon_\perp$, is simply proportional to $\epsilon_\parallel$:

$$\epsilon_\perp = \frac{-\epsilon_\parallel}{\sigma} \qquad (6.37)$$

where the constant σ is known as Poisson's ratio.

Noting that there is *no stress* in the direction of growth it can be simply shown that for the strained layer grown on a (001) substrate (for an *fcc* lattice)

$$\sigma = \frac{c_{11}}{2c_{12}} \tag{6.38}$$

$$\epsilon_{xx} = \epsilon_\parallel$$

$$\epsilon_{yy} = \epsilon_{xx}$$

$$\epsilon_{zz} = \frac{-2c_{12}}{c_{11}}\epsilon_\parallel$$

$$\epsilon_{xy} = 0$$

$$\epsilon_{yz} = 0$$

$$\epsilon_{zx} = 0$$

while in the case of strained layer grown on a (111) substrate

$$\sigma = \frac{c_{11} + 2c_{12} + 4c_{44}}{2c_{11} + 4c_{12} - 4c_{44}}$$

$$\epsilon_{xx} = \left[\frac{2}{3} - \frac{1}{3}\left(\frac{2c_{11} + 4c_{12} - 4c_{44}}{c_{11} + 2c_{12} + 4c_{44}}\right)\right]\epsilon_\parallel$$

$$\epsilon_{yy} = \epsilon_{xx}$$

$$\epsilon_{zz} = \epsilon_{xx}$$

$$\epsilon_{xy} = \left[\frac{-1}{3} - \frac{1}{3}\left(\frac{2c_{11} + 4c_{12} - 4c_{44}}{c_{11} + 2c_{12} + 4c_{44}}\right)\right]\epsilon_\parallel$$

$$\epsilon_{yz} = \epsilon_{xy}$$

$$\epsilon_{zx} = \epsilon_{yz} \tag{6.39}$$

In general, the strained epitaxy causes a distortion of the cubic lattice and, depending upon the growth orientation, the distortions produce a new reduced crystal symmetry. It is important to note that for (001) growth, the strain tensor is diagonal while for (111), and several other directions, the strain tensor has nondiagonal terms. The nondiagonal terms can be exploited to produce built-in electric fields in certain heterostructures as will be discussed in the next section.

An important heterostructure system involves growth of *hcp* lattice-based Al-GaN or InGaN on a GaN substrate along the c-axis. In this case the strain tensor is given by

$$\epsilon_{xx} = \epsilon_{yy} = \frac{a_S - a_L}{a_S}$$

$$\epsilon_{zz} = -2\frac{c_{13}}{c_{33}}\epsilon_{xx} \tag{6.40}$$

This strain is exploited to generate piezoelectric effect based interface charge as discussed in the next section.

As a result of the piezoelectric effect, the built-in strain can produce polarization and interfacial charge between regions with different strain values and polarization. The

6.4. Tailoring polarization: piezoelectric effects

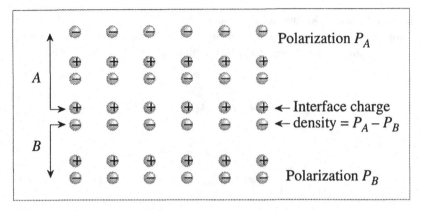

Figure 6.13: A schematic showing how interface charge density can be produced at heterointerfaces of two polar materials.

interfacial charge simply results from the difference of the polarization of the two layers, as shown in Fig. 6.13.

Nitride heterostructures have polarization charges at interfaces because of the strain-related piezoelectric effect as well as from spontaneous polarization. For growth along (0001) orientation, the following relation relates piezoelectric polarization to the strain tensor

$$P_{pz} = e_{33}\epsilon_{zz} + e_{31}(\epsilon_{xx} + \epsilon_{yy}) \tag{6.41}$$

The piezoelectric effect is also present in zinc blende structures. However, the piezoelectric effect only occurs when the strain tensor has off-diagonal components. The polarization values for zinc blende structures (such as GaAs, InAs, etc.) are given by

$$\begin{aligned} P_x &= e_{14}\epsilon_{yz} \\ P_y &= e_{14}\epsilon_{xz} \\ P_z &= e_{14}\epsilon_{xy} \end{aligned} \tag{6.42}$$

The strain tensor is diagonal for growth along (001) direction. As a result there is no piezoelectric effect. However, for other orientations, notably for (111) growth, there is a strong piezoelectric effect.

Piezoelectric effect can be exploited to create interface charge densities as high as 10^{13} cm^{-2} in materials. In Table 6.3 we provide the values of piezoelectric constants for some semiconductors. Also shown in Table 6.4 are elastic constants for some materials.

EXAMPLE 6.1 Consider a piezoelectric ceramic of length 1 cm. A uniform stress of 10 Mpa is applied along the axis. Calculate the potential developed between the two faces. Use these parameters: $d_{33} = 350$ pCN^{-1}; $\epsilon_{33}^x = 700\ \epsilon_0$.

The field developed is

$$E = \frac{d_{33}X}{\epsilon_{33}^x} = \frac{(350 \times 10^{-12}\ \text{CN}^{-1})(10^7\ \text{N})}{700 \times 8.85 \times 10^{-12}\ \text{F/m}}$$

ZINC BLENDE	
MATERIAL	e_{14}(C/m^2)
AlAs	−0.23
GaAs	−0.16
GaSb	−0.13
GaP	−0.10
InAs	−0.05
InP	−0.04

WURTZITE (c-axis growth)			
MATERIAL	e_{31}(C/m^2)	e_{33}(C/m^2)	P_{sp}(C/m^2)
AlN	−0.6	1.46	−0.081
GaN	−0.49	0.73	−0.029
InN	−0.57	0.97	−0.032

Table 6.3: Piezoelectric constants in some important semiconductors. For the nitrides the spontaneous polarization values are also given. (Data for zinc blende materials are from S. Adachi, *J. Appl. Phys.*, vol. 58, **R1** (1985). For nitrides see E. Bernardini, V. Fiorentini, and D. Vanderbilt, *Phys. Rev. B*, vol. 56, **R10024** (1997).)

MATERIAL	C_{11}(N/m^2)	C_{12}(N/m^2)	C_{44}(N/m^2)
Si	1.66 × 10^{11}	0.64 × 10^{11}	0.8 × 10^{11}
Ge	1.29 × 10^{11}	0.48 × 10^{11}	0.67 × 10^{11}
GaAs	1.2 × 10^{11}	0.54 × 10^{11}	0.59 × 10^{11}
C	10.76 × 10^{11}	1.25 × 10^{11}	5.76 × 10^{11}

MATERIAL	C_{13}(N/m^2)	C_{33}(N/m^2)
GaN	10.9 × 10^{11}	35.5 × 10^{11}
AlN	12.0 × 10^{11}	39.5 × 10^{11}

Table 6.4: Elastic constant for some fcc- and hcp-based semiconductors. (For Si, Ge, and GaAs see H. J. McSkimin and P. Andreatch, *J. Appl. Phys.*, **35**, 2161 (1964) and D. I. Bolef and M. Meres, *J. Appl. Phys.*, **31**, 1010 (1960). For nitrides see J. H. Edgar, *Properties of III–V Nitrides*, INSPEC, London (1994) and R. B. Schwarz, K. Khachaturyan, and E. R. Weber, *Appl. Phys. Lett.*, **74**, 1122 (1997).)

6.5. Tailoring polarization: pyroelectric effect

$$= 5.65 \times 10^5 \text{ V/m}$$

The voltage developed is then

$$V = E.L = 5.65 \times 10^3 \text{ V}$$

EXAMPLE 6.2 A thin film of $Al_{0.3}Ga_{0.7}N$ is grown coherently on a GaN substrate. Calculate the polar charge density and electric field at the interface.

The lattice constant of $Al_{0.3}Ga_{0.7}N$ is given by Vegard's law

$$a_{all} = 0.3 a_{AlN} + 0.7 a_{GaN} = 3.111 \text{ Å}$$

The strain tensor components are

$$\epsilon_{xx} = 0.006 = \epsilon_{yy}$$

Using the elastic constant values from Table 6.4

$$\epsilon_{zz} = -0.6 \times 0.006 = 0.0036$$

The piezoelectric effect induced polar charge then becomes

$$P_{pz} = 0.0097 \text{ C/m}^2$$

This corresponds to a density of $6.06 \times 10^{12} \text{ cm}^{-2}$ electronic charges.

In addition to the piezoelectric charge the spontaneous polarization charge is

$$P_{sp} = 0.3(0.089) + 0.7(0.029) - 0.029 = 0.018 \text{ C/m}^2$$

which corresponds to a density of $1.125 \times 10^{13} \text{ cm}^{-2}$ charges. The total charge (fixed) arising at the interface is the sum of the two charges.

6.5 TAILORING POLARIZATION: PYROELECTRIC EFFECT

An important physical effect exploited for thermal imaging applications is the pyroelectric effect. This effect refers to the change in the spontaneous polarization of a material as a function of temperature. Ferroelectric materials (which have a large spontaneous polarization) exhibit it as well as materials, such as AlN, GaN, and InN, which have a fixed spontaneous polarization. As shown in Fig. 6.14 a change in temperature alters the spontaneous polarization (or the surface charge). A current flows over a time interval Δt to neutralize the surface charge and can be used to detect the pyroelectric effect.

To develop an expression for the pyroelectric effect let us examine the relation between displacement, applied electric field, and polarization

$$\begin{aligned} D &= \epsilon_0 E + P_{\text{total}} \\ &= \epsilon_0 E + (P_s + P_{\text{induced}}) = \epsilon E + P_s \end{aligned} \quad (6.43)$$

where ϵ is the dielectric constant. Here we have used Eqs. 6.4 and 6.5 for the derivation.

Assuming a constant electric field, we can write for $\mathbf{P}_g$ the generalized pyroelectric coefficient

$$\mathbf{P}_g = \frac{\partial \mathbf{D}}{\partial T} = \frac{\partial \mathbf{P}_S}{\partial T} + \mathbf{E}\frac{\partial \epsilon}{\partial T} \quad (6.44)$$

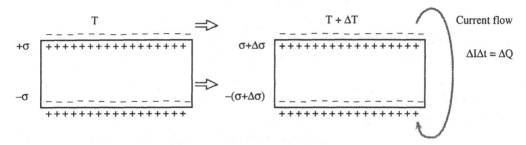

Figure 6.14: A schematic of how a temperature change induced change in spontaneous polarization can be detected by a current flow over a time interval.

This gives

$$\mathbf{p}_g = \mathbf{p} + \mathbf{E}\frac{\partial \epsilon}{\partial T} \quad (6.45)$$

where **p** is the true pyroelectric coefficient ($= \partial \mathbf{P}_s/\partial T$) and, in general, is a vector, but for practical reasons is treated as a scalar. As can be seen from the above equation, the temperature dependence of measured polarization arises from p and $E\partial \epsilon/\partial T$. For non-ferroelectric materials ϵ does not vary much with temperature and the overall pyroelectric effect is very small. In Table 6.5 we provide values for the pyroelectric coefficient for several materials. The pyroelectric effect is quite dependent on temperatures, especially if measured near the Curie temperature.

One of the most sensitive materials used for applications is LiTaO$_3$. Single crystal LiTaO$_3$ can be grown and the material is able to be processed into high sensitivity devices. Lead zirconate (PZ) is another widely used material for devices.

MATERIAL	p (μCm^{-2}K^{-1})	T_c (K)
Triglycine sulphate (TGS, 308 K)	280	322
Deuterated TGS	550	334
LiTaO$_3$ single crystal	230	938
PZT powder	380	500

Table 6.5: Values for the pyroelectric coefficient and Curie temperature of some materials.

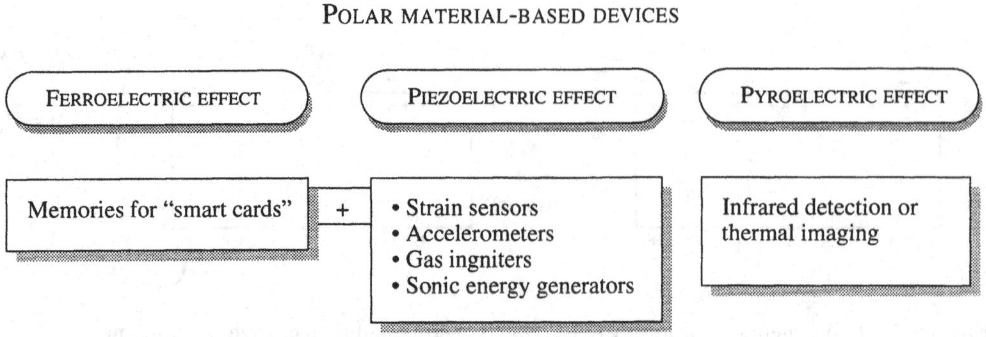

6.6. Device applications of polar materials

POLAR MATERIAL-BASED DEVICES

```
┌─────────────────────┐  ┌─────────────────────┐  ┌─────────────────────┐
│ FERROELECTRIC EFFECT│  │  PIEZOELECTRIC EFFECT│  │ PYROELECTRIC EFFECT │
└─────────────────────┘  └─────────────────────┘  └─────────────────────┘
```

| Memories for "smart cards" | + | • Strain sensors
• Accelerometers
• Gas ingniters
• Sonic energy generators | | Infrared detection or thermal imaging |

Figure 6.15: Polar material based devices.

6.6 DEVICE APPLICATIONS OF POLAR MATERIALS

Traditional semiconductors. e.g. Si, GaAs, etc., are ideally suited for devices such as transistors (switches, amplifiers), light emitters, and detectors. The central reason for the widespread use of semiconductors in such applications is the ease with which free carrier concentration (and, hence, conductivity and optical properties) can be altered by an electrical or electromagnetic perturbation. Traditional semiconductors, however, are not suitable for many applications where the perturbation to be "sensed" is not electrical. This is because usually properties such as piezoelectric constants, pyroelectric constants, etc., are very weak in traditional semiconductors. Clearly there are perturbations, such as stress, temperature change, acceleration, etc., which need to be studied. For such devices the polar materials discussed in this chapter have become materials of choice.

Polarization change based effects, such as the ferroelectric effect, piezoelectric effect, and pyroelectric effects, are exploited in a number of important devices. In Fig. 6.15 we show some of the devices based on polar materials. We will briefly discuss the operation of these devices.

6.6.1 Ferroelectric memory

We have seen that the polarization in a ferroelectric material can be switched in direction by an applied field. As a result there are two stable electrical states of a thin ferroelectric film. This feature has been exploited for memory devices.

A key application of ferroelectric material based memories is the smart card. These cards are traditionally used for banking and retail. Both contact (where the card is inserted into a slot) and contactless smart cards are needed for applications, such as public transportation, entertainment, access control, etc. Ferroelectric devices have proven to be an excellent solution for such applications. The devices consume little power (can be non-volatile, since they depend on atomic motion from one stable state to another) and can store information for up to ten years without power. The content of a card can easily be read by a rf input–output system without any need for physical contact.

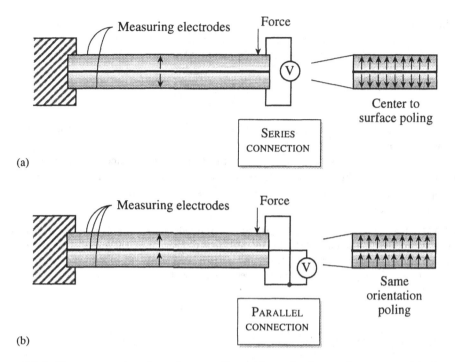

Figure 6.16: Two approaches for using cantilever bimorphs for strain sensors. (a) A serial connection with poling done as shown. (b) A parallel connection.

6.6.2 Strain sensor and accelerometer

An important application of piezoelectric materials is a strain sensor, which allows one to measure the strain or displacement in terms of a voltage signal. For such applications long thin strips or plates are used, since the forces required to create the strain are then smaller.

In Fig. 6.16 we show a typical piezoelectric strain sensor geometry. When a thin plate is bent, half of the plate is compressed, while the other half stretches. As a result, no net voltage is created across the plate. However, if a bimorph, as shown in Fig. 6.16 is used, net voltage can be detected by using an electrode in the middle of the plate. Depending upon the initial poling on the bimorph we can use a serial or parallel connection to obtain the voltage signal across the beam.

It can be shown using simple geometric arguments and the mechanical properties of a cantilever that the voltage developed across a beam of total thickness H and length L is (δz is the displacement of this edge)

$$V = \frac{3}{8}\left(\frac{H}{L}\right)^2 h_{31}\delta z \qquad (6.46)$$

where h_{31} is the voltage per strain for the material. The potential developed for displacements as small as 0.1 μm can be several hundred mV in materials with high piezoelectric coefficients.

It is possible to apply a bias voltage to the cantilever to create a displacement. In this case it can be shown that the displacement is

$$\delta z = \frac{3}{2}\left(\frac{L}{H}\right)^2 d_{31} V \tag{6.47}$$

where V is the applied voltage. For a value of $d_{31} = -79\ pCN^{-1}$ (for BaTiO$_3$) the displacement is 1.2 μm if $L/H = 10$ and applied voltage of 100 V.

The piezoelectric effect can be exploited to design accelerometers as well. A mass is attached to the free end and an acceleration is detected in terms of the voltage developed as a function of time. Series based as shear strain have also been developed for accelerometers.

EXAMPLE 6.3 A cantilever of length 10 mm and thickness 1 mm is used for a strain sensor. A force is applied at the end of the beam causing a deflection of 1.0 μm. Calculate the voltage produced if $h_{31} = -6.2 \times 10^8$ Vm^{-1}.

Using the equation given in this section the voltage is found to be

$$\begin{aligned} V &= \frac{3}{8}\left(\frac{1}{10}\right)^2 (-6.2 \times 10^8\ \text{V/m})\left(10^{-6}\ \text{m}\right) \\ &= 2.3\ \text{V} \end{aligned} \tag{6.48}$$

6.6.3 Ultrasound generation

Important applications of piezoelectric materials are a generation of ultrasonic energy and use as resonators. Under static electric field conditions the piezoelectric strain is small. However, under ac conditions much larger strains can be created if the ac frequency is equal to the mechanical resonant frequency of the bar. In Fig. 6.17a we show a simple bar of length ℓ_1 along with its fundamental resonant frequency given by

$$f_R = \frac{1}{2\ell_1 \sqrt{\rho s_{11}^E}} \tag{6.49}$$

where ρ is the material density and s_{11}^E is the elastic compliance for the applied field. The frequency can be tuned by choice of the dimensions. The piezoelectric bar can be used in conjunction with a microwave cavity to create ultrasound energy, as shown in Fig. 6.17b. The microwave field energy is transferred to ultrasound energy via the piezoelectric field.

6.6.4 Infrared detection using pyroelectric devices

It is well known that objects radiate electromagnetic radiation whose intensity and spectral distribution is controlled by the object's temperature. An important need in technology is the ability to detect temperatures of distant objects. Thermal imaging finds important uses in environmental areas as well as in night vision applications.

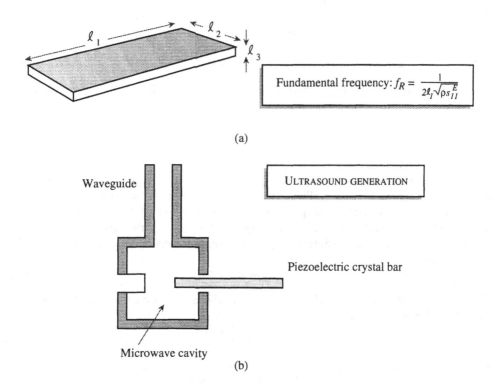

Figure 6.17: A piezoelectric bar, which has a resonant frequency determined by the length, density ρ, and elastic compliance s_{11}^E at constant electric field. (b) A schematic of a system to generate ultrasounds at microwave frequencies.

There are two approaches to detecting long wavelength photons (photons coming from room temperature objects have a peak wavelength at $\sim$ 5–10 μm. By using a narrow bandgap material ($E_g \sim \hbar\omega$) we can use the photons to create e–h pairs that can be used to create a photo signal. This approach was discussed in Chapter 5, Section 5.5.1. In the second approach the temperature change can be used with an effect such as pyroelectricity to create a detectable signal.

Pyroelectric materials offer low-cost alternatives to semiconductor devices for infrared detection. Even though they are not as sensitive as semiconductor devices, they can operate at high temperatures. Thin slices of the material are used as detector elements as shown in Fig. 6.18. If a power density W_i/A is incident on a pixel of area A the energy absorbed in time Δt is

$$\Delta E = \eta W_i \Delta t \qquad (6.50)$$

where η is the emissivity of the material and represents the fraction of energy absorbed. If H is the heat capacity of the element, the change in temperature is given by

$$H \Delta T = \eta W_i \Delta t \qquad (6.51)$$

where $H = \rho c A h$, and ρ is the density, c the specific heat, and h the film thickness. The temperature change results in a voltage signal that can be amplified by an amplifying circuit, as shown in Fig. 6.18. The pyroelectric sensor works when the temperature is changing, since at constant temperature the free internal charge distribution is neutralized by free electrons and surface charges. The pyroelectric capacitor has metallic electrodes and when the radiation impinges a temperature change ΔT develops. The charge associated is

$$\Delta Q = pA\Delta T \qquad (6.52)$$

where p is the pyroelectric coefficient and A is the device area receiving the radiation. The photocurrent is then

$$I_s = pA \frac{d(\Delta T)}{dt} \qquad (6.53)$$

The rms signal voltage is

$$V_s = \frac{I_s(rms) \cdot R}{(1 + \omega^2 R^2 C^2)^{1/2}} \qquad (6.54)$$

where R is the parallel equivalent resistance and C the capacitance.

6.7 SUMMARY

We summarize the topics discussed in this chapter in Tables 6.6 and 6.7.

6.8 PROBLEMS

6.1 The spontaneous polarization in the ferroelectric LiNbO$_3$ is found to be 3 Cm^{-2}. Calculate the dipole moment per unit cell and the relative displacement (from ideal

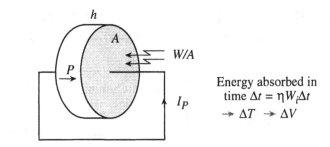
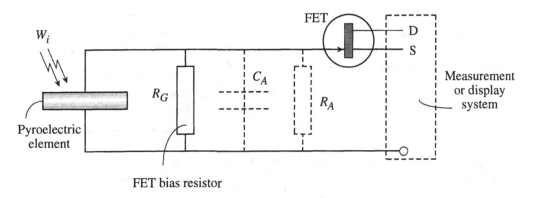

Figure 6.18: A schematic of a pyroelectric detector element and a circuit used to amplify the voltage signal.

6.8. Problems

Topics Studied	Key Findings
Polarization in solids	There are a number of sources for polarization in materials. Atomic, ionic, dipolar, and free charge all contribute. Each source has its own frequency dependence. At high frequency, only electronic contributions are important.
Dielectric response in solids	• Dielectric response describes how an external field's effect is modified inside a solid due to polarization effects. • Both real and imaginary parts of dielectric response are important in describing ac power propagation and dissipation in solids.
Frequency dependence of dielectric response	Due to internal resonances in a solid, the dielectric response shows resonances at certain frequencies.

Table 6.6: Summary table.

Topics studied	Key findings
Ferroelectric materials	Some materials have ionic structure that allows them to have stable structures with net polarization at zero applied field. The polarization versus field relation shows hysterisis in these ferroelectric materials. The ferroelectric effect disappears above a temperature (the Curie temperature) determined by the material properties.
Piezoelectric effect	The polarization in a material can be altered by a strain. Conversely, an electric field can induce strain in a solid. The piezoelectric effect can be exploited for strain sensors and sound generation.
Pyroelectric effect	Temperature changes can alter the polarization in materials. This effect can be exploited for thermal imaging.

Table 6.7: Summary table.

position) of the cations and anions.

6.2 A device is made from a piezoelectric material of area 1 mm × 0.3 mm. The thickness of the device is 0.1 mm. A potential of 10 V is applied by placing electrodes on the two forces of the device. Calculate the dimension of the device after the voltage is applied. Assume that the poling direction is the same as the field direction. The piezoelectric parameters are $d_{33} = 520 \ pCN^{-1}; d_{31} = -200 \ pCN^{-1}$.

6.3 Consider a c-axis grown GaN/Al$_x$Ga$_{1-x}$N interface where the AlGaN layer is under strain to fit the GaN substrate. Calculate the fixed charge density arising from polarization differences at the interface. Consider both piezoelectric and spontaneous polarization effects. Use weighted mean values for alloy parameters.

6.4 Consider an In$_{0.2}$Ga$_{0.8}$N/GaN heterostructure. The InGaN film has biaxial strain and growth is along the c-axis. Calculate the fixed charge density at the interface. Use weighted mean for InGaN piezoelectric coefficients.

6.5 Consider a uniform stress of 5 MPa along the axis of a cylinder of length 1 mm and diameter of 0.5 mm. Calculate the potential across the faces of the cylinder. The parameters for the material are $\epsilon_{33}^x = 700 \ \epsilon_0; d_{33} = 350 \ pCN^{-1}$.

6.6 Consider a pyroelectric detector with the following parameters

Photocurrent	$I_s = 10$ mA
Parallel equivalent resistance	$R = 10 \ \Omega$
Parallel equivalent capacitance	$C = 20 \ pF$
Operating frequency	$\nu = 10$ Mhz

Calculate the signal voltage produced.

6.7 In a pyroelectric capacitor a change in temperature of 1 K is produced in a time of 1 μs. Calculate the current that flows through the device.

$$p = 200 \ \mu Cm^{-2}K^{-1}$$
$$A = 50 \ \mu m \times 50 \ \mu m$$

6.9 FURTHER READING

- **General**

    - A.J. Moulson and J.M. Herbert, *Electroceramics: Materials, Properties, Applications*, Chapman Hall (1992).
    - J.F. Nye, *Physical Properties of Crystals*, Clarendon Press, Oxford (1985).
    - J.C. Burfoot, *Polar Dielectrics and Their Applications*, Macmillan, London (1979).
    - J.M. Herbert, *Ferroelectric Transducers and Sensors*, Gordon & Breach, London (1982).
    - C. Kittel, *Introduction to Solid State Physics*, J. Wiley, New York (1986).

Chapter 7

OPTICAL MODULATION AND SWITCHING

7.1 INTRODUCTION

In Chapter 5 we examined how light detection and emission occurs in devices. In addition to lasers, LEDs, and detectors discussed in Chapter 5, we need devices that can: (i) modulate light, i.e. alter the strength of the light signal; (ii) switch light from one path to another for, say, the purpose of sending a beam to a particular route; and (iii) selectively filter a particular optical wavelength. Such devices are needed for optical communication networks as well as for display technologies and optical sensors.

In Chapter 6 we have seen how the dielectric response of a material is influenced by internal charges. The distribution of these charges and the polarization can be altered by electrical, mechanical, and thermal perturbations. Mechanical stress (strain) leads to devices that can be used as piezoelectric sensors and transducers. Similarly temperature changes can be exploited for infrared sensors. In this chapter we will examine how electrical signals can be used to alter the optical response of a device.

Optical signals are electromagnetic waves, which propagate through free space and solids in accordance with Maxwell's equations. As noted in Chapter 5, in light absorption and emission we need to use quantum mechanics (i.e., treat electromagnetic waves as particles) to understand the physical properties (absorption coefficient, gain, etc.). Once these properties are established the propagation is described by the classical wave equation of Maxwell. The properties of light waves are described by physical parameters, such as polarization, intensity, wavelength, speed of the wave, etc. Most optical modulators and switching devices are based on phenomena that allow altering the polarization or intensity of light.

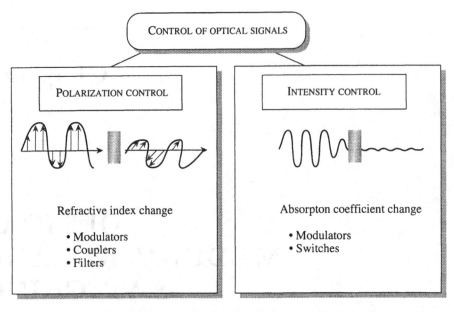

Figure 7.1: An overview of how optical signals can be controlled by altering the optical response of materials.

In Fig. 7.1 we provide an overview of how optical signals can be altered. One approach involves altering the polarization of light. This can be accomplished if the refractive index of a material can be altered. In this Chapter we will examine how this is done and how the results can be used to design optical switches, routers, filters, etc. Another approach to modulate optical signals is by altering the intensity of an optical signal by changing the absorption coefficient of a medium. A change in the absorption coefficient can be used to design optical modulators or a programmable transparency.

In this chapter we will examine how the optical properties of a material are altered to create devices mentioned in Fig. 7.1. Before starting our discussion on materials we will first review some basic properties related to light propagation in materials.

7.2 LIGHT PROPAGATION IN MATERIALS

In Chapter 1 we have discussed the structural properties of crystalline materials. We have seen that unlike the non-crystalline materials, there is a long-range order in the arrangement of atoms in these materials, which leads to anisotropic physical properties. Thus, for example, the light propagation along different directions is not described by the same refractive index. In fact, light polarized along different directions will propagate with different speeds, in general. These anisotropic properties are of great value in designing remarkable optical devices – both passive and active. Among passive devices that use the anisotropy of light propagation in crystals are quarter wave plates to alter polarization, polarizers, birefringent plates, etc. The active devices that use anisotropy of the material are electro-optic devices, liquid crystal devices, acousto-optic devices, etc. We will review the relevant physics of light propagation in anisotropic media in this section. We will not provide detailed derivations of some of the results given, but simply focus on the important physics issues.

In an isotropic medium, the propagation of light waves is described by a direction independent dielectric constant (or refractive index). However, in crystalline materials, the medium is not isotropic. It is useful to describe the properties of a crystal by choosing principal axes determined by the crystal symmetry. The displacement D and the electric field E of the light waves, in general, have a relation given by the dielectric tensor

$$\begin{aligned} D_1 &= \epsilon_{11} E_1 + \epsilon_{12} E_2 + \epsilon_{13} E_3 \\ D_2 &= \epsilon_{21} E_1 + \epsilon_{22} E_2 + \epsilon_{23} E_3 \\ D_3 &= \epsilon_{31} E_1 + \epsilon_{32} E_2 + \epsilon_{33} E_3 \end{aligned} \qquad (7.1)$$

with $\epsilon_{ij} = \epsilon_{ji}$. In general, in anisotropic materials, the vector D and E are not parallel to each other. As a consequence the electrical power given by the Poynting vector S and wave propagation direction k may not be the same. It is possible to define the principle axes of any system, where the E and D vectors are parallel to each other. These axes are found by diagonalizing the dielectric tensor ϵ_{ij}.

To describe the propagation of electromagnetic waves in a solid, we define the energy density in the medium and then examine the constant energy surfaces. The energy density is given by

$$W = \frac{1}{2} E \cdot D = \frac{1}{2} E_i \epsilon_{ij} E_i \qquad (7.2)$$

Using the principle axes system x_1, x_2, x_3 (which need not be the cartesion coordinates), we get

$$2W = \epsilon_1 E_1^2 + \epsilon_2 E_2^2 + \epsilon_3 E_3^2$$

or in terms of the displacement vectors

$$2W = \frac{D_1^2}{\epsilon_1} + \frac{D_2^2}{\epsilon_2} + \frac{D_3^2}{\epsilon_3} \qquad (7.3)$$

We write

$$x_1 = \frac{D_1}{\sqrt{2W \epsilon_0}}; \quad x_2 = \frac{D_2}{\sqrt{2W \epsilon_0}}; \quad x_3 = \frac{D_3}{\sqrt{2W \epsilon_0}} \qquad (7.4)$$

This leads to the equation known as index ellipsoid (or indicatrix)

$$\frac{x_1^2}{n_{r1}^2} + \frac{x_2^2}{n_{r2}^2} + \frac{x_3^2}{n_{r3}^2} = 1 \qquad (7.5)$$

Here n_{ri} represents the refractive index along i. In general we have the ellipsoid (η is called the impermeability tensor)

$$\sum \eta_{ij} x_i x_j = 1 \qquad (7.6)$$

7.2. Light propagation in materials

where $\eta = \epsilon_0/\epsilon$. As noted above, for the principle axes there are no off-diagonal terms in the equation above.

We will briefly describe how the index ellipsoid is used to describe the polarization of light propagating in a crystalline material. In Fig. 7.2a, we show the index ellipsoid of a crystal and a light wave propagating along a direction k. In an isotropic medium, the wave can have an arbitrary polarization in the plane perpendicular to k. However, in an anisotropic medium the wave has either of two linear polarizations and the velocity of the light with each polarization is, in general, different.

To calculate the polarization, we use the construction outlined in Fig. 7.2b. A plane is drawn perpendicular to the **k**-vector and the intersection of this plane with the ellipsoid produces an ellipse. The ellipse produced has principal axes a and b, as shown in Fig. 7.2b. The directions of polarization allowed for the wave are now given by D_a and D_b; i.e., parallel to the principal axes. *The velocities of the light with the two polarizations are inversely proportional to the length of the principal axes.* In particular, if the light is propagating along the axis $i = 3$, the light is polarized along $i = 1$ and $i = 2$ with velocities c/n_{r1} and c/n_{r2}, respectively.

In the analysis discussed above, the indices n_{r1}, n_{r2}, n_{r3} are, in general, different. Their values and their differences depend upon the details of the material structure. We can have the following cases:

Isotropic:
$$n_{r1} = n_{r2} = n_{r3} \tag{7.7}$$

Uniaxial:
$$n_{r1} = n_{r2} \neq n_{r3} \tag{7.8}$$

Biaxial:
$$\begin{aligned} n_{r1} &\neq n_{r2} \\ n_{r2} &\neq n_{r3} \\ n_{r1} &\neq n_{r3} \end{aligned} \tag{7.9}$$

The case of most interest to us is the uniaxial medium, which describes most electro-optic devices used for light modulation and also describes the liquid crystals.

Focusing on the uniaxial crystals, let us denote the axis-3 by z, the 1 and 2 being x and y. If light is propagating along the z-axis, it could have any polarization in the x–y plane, and light with all these polarizations will have the same propagation velocity. The z-axis (also called the c-axis in optics) is then called the optic axis, and $n_{r3}(=n_{rz})$ is denoted by the n_{re}, the extraordinary refractive index, while n_{rx} and n_{ry} are denoted by n_{ro}, the ordinary refractive index. If $n_{ro} < n_{re}$, the crystal is said to be positive, while, if $n_{ro} > n_{re}$, the crystal is said to be negative. Values of some important uniaxial crystals are given in Table 7.1.

In a uniaxial crystal, if light is propagating in a direction other than the optic axis, a phase delay will develop between the two polarizations of light, due to their different propagation velocities. This phase delay is exploited for designing devices that can alter the polarization of light. If an external perturbation can alter the refractive

300   Optical modulation and switching

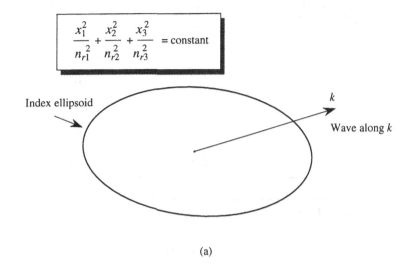

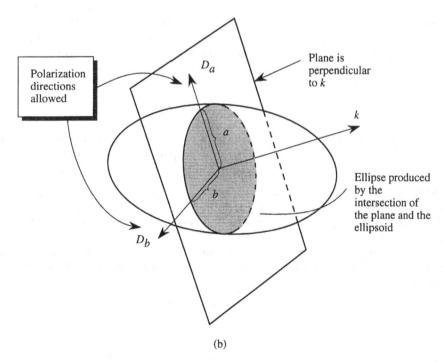

Figure 7.2: (a) An index ellipsoid for a crystal. Shown is a wave along the direction k. (b) The construction used to obtain the polarization of the wave.

7.2. Light propagation in materials

Material	n_{ro}	n_{re}
Quartz	1.544	1.553
ZnS	2.354	2.358
KDP	1.507	1.467
Calcite	1.658	1.486
LiNbO$_3$	2.300	2.208
BaTiO$_3$	2.416	2.364

Table 7.1: Refractive indices of some uniaxial crystals. The refractive indices are wavelength dependent and are given for a wavelength of 0.63 μm.

index, the device can become active and can be used to modulate a light signal as will be discussed later.

Finally consider polarization of a wave propagating along a general direction $\hat{s}$ making an angle θ with the optic axis (the z-axis). The displacement fields are polarized as shown in Fig. 7.3 and as discussed earlier. For the uniaxial crystal, the wave polarized along the x-axis (choosing the $\hat{s}$ direction to be in the y–z plane) is the ordinary wave with index n_{ro}. The wave polarized along the orthogonal direction along the other semi-major axis of the ellipse (see Fig. 7.3) is the extraordinary wave with a refractive index given by

$$\frac{1}{n_{re}^2(\theta)} = \frac{\cos^2\theta}{n_{ro}^2} + \frac{\sin^2\theta}{n_{re}^2} \tag{7.10}$$

If the wave is propagating along the z-axis ($\theta = 0$), i.e., the optic axis, the value of $n_{re}(\theta)$ is simply n_{ro} as expected. If the wave is propagating along the y-axis, $n_{re}(\theta) = n_{re}$.

EXAMPLE 7.1 Consider a quarter-wave plate on which light initially polarized along a direction 45° to the x-axis impinges. What is the polarization of the emerging light if the plate is designed to produce a phase difference of $\pi/2$ between light initially polarized along the x- and y-axes?

The quarter wave plate will create a phase difference of $\pi/2$ between the x- and y-polarized waves after transmission through the plate. Initially the incident wave has the electric fields given by

$$E_x = E_o \cos(\omega t - kz)$$
$$E_y = E_o \cos(\omega t - kz)$$

After transmission we have

$$E_x = E_o \cos(\omega t - kz + \phi_o)$$
$$E_y = E_o \cos(\omega t - kz + \phi_o + \pi/2)$$

The light is thus circularly polarized.

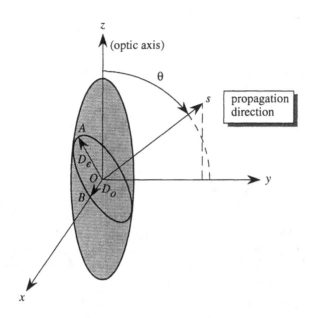

Figure 7.3: The polarization of an optical beam propagating along a direction s, making an angle θ with the optic axis.

7.3 MODULATION OF OPTICAL PROPERTIES

We will now discuss the physical effects that form the basis of intelligent optoelectronic devices, such as switches, routers, tunable filter, etc. To alter optical response of a material, we need to alter the refractive index or absorption coefficient (or dielectric response). As we saw in Chapter 6, a variety of perturbations can alter the dielectric response, although we are interested in changes in dielectric response ϵ at frequencies around 10^{15} Hz, i.e., at optical frequencies. At such high frequencies only the electrons (rather than ions or molecules) are able to respond and contribute to the changes in the dielectric response. The perturbation we will consider is electric field. As discussed in Chapter 6 other perturbations, such as mechanical stress, temperature changes, etc., can also be used to modulate an optical signal. However, an electrical signal can be operated at high speeds ($\sim$ up to 50 GHz) and is widely used in optical systems.

When the optical properties of a material are modified, the effect on a light beam propagating in the material can be classified into two categories, depending upon the photon energy. As shown in Fig. 7.4, if the photon energy is in a region where the absorption coefficient is zero (beam with frequency ω_1), the effect of the modification of the refractive index is to alter the velocity and polarization of propagation of light. However, if the photon energy is in a region where the absorption coefficient is altered (beam with frequency ω_2), the intensity of light emerging from the sample will be altered. These two approaches for the modification of the optical properties by a applied electric field are called the electro-optic and the electro-absorption approaches, respectively.

In the electro-optic effect, an applied electric field is used to alter the phase

7.3. Modulation of optical properties

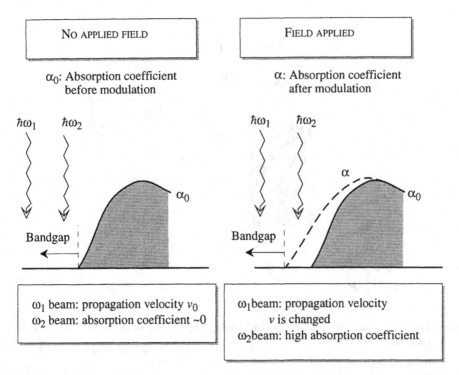

Figure 7.4: A schematic of the effect of a change in optical properties of a material on an optical beam. For energy $\hbar\omega_1$, the main effect of the change in the optical properties is a change in propagation velocity. For $\hbar\omega_2$, the effect is a change in intensity.

velocity of a propagating signal and this effect can be exploited in an interference scheme to alter the polarization or intensity of the light. We will first discuss this approach.

7.3.1 Electro-optic effect

The electro-optic effect depends upon the modification of the refractive index of a material by an applied electric field and is the basis of numerous important devices used for optical routers for optical communication, color filters, display, image storage, etc. The effect is based on the modification of the refractive index of a material by an electric field. At small electric fields ($\leq 10^4$ V/cm) the change in a material's refractive index is small and we have a linear relation between the polarization P and applied field E. However, at higher fields the relationship between P and E is non-linear and the change in the dielectric constant or refractive index is more discernible and can be exploited for device design.

In the previous section, we have reviewed some important principles of light propagation in solids. Changes in refractive index can be exploited to alter the nature of the optical signal. Consider a situation where an electric field is applied to the crystal. The applied electric field modifies the polarization and bandstructure of the semiconductor through a number of interactions. These interactions may involve:

(i) Strain: In a piezoelectric materials the electric field may cause a distortion in the lattice and, as a result, the dielectric response may change. This may cause a change in the refractive index.

(ii) Distortion of the excitonic features: In Chapter 5 we discussed the optical properties of the exciton. The presence of an electric field can modify the excitonic spectra, thus altering the electronic spectra and, hence, the optical spectra of the material.

In general, the change in the impermeability tensor may be written as

$$\eta_{ij}(E) - \eta_{ij}(0) = \Delta\eta_{ij} = r_{ijk}E_k + s_{ijk\ell}E_k E_\ell \quad (7.11)$$

where E_i is the applied electric field component along the direction i, and r_{ijk} and $s_{ijk\ell}$ are the components of the electro-optic tensor. In materials like GaAs where the inversion symmetry is missing, r_{ijk} is non-zero and we have a linear term in the electro-optic effect. The linear effect is called the Pockel effect. In materials like Si, where we have inversion symmetry, $r_{ijk} = 0$ and the lowest-order effect is due to the quadratic effect (known as the Kerr effect).

In general, r_{ijk} has 27 elements, but, since the tensor is invariant under the exchange of i and j, there are only 18 independent terms. It is common to use the contracted notation $r_{\ell m}$, where $\ell = 1, \ldots 6$ and $m = 1, 2, 3$. The standard contraction arises from the identification of $i,j = 1,1; 2,2; 3,3; 2,3; 3,1;1,2$ by $\ell = 1, 2, 3, 4, 5, 6$, respectively. The 18 coefficients are further reduced by the symmetry of the crystals. In semiconductors such as GaAs, it turns out that the only non-zero coefficients are

$$r_{41}$$
$$r_{52} = r_{41}$$
$$r_{63} = r_{41} \quad (7.12)$$

Thus, a single parameter describes the linear electro-optic effect. In Table 7.2, we give the values of the electro-optic coefficients for some materials.

The second-order electro-optic coefficients $s_{ijk\ell}$ are usually not important for materials, unless the optical energy $\hbar\omega$ is very close to the bandgap. In materials such as GaAs, the second-order coefficients that are non-zero from symmetry considerations are in the contracted form $s_{pq}, p = 1\ldots 6, q = 1\ldots 6$,

$$s_{11} = s_{22} = s_{33}$$
$$s_{12} = s_{13}$$
$$s_{44} = s_{55} = s_{66} \quad (7.13)$$

The electro-optic effect is used to create a modulation in the frequency, intensity, or polarization of an optical beam.

Pockels effect

To see how the electro-optic effect influences optical properties of materials we consider

7.3. Modulation of optical properties

MATERIAL	WAVELENGTH (μm)	ELECTROOPTIC COEFFICIENT (10^{-12} m/V)	INDEX OF REFRACTION
LiNbO$_3$	0.623	$r_{13} = 9.6$ $r_{22} = 6.8$ $r_{33} = 30.9$ $r_{51} = 32.6$	$n_0 = 1.8830$ $n_e = 1.7367$
GaAs	0.9 1.15	$r_{41} = 1.1$ $r_{41} = 1.43$	$n = 3.60$
KDP	0.633	$r_{63} = 11$ $r_{41} = 8$	$n_0 = 1.5074$ $n_e = 1.4669$
ADP	0.633	$r_{63} = 8.5$ $r_{41} = 28$	$n_0 = 1.52$ $n_e = 1.48$
Quartz	~0.632	$r_{41} = 0.2$ $r_{63} = 0.93$	$n_0 = 1.54$ $n_e = 1.55$
BaTiO$_3$	~0.632	$r_{33} = 23$ $r_{13} = 8$ $r_{42} = 820$	$n_0 = 2.437$ $n_e = 2.180$
LaTiO$_3$	~0.632	$r_{33} = 30.3$ $r_{13} = 5.7$	$n_0 = 2.175$ $n_e = 2.365$

Table 7.2: Electro-optic coefficients for some materials.

the linear effect, known as Pockels effect, in a material such as GaAs (results for a tetragonal system will also be summarized) when a field E is applied to the crystal. The index ellipsoid is (see Eq. 7.6)

$$\left(\frac{1}{n_x^2} + r_{1k}E_k\right)x^2 + \left(\frac{1}{n_y^2} + r_{2k}E_k\right)y^2 + \left(\frac{1}{n_z^2} + r_{3k}E_k\right)z^2$$
$$+ \; 2yz r_{4k}E_k + 2zx r_{5k}E_k + 2xy r_{6k}E_k = 1 \quad (7.14)$$

where $E_k (k = 1, 2, 3)$ is the component of the electric field in the x, y, and z directions. Using the elements of the electro-optic tensor for GaAs, we get

$$\frac{x_1^2}{n_x^2} + \frac{x_2^2}{n_y^2} + \frac{x_3^2}{n_z^2} + 2x_2 x_3 r_{41} E_1 + 2zx r_{41} E_2 + 2xy r_{41} E_3 = 1 \quad (7.15)$$

Let us now simplify the problem by assuming that the electric field is along the $\langle 001 \rangle$ direction, as shown in Fig. 7.5

$$E_x = E_y = 0, \quad E_z = E \quad (7.16)$$

We now rotate the axes by 45° so that the new principal axes are (see Fig. 7.5b)

$$x' = \frac{x}{\sqrt{2}} - \frac{y}{\sqrt{2}}$$
$$y' = \frac{x}{\sqrt{2}} + \frac{y}{\sqrt{2}}$$
$$z' = z \quad (7.17)$$

In terms of this new set of axes, the index of ellipsoid is written as

$$\frac{x'^2}{n_x'^2} + \frac{y'^2}{n_y'^2} + \frac{z'^2}{n_z'^2} = 1 \quad (7.18)$$

where the new indices are

$$n_x' = n_o + \frac{1}{2}n_o^3 r_{41} E$$
$$n_y' = n_o - \frac{1}{2}n_o^3 r_{41} E$$
$$n_z' = n_o \quad (7.19)$$

where n_o is the index in the absence of the field ($= n_x = n_y = n_z$). As a result of this change in the indices along the x' and y' axes, for light along $\langle 01\bar{1} \rangle$ (x') and $\langle 011 \rangle$ (y') directions, a phase retardation occurs due to the change in index ellipsoid as shown in Fig. 7.5c. The phase retardation for a wave that travels a distance L is ($n_z' = n_o$)

$$\Delta\phi(x') = \frac{\omega}{c}\left(n_z' - n_x'\right)L_1 = -\frac{\pi}{\lambda}n_o^3 r_{41} E L$$
$$\Delta\phi(y') = \frac{\omega}{c}\left(n_z' - n_y'\right)L_1 = \frac{\pi}{\lambda}n_o^3 r_{41} E L \quad (7.20)$$

7.3. Modulation of optical properties

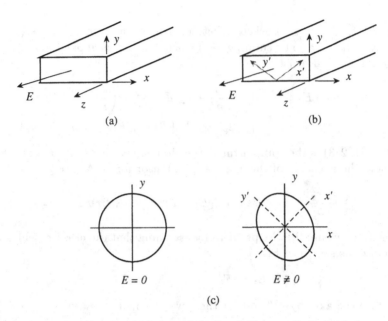

Figure 7.5: (a) A waveguide in which a modulating field is applied along the z-axis. (b) A rotated axis x', y' used to understand wave propagation. (c) The index ellipsoid in the absence and presence of the electric field.

As another class of materials let us consider $BaTiO_3$ which is tetragonal below 400 K. For $BaTiO_3$, the optic axis is along the z-direction and $n_0 = 2.416$ and $n_e = 2.364$. In the presence of an electric field, following the discussion above, we get (again assuming that the field is along the z-direction or the optic axis)

$$\left(\frac{1}{n_0^2} + r_{13}E\right) x^2 + \left(\frac{1}{n_0^2} + r_{13}E\right) y^2 + \left(\frac{1}{n_0^2} + r_{33}E\right) z^2 = 1 \quad (7.21)$$

This leads to

$$\Delta\left(\frac{1}{n_0^2}\right) = r_{13}E \quad \text{and} \quad \Delta\left(\frac{1}{n_e^2}\right) = r_{33}E \quad (7.22)$$

or, since $\Delta n_0 \ll n_0, \Delta n_e \ll n_e$

$$\Delta n_0 = -\frac{1}{2}n_0^3 r_{13}E$$

$$\Delta n_e = -\frac{1}{2}n_e^3 r_{33}E \quad (7.23)$$

As a result of the changes in the ordinary and extraordinary refractive index we have an induced bifringence Δn

$$\Delta n = \Delta(n_e - n_0) = -\frac{1}{2}n_e^3 \left(r_{33} - \frac{n_0^3}{n_e^3}r_{13}\right) E$$

$$\equiv -\frac{1}{2}n^3 r_c E \quad (7.24)$$

where $n \sim n_0 \sim n_e$ and

$$r_c = r_{33} - \frac{n_0^3}{n_e^3} r_{13}$$

$$\sim r_{33} - r_{13} \qquad (7.25)$$

Kerr effect

In cubic materials with inversion symmetry the linear electro-optic effect is absent and the second-order effect, i.e. the Kerr effect, becomes important. We will now discuss this second-order effect.

Using the contracted notation, for quadratic electro-optic coefficients, the index ellipsoid can be written as

$$\left(\frac{1}{n_x^2} + s_{11}E_x^2 + s_{12}E_y^2 + s_{12}E_z^2\right)x^2$$
$$+ \left(\frac{1}{n_y^2} + s_{12}E_x^2 + s_{11}E_y^2 + s_{12}E_z^2\right)y^2$$
$$+ \left(\frac{1}{n_z^2} + s_{12}E_x^2 + s_{12}E_y^2 + s_{11}E_z^2\right)z^2$$
$$+ 2yz(2s_{44}E_yE_z) + 2zx(2s_{44}E_xE_z) + 2xy(2s_{44}E_yE_x) = 1 \qquad (7.26)$$

In the presence of an electric field, E, in the z direction, we have

$$\left(\frac{1}{n_x^2} + s_{12}E^2\right)x^2 + \left(\frac{1}{n_y^2} + s_{12}E^2\right)y^2$$
$$+ \left(\frac{1}{n_z^2} + s_{11}E^2\right)z^2 = 1 \qquad (7.27)$$

This index ellipsoid can be rewritten as

$$\frac{x^2 + y^2}{n_o^2} + \frac{z^2}{n_e^2} = 1 \qquad (7.28)$$

with

$$n_o = n - \frac{1}{2}n^3 s_{12} E^2 \qquad (7.29)$$

and

$$n_e = n - \frac{1}{2}n^3 s_{11} E^2 \qquad (7.30)$$

The phase retardation due to the applied field is thus given by

$$\Delta\Phi = \frac{\omega}{c}(n_e - n_o)L = \frac{\pi}{\lambda}n^3(s_{12} - s_{11})E^2 L$$

The total phase change between waves travelling along x' and y' then becomes after adding the effects of the linear and quadratic electro-optic effects

$$\Delta\phi(x') = -\frac{\pi L}{\lambda}n_o^3\left[r_{41}E_1 + (s_{12} - s_{11})E^2\right]$$

7.3. Modulation of optical properties

$$\Delta\phi(y') = \frac{\pi L}{\lambda} n_o^3 \left[r_{41} E_1 + (s_{12} - s_{11}) E^2 \right] \quad (7.31)$$

The phase changes produced by the electric field can be exploited for a number of important switching or modulation devices. From Table 7.2, we see that the electro-optic coefficients in materials such as LiNbO$_3$ are much larger than those in traditional semiconductors. As a result LiNbO$_3$ is widely used in optical directional couplers and switches.

EXAMPLE 7.2 A bulk GaAs device is used as an electro-optic modulator. The device dimension is 1 mm and a phase change of 90° is obtained between light polarized along $<01\bar{1}>$ and $<011>$. The wavelength of the light is 1.5 μm. Calculate the electric field needed.

The phase change produced is

$$\Delta\phi = \frac{2\pi}{\lambda} n_o^3 r_{41} E L = \frac{\pi}{2}$$

$$E = \frac{\lambda}{4 n_o^3 r_{41} L}$$

$$= \frac{(1.5 \times 10^{-6} \text{ m})}{4(3.3)^3 (1.2 \times 10^{-12} \text{ m/V})(10^{-3} \text{ m})}$$

$$= 8.7 \times 10^6 \text{ V/m}$$

If the field is across a 1.0 μm thickness, the voltage needed is 8.7 V.

7.3.2 Electro-absorption modulation

In Fig. 7.4 we show how an optical signal can be modulated using two different approaches. We have already discussed the electro-optic modulation in which changes in refractive index can be used to alter light propagation. In this mode the wavelength of light is such that there is no absorption (at least intentional) of light. As we will see later, intensity modulation can also be done using the electro-optic effect using interference techniques.

In the electro-absorption modulation scheme (right-hand side of Fig. 7.4) the photon energy is near the bandgap of a semiconductor. The application of an electric field alters the absorption spectrum and thus influences the intensity of the optical signal.

We have seen in Chapter 5 that if the photon energy exceeds the bandgap of the material, absorption occurs due to an electron moving from the valence band to the conduction band. In fact, as discussed in Chapter 5, absorption starts when the photon energy reaches

$$\hbar\omega = E_g - E_{ex} \quad (7.32)$$

where E_{ex} is the exciton binding energy; i.e., the Coulombic binding energy of the electron–hole pair. In Fig. 7.6 we show a typical absorption spectra for a semiconductor. Notice the excitonic transition. In most bulk semiconductors, the exciton transition is clearly observable only at low temperatures and in very high purity materials.

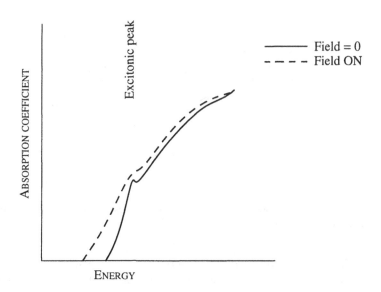

Figure 7.6: A schematic of how absorption coefficient changes with applied field.

When an electric field is applied to a semiconductor, the absorption spectrum shifts towards lower energy, as shown in Fig. 7.6. The shift is due to the shrinking of the effective bandgap of the material. The change in absorption coefficient at a particular photon energy can be used for intensity modulation. The modulation achieved in bulk semiconductors is quite small, and not very useful for devices. However, if quantum wells are used (see Chapter 3), the change in absorption can be quite large. The effect is called quantum confined Stark effect.

Quantum confined stark effect (QCSE) refers to the changes that occur in the electronic and optical spectra of a quantum well when an electric field is applied. In Fig. 7.7 we show schematically a quantum well without and with an electric field in the confinement direction. The field pushes the electron and hole functions to opposite sides making the ground state intersubband separation smaller. This effect is the dominant term in changing the exciton resonance energy.

While the exact calculation of the intersubband separation requires numerical techniques, we can estimate these changes by using perturbation theory. This approach gives reasonable results for low electric fields. If the field is small enough such that

$$|eEW| \ll \frac{\hbar^2 \pi^2}{2m^* W^2} \tag{7.33}$$

i.e., the perturbation is small compared to the ground state energy, then it can be shown that the ground state energy changes by

$$\Delta E_1^{(2)} = \frac{1}{24\pi^2}\left(\frac{15}{\pi^2} - 1\right)\frac{m^* e^2 E^2 W^4}{\hbar^2} \tag{7.34}$$

We see that the second-order effect increases with m^* and has a strong well size dependence. This would suggest that for the best modulation we should use a wide well.

7.3. Modulation of optical properties

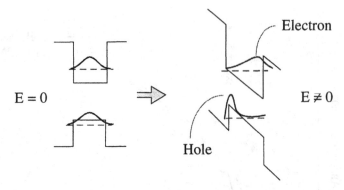

Figure 7.7: A schematic showing how an electric field alters the quantum well shape and the electron and hole wavefunctions.

However, in wide wells the exciton absorption decreases and also the HH, LH separation becomes small. Optimum well sizes are of the order of ~100 Å for most quantum well structures.

The absorption and electro-absorption is strongly dependent on polarization. Denoting the transverse electric (TE) mode for polarization where E-field is in the quantum well plane and transverse magnetic (TM) for the case where the E-field is normal to the quantum well, in Fig. 7.8 we show measured results for electro-absorption in GaAs/AlGaAs quantum wells.

EXAMPLE 7.3 A 100 Å GaAs/Al$_{0.3}$Ga$_{0.5}$As MQW structure has the HH exciton energy peak at 1.51 eV. A transverse bias of 80 kV/cm is applied to the MQW. Calculate the change in the transmitted beam intensity (there is no substrate absorption) if the total width of the wells is 1.0 μm. The photon energy is $\hbar\omega = 1.49$ eV. Peak absorption is 1.16×10^4 cm^{-1}.

The transmitted light is
$$I = I_o \exp(-\alpha d)$$

At zero bias, we have
$$\alpha(V = 0) = 1.16 \times 10^4 \exp\left(\frac{-(1.49 - 1.51)^2}{1.44(2.5 \times 10^{-3})^2}\right)$$
$$\sim 0$$

At a bias of 80 kV/cm, the exciton peak shifts by ~20 meV. The absorption coefficient is
$$\alpha(E = 80 \text{ kV/cm}) = 1.16 \times 10^4 \exp\left(\frac{-(1.49 - 1.49)^2}{1.44(2.5 \times 10^{-3})^2}\right)$$
$$= 1.2 \times 10^4 \text{ cm}^{-1}$$

The ratio of the transmitted intensity is
$$\frac{I(E = 80 \text{ kV/cm})}{I(E = 0)} = 0.3$$

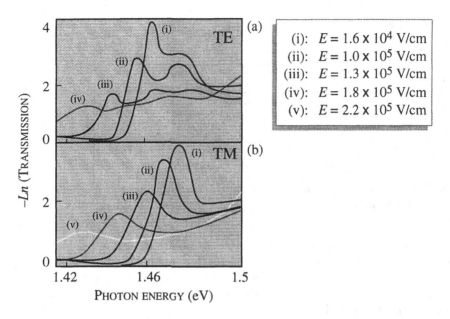

Figure 7.8: Measured polarization dependent transmittances in GaAs/AlGaAs (100 Å) multiquantum well structures when light is coming in the waveguide geometry. (a) Incident polarization parallel to the plane of the layers. (b) Incident polarization perpendicular to the plane of the layers. (After D.A.B. Miller et al., *IEEE J. Quantum Electronics*, QE-22, 1816, 1986.)

7.4 OPTICAL MODULATION DEVICES

The electro-optic and electro-absorption effects discussed above form the basis of numerous optical devices. These devices are used to switch optical signals, modulate their intensity, couple signals from one point to another, display optical images, etc. In addition to using an electric field to alter optical properties other perturbations such as strain (arising from pressure, ultrasonic waves, etc.) or magnetic effects can also alter the dielectric response and be exploited for acousto-optic or magneto-optic devices. In this section we will briefly discuss some of the important modulation/sensing devices.

A most useful technique to modulate an optical signal is through the use of polarizers and an active device that can change the polarization of light. The general approach is illustrated in Fig. 7.9. In this particular geometry (other geometries are also possible) two polarizers aligned in the cross-polarized configuration are placed on each side of the device. The device consists of a crystal (or liquid crystal) in which the two refractive indices n_e and n_o are different. As discussed in the previous section it is possible to alter the difference between n_{re} and n_{ro} by using an external perturbation. This alteration can be done by applying an electric field and utilizing the electro-optic effect.

Let us first consider the case of an electro-optic modulator based on crystals such as lithium niobate. Later we will consider the case of a liquid crystal such as a twisted nematic. Let us assume that a linearly polarized light enters the device. As

7.4. Optical modulation devices

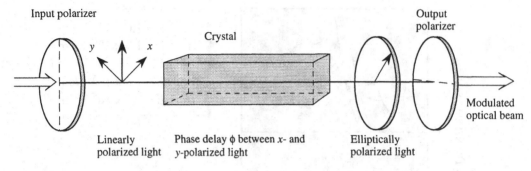

Figure 7.9: A schematic of how a polarization charge produced by a crystal device can alter the intensity of an optical beam.

shown in Fig. 7.9, let us assume that a linearly polarized light is incident on the crystal and the x-axis and the y-axis represent the two polarization axes for the crystal. In general, the two directions have different refractive indices and, as the wave propagates, a phase difference develops between the two polarizations. Consider an input signal that is linearly polarized and given by

$$E_x = \frac{E_o}{\sqrt{2}} \exp(i\omega t) \tag{7.35}$$

$$E_y = \frac{E_o}{\sqrt{2}} \exp(i\omega t) \tag{7.36}$$

After transmission through the modulator, the wave emerges with a general polarization given by

$$E_x = \frac{E_o}{\sqrt{2}} \exp(i\omega t + i\theta_1) \tag{7.37}$$

$$E_y = \frac{E_o}{\sqrt{2}} \exp(i\omega t + i\theta_2) \tag{7.38}$$

with the phase difference given by $\phi = \theta_2 - \theta_1$. If ϕ is $\pi/2$, the output beam is circularly polarized, and, if it is π, it is linearly polarized with polarization 90° with respect to the input beam. If the output beam passes through a polarizer at 90° with respect to the input beam polarizer, as shown in Fig. 7.9, the modulation ratio is given by (assuming no absorption losses)

$$\frac{I_{\text{out}}}{I_{\text{in}}} = \sin^2 \frac{\phi}{2} \tag{7.39}$$

Thus, if ϕ can be controlled by an electric field, the intensity can be modulated.

Polarization and modulation properties of a twisted nematic

Over the last decade there has been an explosion in flat panel displays for applications in laptop computers, televisions, etc. A considerable part of this technology arises from improvement in liquid crystal displays. As we have noted in Chapter 1, the liquid crystal is a material which has good long-range order along some direction. Also, since the material is made up of rod-like or disc-like molecules, it has a very strong anisotropy between n_e and n_o. In a nematic liquid crystal, we can introduce a twist in the order in which molecules are arranged by using two glass plates that have been rubbed in a particular orientation. As a result, the optic axis of the liquid crystal changes from point to point, consequently, the direction of the polarizations corresponding to the ordinary and extraordinary rays changes from point to point (as shown in Fig. 7.10).

An important approximation that is used to describe how light propagates (i.e., how the polarization changes) through a twisted nematic crystal is called the adiabatic approximation. The adiabatic approximation depends upon the fact that the twist in the crystal is "slowly varying." This is a good approximation for liquid crystals, since a twist of $\pi/2$ is produced over several microns (say $\sim$ 10-20 μm). As a result, the light responds according to the local refractive indices and the local polarization axes. Thus, if light enters the crystal along the "slow polarization" direction, it remains along this polarization as it travels down the liquid crystal.

From the adiabatic approximation discussed above, we can see that there are two sources for the polarization change in a twisted nematic liquid crystal: (i) As a result of the difference between n_e and n_o, the phase difference between the two rays, states changes, thus the polarization changes. This is the effect discussed above and produces a modulation of light as given by Eq. 7.39. (ii) Additionally, due to the twist in the crystal, the polarization is rotated. This effect is exploited in most liquid crystal displays.

According to the adiabatic approximation, if the twist angle is 90° (or 270°) from the top plate to the lower one, and light polarized as one of the ordinary or extraordinary waves is sent in, we have the following possibilities: (i) If the output polarizer is oriented along the input polarizer, the transmitted intensity is zero. (ii) If the output and output polarizers are cross-polarized, the light passes through. A more accurate treatment of the problem shows that the transmittance in the first case (i.e., the polarizers having the same orientation) is given by the following relation for a $\pi/2$ twist

$$T = \frac{\sin^2\left(\frac{\pi}{2}\sqrt{1+(\phi/\pi)^2}\right)}{1+(\phi/\pi)^2} \tag{7.40}$$

where ϕ is the phase difference produced due to the difference in the values of n_e and n_o and is for a device of thickness d

$$\phi = \frac{2\pi}{\lambda}(n_e - n_o)d \tag{7.41}$$

If ϕ is much larger than π, we see that T approaches zero as is the case where the adiabatic approximation is valid.

We have noted earlier that in a uniaxial crystal there is one orientation (the c-axis) which defines the optic axis along which light propagates with the same speed regardless of its polarization. The liquid crystal display devices depend upon the ability to change the c-axis (also known as the director for liquid crystals) by an external perturbation such as an applied field. Consider the following situations: (a) The c-axis

7.4. Optical modulation devices

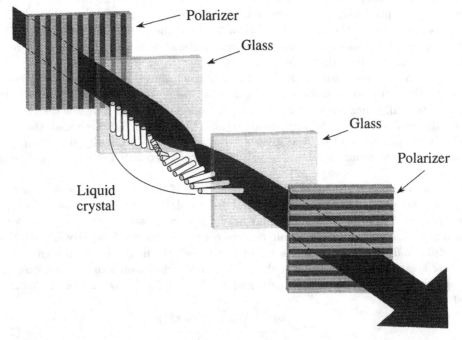

Figure 7.10: A schematic of a twisted nematic crystal. In the adiabatic approximation, if the twist is "slow," the polarization of light simply follows the twist as shown.

is parallel to the input polarizer (the refractive index is n_e for light polarized parallel to the c-axis). In this case the value of ϕ is maximum and the transmittance for the case where the output polarizer is parallel to the input polarizer is minimum. (b) An applied external perturbation forces the c-axis to be oriented along the propagation direction, so that there is no propagation delay for light polarized in different orientations. When this happens, the liquid crystal becomes transparent since light simply propagates at its original polarization. This can also be seen by putting $\phi = 0$ in Eq. 7.40. Thus, if the c-axis is altered, the device can change from opaque to transparent, which is what liquid crystal display devices are supposed to do.

7.4.1 Electro-optic modulators

Electro-optic modulators can produce amplitude, frequency, or phase modulation in an optical signal by exploiting the electro-optic effect in which the optical properties of a crystal can be altered by an electric field. A number of crystals exist which have desirable response behavior. These include potassium dihydrogen phosphate (KDP), ferroelectric peroskites such as $LiNbO_3$ and $LiTaO_3$, as well as semiconductors, such as GaAs and CdTe. We have discussed the basis of the electro-optic effect in Section 7.3.

We have discussed above (see Eqs. 7.35 to 7.39) how a phase change of ϕ can cause modulation of light. If the phase ϕ can be controlled by an electric field, the intensity can be modulated. For GaAs, the electric field dependent phase is given by (see Section 7.3.1)

$$\phi = \frac{2\pi}{\lambda} L n_o^3 r_{41} \frac{V}{d} \tag{7.42}$$

where λ is the wavelength of light, L the device length, n_{ro} the GaAs refractive index, r_{41} the electro-optic coefficient for GaAs, V the transverse applied bias, and d the thickness of the modulator. A similar analysis for materials like KDP shows that the phase change between the two polarized waves at the output is given by

$$\phi = \frac{2\pi n_{ro}^3 r_{63} E L}{\lambda_o} \tag{7.43}$$

where r_{63} is the electro-optic coefficient and E is the electric field $(= V/d)$. If cross-polarized polarizers are used, the maximum transmittance occurs when the phase change is π.

It may be noted that the electric field can be applied in a transverse or longitudinal way to the modulator.

It is clear from Eqs. 7.42 and 7.43 that a high electro-optic coefficient can allow us to achieve a modulation using a smaller interaction length for the same applied field. However, the electro-optic coefficients of most materials are rather small ($\sim 10^{-12}$ m/V) as can be seen from Table 7.2, so that for realistic bias values the length required is quite long (millimeters or more).

Electro-optic materials and image recording
The electro-optic effect is useful not only for optical modulation but also for optical image recording. Ferrroelectric materials, which have a strong electro-optic effect, find

7.4. Optical modulation devices

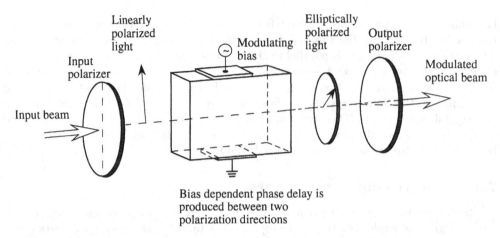

Figure 7.11: The use of an electro-optic device to modulate an optical signal. The applied bias introduces a phase change between light travelling in two polarization directions and the output light is modulated.

important uses in applications involving image recording, image contrasting etc. The material most widely used for this application is the polycrystalline ceramic PLZT, discussed in Chapter 6. Being polycrystalline, this material can be fabricated in large dimensions at low cost.

We recall from Chapter 6 that ferroelectric materials have a non-zero spontaneous polarization. When an external electric field is applied to them, the polarization can be altered. In particular the polarization can be reduced to zero by applying a field called the coercive field. The coercive field that is needed can be altered if a built-in field can be created by photo-generated carriers. This is the principle behind optical image recording.

The photoferroelectric (PFE) imaging device consists of a thin (0.1–0.3 mm) plate of PLZT ceramic with transparent electrodes applied to the major faces. The image to be stored is made to illuminate the face of the plate using near-ultraviolet illumination. Simultaneously a voltage is applied to the device. When light shines on the photo-sensitive PLZT, photo-generated carriers are produced with a local concentration proportional to the local image intensity. The carriers (electrons and holes) are separated by the applied field and trapped at defect sites. A local field is superimposed on the external field, changing the coercive field and therefore the local polarization of the material. This in turn results in local strain variations on the PLZT plate. The image is thus faithfully recorded on the plate and can be read through a projection device.

By using proper fabrication techniques, PLZT ceramics can be made with grain sizes of ~ 2 micrometers. Such plates can store images with resolution of up to 100 lines per centimeter. The stored images can be erased by shining a uniform beam of near-ultraviolet light on the PLZT plate and simultaneously applying a voltage pulse to switch the ferroelectric polarization to its initial remnant state.

EXAMPLE 7.4 A bulk GaAs device is used as an electro-optic modulator. The device dimension is 1 mm and a phase change of 90° is obtained between light polarized along $<01\bar{1}>$ and $<011>$. The wavelength of the light is 1.5 μm. Calculate the electric field needed.

The phase change produced is ($\xi = 1$)

$$\Delta\phi = \frac{2\pi}{\lambda} n_{ro}^3 r_{41} EL = \frac{\pi}{4}$$

$$E = \frac{\lambda}{8 n_{ro}^3 r_{41} L}$$

$$= \frac{(1.5 \times 10^{-6} \text{ m})}{8(3.3)^3 (1.2 \times 10^{-12} \text{ m/V})(10^{-3} \text{ m})}$$

$$= 4.35 \times 10^6 \text{ V/m}$$

EXAMPLE 7.5 The crystal KD*P (potassium dideuterium phosphate) is an important material for optoelectronics. Calculate the voltage needed to produce a phase change of π in a KDP device. This voltage is called the half-wave voltage. The wavelength of light is 1.064 μm. The refractive index is 1.52.

The half-wave voltage is ($r_{63} = 26.4 \times 10^{-12}$ m/V)

$$V(\lambda/2) = EL = \frac{(1.064 \times 10^{-6} \text{ m})\pi}{2(1.52)^3 (26.4 \times 10^{-12} \text{ m/V})}$$

$$= 5.74 \text{ kV}$$

7.4.2 Interferroelectric modulators

In the previous section, we have seen how the phase change produced by the electro-optic effect can be used to modulate a signal. However, the device configuration shown in Fig. 7.11 is not the only configuration that is used to design modulators. There are a number of other configurations that do not require polarizers, but modulate a signal through interferometric effects.

Fabry-Perot Modulators

The Fabry-Perot modulator (often called the *etalon*) consists of two partially transmitting mirrors enclosing an electro-optic material as shown in Fig. 7.12. If n_r is the refractive index of the electro-optic material and L is its length, the transmission through the etalon is maximum when (λ is the free space wavelength)

$$L = \frac{m\lambda}{2n_r} \tag{7.44}$$

The transmission coefficient for an etalon with mirror reflectivity R is given by

$$T = \frac{1}{1 + \frac{4R}{(1-R)^2} \sin^2\left(\frac{2\pi n_r L}{\lambda}\right)} \tag{7.45}$$

As can be seen from this expression and as is illustrated in Fig. 7.12, the selectivity of the etalon increases as R increases. In Fig. 7.12, the transmission coefficient is also shown as a function of round trip phase change of a wave, $2\pi n_r L/\lambda$, as it crosses the etalon.

7.4. Optical modulation devices

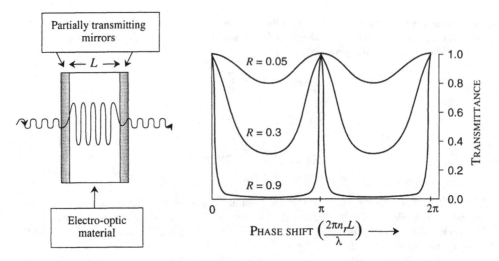

Figure 7.12: A schematic of a Fabry–Perot modulator. In epitaxially grown devices, the partially transmitting structures are distributed Bragg reflectors. The transmission coefficient of a Fabry–Perot structure as a function of the phase shift is shown at the right of the figure. R is the mirror reflectivity.

The Fabry–Perot modulator operates on the basis that, if n_r can be altered by an electric field, the phase change will alter and, as a result, the transmission of the optical signal will change. The goal of the device design is to be able to switch between T_{max} and T_{min} by applying the field.

The frequency difference between two successive maximas in the transmission of the Fabry–Perot structure is denoted as the free spectral range (FSR) and is given by

$$FSR = \frac{c}{2n_r L} \qquad (7.46)$$

An important parameter for an etalon is its finesse, which gives the ratio of the FSR and the full width at half maximum of any transmission peak. It has a value

$$F = \frac{\pi (R_1 R_2)^{1/4}}{1 - (R_1 R_2)^{1/2}} \qquad (7.47)$$

where R_1 and R_2 are the reflection coefficients of the front and back mirrors of the etalon. Fabry–Perot modulators have been demonstrated to operate up to 10 GHz, with contrast ratios up to 10 dB.

Mach–Zender modulators

The phase modulation produced by electro-optic effect can be used to create intensity modulation in a Mach–Zender interferometer. In Fig. 7.13, we show a schematic of the modulator. An optical signal coming form a single-mode waveguide is split by a coupler. The two split beams travel through two different guides (in general of different length) and then recombine to produce the output.

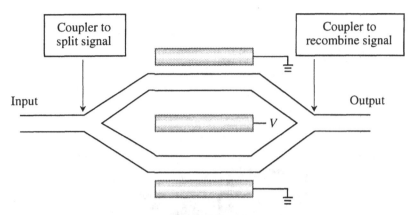

Figure 7.13: A schematic of the Mach-Zender interferometer used as a modulator. Two 3 dB couplers are used to split and recombine an incoming signal.

If the optical paths in the two arms of the interferometer is an integral number of optical wavelengths, the two waves will arrive at the second coupler in phase, and interface constructively to produce a high intensity. If an electric field is now used to create a relative phase difference between the two waves, the intensity can be reduced. If the overall phase difference between beams traveling in the two paths is π, a minimum intensity will be produced.

Due to the loss in optical signal suffered as a result of the couplers, the Mach-Zender modulators do not have very high efficiencies.

Liquid crystal display devices

Liquid crystal-based displays are used widely in laptop computers, cockpit displays, flat screen TVs, etc. The technology used is based on active matrix display where each pixel is controlled by a thin film transistor (TFT) switch. This switch allows the signal voltage to be applied to the liquid crystal cell for the entire cycle time between refreshes.

A schematic of the active matrix liquid crystal display (AMLCD) is shown in Fig. 7.14. A sample pulse is applied to the gate of a transistor pulling the device into inversion. A data voltage is applied via the column line to the drain (or the source) of the TFT. The source (or the drain) is connected to a storage capacitor, which holds the applied voltage once the gate pulse is removed.

The TFT must have the properties that its resistance is very low when the gate bias is ON and very high when the gate bias is OFF. This allows the storage capacitor to charge to the applied potential during the time the signal is on. Also, the capacitances (the gate source, gate drain, storage, liquid crystal cell, and $C_{parasite}$) should be such as to allow minimum charge leakage during a cycle time. The success of the AMLCD depends critically upon the TFT discussed in Chapter 5.

7.4. Optical modulation devices

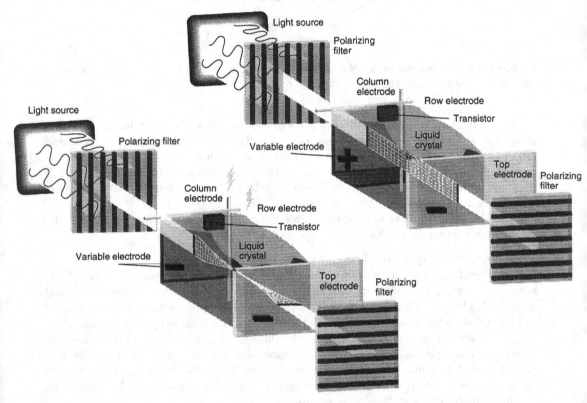

Figure 7.14: A schematic of an active matrix liquid crystal display.

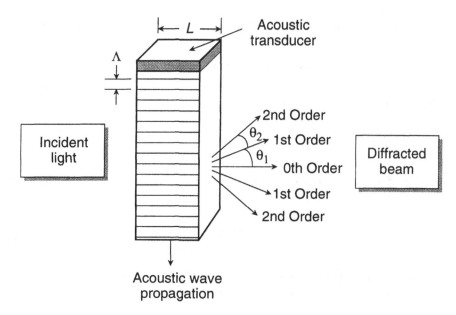

Figure 7.15: A schematic of an acousto-optic device where a coherent light beam is diffracted form the index modulation created by an acoustic wave.

Acousto-optic devices

In Chapter 6 we have seen that strain in a material can alter the dielectric response of a material. Consequently the propagation of an optical beam can be influenced by strain. An important area where this effect is exploited (in addition to electrical impulse based devices discussed in Chapter 6) is the acousto-optic technology. In acousto-optic devices a sound wave propagating in a medium creates local modulation of dielectric constant or refractive index. An optical beam can then be used to "read" this modulator. In Fig. 7.15 we show a schematic of a typical acousto-optic device structure. This device called the Raman–Nath modulator embodies the basic principles which are used for other acousto-optic devices as well.

In the Raman–Nath device shown in Fig. 7.15 an acoustic wave propagates through a thin region and creates a modulation in the refractive index with a wavelength Λ the same as the sound wavelength. The material now acts like a diffraction grating and an optical beam passing through the region suffers diffraction as shown. If Δn is the amplitude of the refractive index modulation, the phase difference arising from the modulation is

$$\Delta \Phi = \frac{\Delta n 2\pi L}{\lambda_0} \sin \frac{2\pi y L}{\Lambda} \tag{7.48}$$

Here L is the interaction distance. If the interaction distance is small the device acts as a diffraction grating, leading to the Raman–Nath modulator. If the interaction distance is large the optical beam suffers multiple refractions from the periodic structure and the device suffers Bragg scattering.

7.5. Summary

According to standard optics of gratings the diffraction orders are given by

$$\sin \theta_m = \frac{m\lambda}{\Lambda}, \quad m = 0, \pm 1, \pm 2 \cdots \quad (7.49)$$

If the interaction distance is small, i.e.

$$L \ll \frac{\Lambda^2}{(\lambda/n)} \quad (7.50)$$

the intensity of the diffracted orders is given by Bessel function (I_0 is the intensity without the acoustic modulation)

$$\frac{I_m}{I_0} = \frac{[J_m(\Delta\Phi_{max})]^2}{2} \quad (7.51)$$

where $\Delta\Phi_{max}$ is the maximum value of $\Delta\Phi$ given by Eq. 7.48. The device is used as a modulator using the $m = 0$ beam. The modulator index is then

$$\frac{I_0 - I_0(m=0)}{I_0} = 1 - [J_0(\Delta\Phi_{max})]^2 \quad (7.52)$$

The Raman–Nath modulation is not able to get a high ON/OFF ratio because of the short interaction length. By increasing the interaction length, the device acts as a Bragg reflector and the optical beam satisfies the Bragg law

$$\sin \theta_B = \frac{\lambda}{2\Lambda} \quad (7.53)$$

In this case the modulator is given by (in the ideal case)

$$\frac{I_0 - I}{I_0} = \sin^2 \frac{\Delta\Phi}{2} \quad (7.54)$$

As we can see the Bragg reflection-based acousto-optic modulators can reach very high ON/OFF ratios.

In addition to modulation of an optical signal, acousto-optic devices are used to carry out spectral analysis of rf signals. For this application an incoming microwave signal is used to launch a surface acoustic wave (SAW) into a piezoelectric material. The diffraction of a laser beam by this SAW is then used to analyze the content of the incoming signal.

7.5 SUMMARY

Issues discussed in this chapter are summarized in Table 7.3.

Topics studied	Key findings
Wave propagation in crystals	Crystals are anisotropic and light propagates with electric field polarized along well defined directions. The velocity of the light polarized along the two directions are different. Phase difference develops between the light polarized in different directions.
Polarization of light in crystals	In uniaxial crystals if light is propagating along the optic axis, the light can be polarized in any direction normal to the axis. However, for a general direction, light can be polarized only along two directions. The two polarization directions are the axes of the ellipse produced by the intersection of the normal plane and the indicatrix ellipse.
Light modulation through polarization control	In general, light traveling in a crystal has different velocities for different polarizations. If the refractive index for one polarization can be altered, the output light intensity can be modulated.
Light modulation through liquid crystals	The polarization direction of light propagating in liquid crystal can be altered by twisting the liquid crystal. By using polarizers, this effect can be used to modulate light.
High speed electro-optic modulators	Solid crystals are used for light modulation or switching. An electric field alters the refractive index of the crystal anisotropically, resulting in an extra phase difference between light polarized along different directions. The phase difference can be exploited (with polarizers) to modulate the light beam.
Advanced modulation/ switching devices	Semiconductor quantum well structures are increasingly being exploited to improve device response. The photon-electron interaction in these structures can be tuned to improve the desired performance.

Table 7.3: Summary table.

7.6 PROBLEMS

7.1 The electric fields of an optical beam are represented by the following

$$E_x = \frac{1}{2} \cos(\omega t - kz)$$
$$E_y = \cos(\omega t - kz + \delta)$$

Sketch the polarization ellipses for the values given by $0, \pi/4, \pi/2,$ and $3\pi/4$.

7.2 An optical beam is traveling along the z-axis, and its fields are given by $E_x = 0.1 \cos(\omega t - kz); E_y = \cos(\omega t - kz + \pi/2)$. Calculate the major and minor axis of the polarization ellipse.

7.3 Sketch a diagram similar to Fig. 7.3 of the text for various δ values when the magnitude of the electric field in the x-direction is twice the magnitude of that in the y-direction.

7.4 A GaAs electro-optic modulator is needed in an optical communication system. The maximum voltage available is 10 V. A device of length no more than 1 mm is needed. Calculate the thickness of the device. How long does it take for an optical signal to pass the device? Use data in Table 7.1 with $n_r = 3.6$ and $\lambda = 1.0$ μm.

7.5 A lithium niobate modulator is to be designed for a 1.06 μm system. The device length is 1 mm and the thickness is 10 μm. Calculate the voltage needed for the modulator.

7.6 Discuss the incompatibility of modern microelectronic devices with bulk electro-optic modulator. Note that the dimensions of most microelectronic devices are ~ 1 μm.

7.7 Consider a liquid crystal cell using the parallel configuration (Fig. 7.10). The splay elastic constant is found to be $K_1 = 2 \times 10^{10}$ N for a wide range of liquid crystals. If the maximum voltage available is 2.0 V, calculate the minimum dielectric anisotropy needed to use the cell in a display application.

7.8 In a twisted orientation liquid crystal cell (Fig. 7.10), what should be the polarization of the incoming light for optimum performance?

7.9 Calculate approximately the maximum switching time acceptable in a liquid crystal cell that is to be used in a 500 × 500 display.

7.7 FURTHER READING

- General

    - J.F. Nye, *Physical Properties of Crystals*, Oxford, Clarendon Press (1957).
    - C.R. Pollock, *Fundamentals of Optoelectronics*, Irwin (1995).
    - P. Yariv and P. Yeh, *Optical Waves in Crystals*, Wiley-Interscience (1984).

Chapter 8

MAGNETIC EFFECTS IN SOLIDS

8.1 INTRODUCTION

Magnets are one of the most fascinating materials. Children use them as toys and their parents use them to fasten images of their cute faces on to refrigerators. However, magnetic materials also find all kinds of important roles in technology. Magnets based on the traditional metals (iron, nickel, and manganese) have been used for transformers and motors. These magnets have high conductivities and, as a result, carry large (unwanted) Eddy currents at high frequencies. With advances in ceramic magnets (containing iron or other magnetic elements) the applications of magnetic materials has greatly expanded. Ceramic magnets have become important in numerous information processing technologies, although their most dominant impact has been on information storage. Also, propagation of electromagnetic waves in magnetic materials allows for a variety of interesting devices that find use in microwave technology. In recent years, traditional semiconductors, such as GaAs, have been doped with Mn to create magnetic semiconductors. There have been suggestions that devices based on electron spin in such materials can lead to a new field of spintronics with applications in high-performance information processing. However, spintronics has not yet made any impact on technology, primarily because of the very low temperatures needed for such devices to operate.

In this chapter we will first examine some basic physics of magnetic materials and the interaction of electrons with magnetic fields. We will then examine how magnetic materials can be exploited for device applications.

8.2 MAGNETIC MATERIALS

The magnetic properties of a material are described through the magnetization M (magnetic moment per volume). There are several classes of magnetic materials determined by how magnetization responds to an applied magnetic field H. The most common and well-known magnetic materials are iron, nickel, and manganese, but almost every material has some response to a magnetic field. The magnetic susceptibility χ, of a material is defined by

$$\chi = \frac{M}{H} = \frac{\mu_0 M}{B} \tag{8.1}$$

Most materials are diamagnetic and have a small ($\sim 10^{-6}$) negative susceptibility. In such materials the electron spins are aligned in such a manner so that the net spin is zero. When a magnetic field is applied the electron orbit is modified to produce magnetization opposite to the applied field. Insulators and many organic compounds are diamagnetic.

In paramagnetic materials the applications of a magnetic field tends to orient the magnetic moments in the material, so that there is a net magnetization in the field direction. The susceptibilities are positive, but small ($\sim 10^{-3}$–10^{-6}). Paramagnetic materials obey Curie's law; i.e., $\chi \propto 1/T$.

Ferromagnetic materials have spontaneous magnetization below a certain temperature (Curie temperature). A magnetic field is needed to align the domains of magnetic field in a ferromagnetic material, since in the as grown materials the magnetization of different domains cancels out. As noted in the introduction the use of ceramic magnetic materials have allowed suppression of the Eddy current, making these materials very useful for high frequency operation.

One of the earliest known ferrites is magnetite (lodestone) Fe_3O_4, a naturally occurring material. The structure can be written as $FeOFe_2O_3$, which can be generalized as $MOFe_2O_3$ where M represents divalent ions, such as Mn^{2+}, Fe^{2+}, Co^{2+}, Ni^{2+}, as well as Cu^{2+} or Zn^{2+}. M could also represent a combination of ions with average valence of 2. Such general configurations are termed spinel ferrites after the mineral $MgOAl_2O_3$ (spinel). Another class of ferrites have the structure of hexaferrite, a model being $BaFe_{12}O_{19}$. Finally we have garnets, a model being $Y_3Fe_5O_{12}$ (YIG). Yttrium iron garnets or YIG is widely used in microwave device applications. The general garnet can be written as $R_3Fe_5O_{12}$, where R can be Y or can be totally or partially replaced by lanthanum, cerium, gallium, etc.

8.3 ELECTROMAGNETIC FIELD MAGNETIC MATERIALS

In Chapters 5 and 6 we have considered the propagation of electromagnetic waves in non-magnetic materials. The propagation (intensity, polarization, speed, absorption, or gain) are described via the complex dielectric response. Similar concepts can be developed for wave propagation in magnetic materials and the response is described via the complex permeability, μ.

The appropriate equations for the magnetic induction (or magnetic flux density), **B**, and magnetic field, **H**, are

$$\mathbf{B} = \mu \mathbf{H} = \mu_0 \mathbf{H} + \mathbf{M} \tag{8.2}$$

where μ_0 is the free space permeability and M is the magnetization (compare this with $\mathbf{D} = \epsilon_0 \mathbf{E} + \mathbf{P}$). As noted above, the magnetic susceptibility (χ) is defined to describe

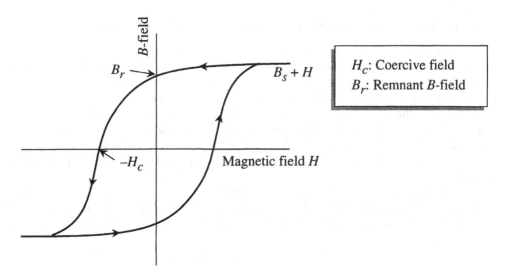

Figure 8.1: A typical hysterisis curve for the B-H relation in a magnetic field. The coercive field is H_c. At this field the magnetic induction B is brought back to zero. At $H = 0$, the remnant B field is B_r.

the relation between magnetization and magnetic field

$$M = \chi H \tag{8.3}$$

This gives

$$\mu = \mu_0(1+\chi) \tag{8.4}$$

In most materials the permeability is very close to unity and typically differs from unity by a few parts per 10^5. Materials where $\mu > 1$ are called paramangetic and where $\mu < 1$ are called diamangetic. For ferromagnetic materials, the value of μ can be very large and, in fact, we have to replace the linear relation by a function

$$B = F(H) \tag{8.5}$$

As discussed for the ferroelectric materials in Chapter 6, there is a hysterisis curve that represents the B–H relationship in ferromagnetic materials. Permeability defined by the slope of the curve can reach a million in ferromagnetic materials. In Fig. 8.1 we show a typical hysterisis curve for ferromagnets.

The use of magnetic materials for devices is based on several phenomena:

- Propagation of electromagnetic waves in magnetic material. As in materials with anisotropic dielectric response, waves with different polarization propagate differently and this can be exploited to make magneto-optic devices.

- Use of external current/magnetic field to alter the magnetization of a material to make switches, memories, etc.

8.3. Electromagnetic field magnetic materials

- Use of high permeability materials to make transformers, antennas, etc.

We will briefly review how electromagnetic waves propagate in magnetic materials. We will also discuss the physics behind magnetic properties and the underlying reasons for diamagnetic, paramagnetic, and ferromagnetic materials.

An important and useful phenomena that occurs when electromagnetic waves propagate in a medium with magnetization and a dc magnetic field (as in the ionosphere and in magnetic materials) is the difference in propagation vectors for different polarizations. This effect is exploited for microwave device (phase shifters) applications. To understand this phenomena we will develop a simple model for the propagation of em waves in a medium with a dc magnetic field and certain electron density.

We assume that an electromagnetic wave is propagating along the dc magnetic field B_0 direction (z-direction). The ac magnetic field will be assumed to be small compared to B_0. The equation of motion for the electrons in the medium is (electron charge is $-e$)

$$m\frac{d^2x}{dt^2} - eB_0 \times \frac{dx}{dt} = -eEe^{-i\omega t} \tag{8.6}$$

The transverse electric field may be written as circularly polarized waves

$$\mathbf{E} = (\mathbf{a}_1 \pm i\mathbf{a}_2)E \tag{8.7}$$

where $\mathbf{a}_1, \mathbf{a}_2$ are unit vectors. Assuming a similar time-dependent expression may be written for the electron displacement, the steady state solution for the electronic coordinate is

$$\mathbf{x} = \frac{e}{m\omega(\omega \mp \omega_B)}\mathbf{E} \tag{8.8}$$

where ω_B is the frequency of precession of electrons in a B-field

$$\omega_B = \frac{eB_0}{m} \tag{8.9}$$

The oscillating charge creates a dipole moment, which in turn influences the dielectric response of the medium. The dielectric response (form our discussion in Chapter 7) is

$$\epsilon = 1 - \frac{\omega_p^2}{\omega(\omega \mp \omega_B)} \tag{8.10}$$

where

$$\omega_p^2 = \frac{Ne^2}{\epsilon_0 m} \tag{8.11}$$

where N is the carrier density.

The upper sign corresponds to left-handed circular polarization and the lower sign corresponds to right-handed circular polarization. We see that the dielectric response and, hence, the propagation wavevector is different, depending upon the polarization.

For propagation of a wave in a direction other than that given by the dc magnetic field, the frequency ω_B is given by the component of B_0 in the propagation direction, making the medium not only binefringent, but also anisotropic.

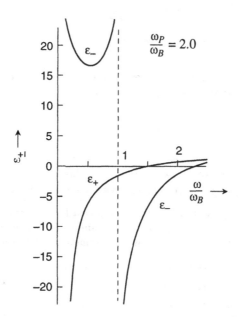

Figure 8.2: The dielectric response for waves propagating in a medium with dc magnetic field. ϵ_+ and ϵ_- are the responses for right- and left-circular polarized waves. The ratio of ω_p/ω_B is 2 for the plots.

In Fig. 8.2 we show a typical plot for ϵ_+ and ϵ_- as a function of frequency for a ω_p/ω_B value of 2.0. The k-vector of the traveling depends upon the dielectric constant and, as we can see, can have very different values for different polarizations.

The simple theory given above can be extended to em wave propagation in magnetic materials with dc mangetization and dc H-field. We will summarize the results for isotropic ferrites, which are materials used for microwave applications. The material is described on internal magnetization M_0 and internal field H_0. We define the following frequencies

$$\omega_m = g\frac{eM_0}{2m} = \gamma M_0 \qquad (8.12)$$

$$\omega_0 = g\frac{eH_0}{2m} = \gamma H_0 \qquad (8.13)$$

where g is called the spectroscopic splitting factor or the Lande g factor. It has a value of 2 for free electrons. The parameter γ is called the gyromagnetic ratio. If an em wave of frequency ω propagates in the material, the propagation constants for right-circular and left-circular polarized waves is given by

$$\beta_+ = \omega\sqrt{\epsilon\mu_0\left(1 + \frac{\omega_m}{\omega_0 - \omega}\right)} \qquad (8.14)$$

8.4. Physical basis for magnetic properties

$$\beta_- = \omega\sqrt{\epsilon\mu_0\left(1+\frac{\omega_m}{\omega_0+\omega}\right)} \qquad (8.15)$$

If right- and left-circular polarized waves propagate a distance z through the material, the Faraday notation for the waves is defined as

$$\theta = \frac{1}{2}(\beta_+ - \beta_-)z \qquad (8.16)$$

The results above are for the case where the wave propagates in the direction of the dc magnetic field. The change in phase for the two polariztions is the basis for devices used in microwave technology. We will discuss these devices in Section 8.8.3.

EXAMPLE 8.1 An isotropic ferrite is used to make a Faraday rotator. Calculate the rotated angle when an electromagnetic wave passes over 1 mm of the device with propagation direction parallel to the dc magnetic field. The following parameters define the problem:

Operating frequency	$f = 10$ Ghz
Gyromagnetic ratio	$\gamma = 2.2 \times 10^5$ rad/s/A/m
Relative dielectric	$\epsilon_r = 10$
Dc Magnetic field	$H_0 = 1000$ A/m
Dc Magnetization	$M_0 = 10^5$ A/m

We have

$$\omega_m = \gamma M_0 = 2.21 \times 10^9 \text{ rad/s}$$
$$\omega_0 = \gamma H_0 = 0.221 \times 10^9 \text{ rad/s}$$
$$1 + \frac{\omega_m}{\omega_0 - \omega} = 0.96$$
$$1 + \frac{\omega_m}{\omega_0 + \omega} = 1.04$$
$$\beta_+ = 649.3 \text{ rad/m}$$
$$\beta_- = 675.4 \text{ rad/m}$$

The rotated angle is

$$\theta = \frac{1}{2}(\beta_+ - \beta_-)z = -0.131 \text{ rad} = 7.5 \text{ degrees}$$

8.4 PHYSICAL BASIS FOR MAGNETIC PROPERTIES

To understand the magnetic properties of materials we first need to examine how a magnetic field influences electronic properties. In classical physics we know that a charged particle interacts with a magnetic field via a force given by the Lorentz equation. This interaction leads to a number of phenomena, which are adequately described by classical physics. The Lenz law is one example, where the current induced by a magnetic field is accurately described by classical physics.

The classical description can be extended to the quantum description by the usual procedure where energy is treated as an operator. In general, the hamiltonian for an electron in the absence of a magnetic field is

$$H_0 = \frac{p^2}{2m} + U(r) = -\frac{\hbar^2}{2m}\nabla^2 + U(\mathbf{r}) \tag{8.17}$$

In the presence of a magnetic field we know that a charged particle feels the Lorentz force ($q\mathbf{v} \times \mathbf{B}$). The energy function describing the particle energy can be shown to be given by ($q = -e$ for electrons)

$$\begin{aligned} H &= \frac{1}{2m}(\mathbf{p} - q\mathbf{A})^2 + U(\mathbf{r}) \\ &= \frac{1}{2m}\left(\frac{\hbar}{i}\nabla - q\mathbf{A}\right)^2 + U(\mathbf{r}) \end{aligned} \tag{8.18}$$

Here $\mathbf{A}$ is the vector potential characterizing the magnetic field. The hamiltonian adequately describes the spatial motion of an electron in a magnetic field. However, an important effect is still missing and is related to the spin of the electron.

The spin of the electron (or other charged elementary particles) interacts with a magnetic field to contribute an important term to the hamiltonian. To motivate this term, we assign a magnetic moment to the spin part of the electron angular momentum. From symmetry, this moment is parallel or antiparallel to the spin. We write it as

$$\mu_s = -g\mu_B \mathbf{S} = \gamma\hbar\mathbf{S} \tag{8.19}$$

where $\mathbf{S}$ is the spin of the particle; g is known as the g-factor and characterizes the particle. The constant μ_B is known as the Bohr magneton and has a value

$$\mu_B = \frac{e\hbar}{2m} \tag{8.20}$$

or, in cgs units

$$\mu_B = \frac{e\hbar}{2mc}$$

The constant γ is called the gyromagnetic or magnetogyric ratio. The magnetic moment associated with the spin then gives the usual term in the hamiltonian

$$H_{\text{spin}} = -\mu_s \cdot \mathbf{B} \tag{8.21}$$

This term is now added to the hamiltonian to give the full hamiltonian describing a charged particle in a magnetic field

$$H = \frac{1}{2m}(\mathbf{p} - q\mathbf{A})^2 + U(\mathbf{r}) - \mu_s \cdot \mathbf{B} \tag{8.22}$$

The approach used to solve the Schrödinger equation resulting from this hamiltonian depends on the form of the background potential, $U(\mathbf{r})$. The spatial part of the problem can be solved exactly for the free electron problem ($U = 0$) or for electrons in

8.4. Physical basis for magnetic properties

a periodic potential where the effective mass approach can be used. The effective mass theory allows us to describe the effect of the background potential by including it via an equivalent hamiltonian

$$-\frac{\hbar^2}{2m}\nabla^2 + U(\mathbf{r}) \Longrightarrow \frac{-\hbar^2}{2m^*}\nabla^2 \tag{8.23}$$

where m^* is the effective mass of the electron in the band of interest. Thus, in the free electron case, and in the case of electrons in crystals where the effective mass theory is adequate, we can write (m is the effective mass for crystalline materials)

$$H = \frac{1}{2m}(\mathbf{p} - q\mathbf{A})^2 - \boldsymbol{\mu}_s \cdot \mathbf{B} \tag{8.24}$$

If we ignore the spin effects, the Schrödinger equation becomes

$$\frac{1}{2m}\left(\frac{\hbar}{i}\nabla - q\mathbf{A}\right)^2 \psi = E\psi \tag{8.25}$$

This problem can be solved exactly, as will be seen in the next section.

For the general background potential, the problem cannot be solved exactly. The problems of special interest are electrons in atoms and molecules where the electronic states are bound. The behavior of such electrons in a magnetic field leads to very important physical phenomena, as indicated in Fig. 8.3. In the previous section we have mentioned diamagnetic, paramagnetic, and ferromagnetic materials. These effects can be understood on the basis of quantum mechanics. Additionally, as shown in Fig. 8.3, electrons in semiconductors (free electrons) show interesting physical properties.

To address the general problem in a constant magnetic field we write (remember $\mathbf{B} = \nabla \times \mathbf{A}$)

$$\mathbf{A} = \frac{1}{2}\mathbf{B} \times \mathbf{r} \tag{8.26}$$

Writing the hamiltonian in the absence of the field as H_0, we get, for an electronic system in a magnetic field

$$\begin{aligned} H &= H_0 + \frac{e}{m}\mathbf{A}\cdot\mathbf{p} + \frac{e^2}{2m}A^2 - \boldsymbol{\mu}_s\cdot\mathbf{B} \\ &= H_0 + \frac{e}{2m}\mathbf{B}\cdot(\mathbf{r}\times\mathbf{p}) + \frac{e^2(\mathbf{B}\times\mathbf{r})^2}{8m} - \boldsymbol{\mu}_s\cdot\mathbf{B} \end{aligned} \tag{8.27}$$

Since $\mathbf{r}\times\mathbf{p}$ is the orbital angular momentum operator $\mathbf{L}$, we have

$$H = H_0 - \frac{e}{2m}\mathbf{L}\cdot\mathbf{B} + g\mu_B\mathbf{S}\cdot\mathbf{B} + \frac{e^2}{8m}(\mathbf{B}\times\mathbf{r})^2 \tag{8.28}$$

To proceed, we need to know the value of g which defines the relation between the spin and the magnetic moment. In analogy to the second term in the hamiltonian which comes from the spatial part of the electron angular momentum, it would appear that $g = 1$. However, this is not the case. It can be shown from Dirac's relativistic formalism

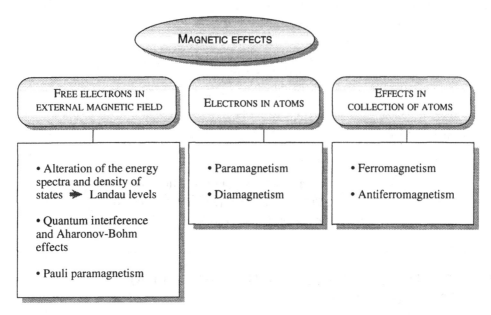

Figure 8.3: An overview of magnetic effects in physical systems studied in this chapter.

of the electron that $g = 2.0023$ (taken as 2.0 for most cases) for the electron spin. This gives us the final hamiltonian

$$H = H_0 - \frac{\mu_B}{\hbar}(\mathbf{L} + 2\mathbf{S}) \cdot \mathbf{B} + \frac{e^2}{8m}(\mathbf{B} \times \mathbf{r})^2 \qquad (8.29)$$

Before starting to solve specific problems of interest, we remind the reader that the response of a system to an external magnetic field is represented via the magnetic susceptibility χ. As noted earlier, $\mathbf{M}$ is the magnetization of a system (magnetic moment per unit volume); the magnetic susceptibility per unit volume is defined by

$$\chi = \frac{\mathbf{M}}{\mathbf{H}} = \frac{\mu_0 \mathbf{M}}{\mathbf{B}}$$

where μ_0 is the free space permeability ($4\pi \times 10^{-7}$ H/m). In cgs units we have

$$\chi = \frac{\mathbf{M}}{\mathbf{B}}$$

As noted in Section 8.2, systems where χ is positive are known as paramagnetic, while those where it is negative are known as diamagnetic. In ferromagnetic materials the magnetization can be finite, even in the absence of a magnetic field. As will be seen, the second term in the hamiltonian given by Eq. 8.28 leads to paramagnetic effects and the Zeeman effect. The third term leads to diamagnetic effects.

EXAMPLE 8.2 Calculate the value of the Bohr magneton.

In SI units the Bohr magneton is

$$\mu_B = \frac{e\hbar}{2m_e}$$
$$= \frac{(1.602 \times 10^{-19} \text{ C})(1.055 \times 10^{-34} \text{ J s})}{2(9.1096 \times 10^{-31} \text{ kg})}$$
$$= 9.274 \times 10^{-24} \text{ JT}^{-1}$$

In cgs units we have

$$\mu_B = \frac{e\hbar}{2mc}$$
$$= \frac{(4.803 \times 10^{-10} \text{ esu})(1.055 \times 10^{-27} \text{ erg/s})}{2(9.1096 \times 10^{-28} \text{ g})(2.9979 \times 10^{10} \text{ cm/s})}$$
$$= 9.277 \times 10^{-21} \text{ erg/gauss}$$

8.5 COHERENT TRANSPORT: QUANTUM INTERFERENCE

We will start our discussion of magnetic phenomena by examining electrons propagating in a solid without scattering. These electrons are described by plane waves, as discussed in Chapter 3, and propagate without scattering in high-quality materials at low temperatures.

Recently there has been increasing interest in devices based upon *quantum interference* of electrons. Much as optical diffraction experiments, these devices use electron wave interference to create constructive or destructive interference at a certain point in space. If the interference pattern can be controlled by an external stimulus, one can design a device that can switch between a conducting and non-conducting state. Such devices can be made from semiconductor or metallic materials, unlike traditional devices that can only be made from semiconductors. They can also be made from superconducting materials. An interesting way to control the interference pattern of electrons is via a magnetic field. To understand this, let us consider the effect of the magnetic field on the wavefunction of an electron.

8.5.1 Aharonov Bohm effect

We consider the Schrödinger equation for an electron in the presence of an electromagnetic potential described by the vector potential $\mathbf{A}$ and scalar potential ϕ

$$\frac{1}{2m}(-i\hbar\nabla - e\mathbf{A})^2 \psi + V\psi = E\psi \qquad (8.30)$$

where $V = e\phi$. We assume that $\mathbf{A}$ and ϕ are time-independent. In a region where the magnetic field is zero, it can be shown that the solution of the problem has the form

$$\psi(x) = \psi^0(x) \exp\left[\frac{ie}{\hbar}\int^{S(x)} \mathbf{A}(x') \cdot d\mathbf{s}'\right] \qquad (8.31)$$

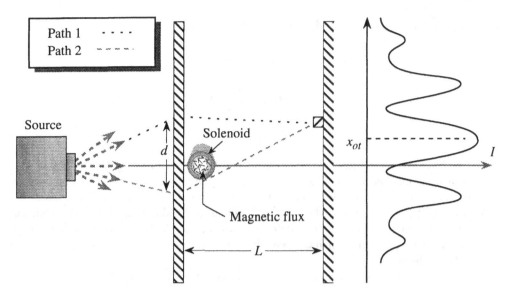

Figure 8.4: A magnetic field can influence the motion of electrons even though it exists only in regions where there is an arbitrarily small probability of finding the electrons. The interference pattern of the electrons can be shifted by altering the magnetic field.

where $\psi^0(x)$ satisfies the Schrödinger equation with the same value of ϕ but with $\mathbf{A}(x) = 0$. The line integral in Eq. 8.31 can be along any path as long as the end point $S(x)$ is the point x and $\nabla \times \mathbf{A}$ is zero along the integral. Notice that this is essentially equivalent to making the change

$$\mathbf{k} \to \mathbf{k} - \frac{e}{\hbar}\mathbf{A} \tag{8.32}$$

in the usual free electron wavefunction $\exp(i\mathbf{k}\cdot\mathbf{r})$.

An interesting effect occurs if electrons can travel through a material without suffering scattering so that the phase relation given above is maintained. Consider the case shown schematically in Fig. 8.4. Here a beam of coherent electrons is separated into two parts and made to recombine at an interference region. This is similar to the double-slit experiment from optics, except now we have a region of magnetic field enclosed by the electron paths as shown. The wavefunction of the electrons at the point where the two beams interfere is given by (we assume that phase coherence is maintained)

$$\psi(x) = \psi_1^0 \exp\left[\frac{ie}{\hbar}\int_{\text{path 1}}^{S(x)} \mathbf{A}(x')\cdot d\mathbf{s}'\right] + \psi_2^0 \exp\left[\frac{ie}{\hbar}\int_{\text{path 2}}^{S(x)} \mathbf{A}(x')\cdot d\mathbf{s}'\right] \tag{8.33}$$

The intensity or the electron density is given by

$$I(x) = \{\psi_1(x) + \psi_2(x)\}\{\psi_1(x) + \psi_2(x)\}^* \tag{8.34}$$

If we assume that $\psi_1^0 = \psi_2^0$, i.e. the initial electron beam has been divided equally along

8.5. Coherent transport: quantum interference

the two paths, the intensity produced after interference is

$$I(x) \propto \cos\left[\frac{e}{\hbar} \oint \mathbf{A} \cdot d\mathbf{s}\right]$$
$$= \cos\left[\frac{e}{\hbar} \int_{\text{area}} \mathbf{B} \cdot \mathbf{n}\, da\right]$$
$$= \cos\frac{e\Phi}{\hbar} \qquad (8.35)$$

where we have converted the line integral over the path enclosed by the electrons to a surface integral and used $\mathbf{B} = \nabla \times \mathbf{A}$. The quantity Φ is the magnetic flux enclosed by the two electron paths. It is interesting to note that, even though the electrons never pass through the $\mathbf{B} \neq 0$ region, they are still influenced by the magnetic field. From Eq. 8.35 it is clear that, if the magnetic field is changed, the electron density will undergo modulation. This phenomenon has been observed in semiconductor structures as well as metallic structures.

In cgs units, the equation for current interference takes the form

$$I(x) \propto \cos\frac{e\Phi}{\hbar c}$$

For completeness, it is illustrative to examine the implications of our results in a superconductor. In superconductors, the electrons find it energetically favorable to form pairs mediated by the electron–lattice interactions. These pairs, called *Cooper pairs*, do not suffer collisions because of the existence of an energy gap between their energy and the energies of state where they could scatter into. We use $2e$ instead of e to describe the wavefunction of the Cooper pairs

$$\psi(x) = \psi^0 \exp\left[\frac{2ie}{\hbar} \int^{S(x)} \mathbf{A}(x') \cdot d\mathbf{s}'\right] \qquad (8.36)$$

If we consider a superconducting ring as shown in Fig. 8.5 enclosing a magnetic field region, the fact that the electron wavefunction should not be multi-valued if we go around the ring gives us the condition

$$\frac{2e}{\hbar} \oint \mathbf{A} \cdot d\mathbf{s} = 2n\pi \qquad (8.37)$$

or

$$\frac{2e\Phi}{\hbar} = 2n\pi \qquad (8.38)$$

The flux enclosed by the superconducting ring is thus quantized

$$\Phi = \frac{n\pi\hbar}{e} \qquad (8.39)$$

This effect was used to confirm that the current in superconductors is carried by a pair of electrons rather than individual electrons.

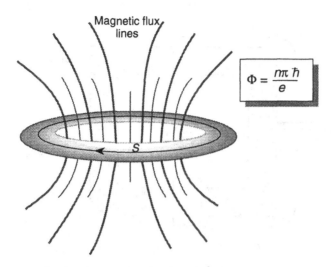

Figure 8.5: Flux quantization in superconducting rings.

EXAMPLE 8.3 A key attraction of the Aharonov–Bohm effect is the possibility of its use in switching devices, where a change in magnetic field can change the state of a device from 0 to 1. The magnetic field (and, hence, switching energy) needed for switching the device would be very small. Consider a device of area 20 μm × 20 μm through which a magnetic field is varied. Calculate the field variation needed to switch the devices.

The flux "quanta" needed to switch from a constructive to destructive electron interference is

$$\Delta \Phi = \frac{\pi \hbar}{e}$$

The magnetic field change needed for our device is

$$\begin{aligned}
\Delta B &= \frac{\pi \hbar}{eA} \\
&= \frac{\pi (1.05 \times 10^{-34} \text{ J s})}{(1.6 \times 10^{-19} \text{ C})(20 \times 10^{-6} \text{ m})^2} \\
&= 5.15 \times 10^{-6} \text{ T}
\end{aligned}$$

Considering that the earth's magnetic field is about eight times larger (40 μT), we can see the small field changes needed. However, from a practical standpoint, it is difficult to incorporate magnetic fields into a circuit element. Unlike electric field or voltage, there is no simple source of magnetic field.

8.5.2 Quantum interference in superconducting materials

The observation in this section that the phase difference around a superconducting ring (or a loop) encompasses a magnetic flux that is an integral product of $2e\phi/\hbar$ is used for important devices.

Let us consider a superconducting loop, as shown in Fig. 8.6. In the absence of any voltage, the phase difference between points 1 and 2, taken through the junction

8.5. Coherent transport: quantum interference

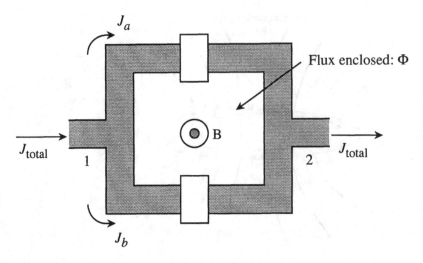

Figure 8.6: Quantum interference in a superconducting loop.

with insulator a is δ_a and that through the other junction, is δ_b. For identical junctions these are equal ($\delta_a = \delta_b = \delta_0$). Let us introduce a flux Φ through the interior of the loop. We now have

$$\delta_b - \delta_a = \frac{2e\Phi}{\hbar} \qquad (8.40)$$

We may write this equation as

$$\delta_a = \delta_0 - \frac{e\Phi}{\hbar}$$
$$\delta_b = \delta_0 + \frac{e\Phi}{\hbar} \qquad (8.41)$$

The total current through the loop is now

$$\begin{aligned} J_{total} &= J_a + J_b \\ &= J_0 \left\{ \sin\left(\delta_0 + \frac{e\Phi}{\hbar}\right) + \sin\left(\delta_0 - \frac{e\Phi}{\hbar}\right) \right\} \\ &= 2J_0 \sin\delta_0 \cos\frac{e\Phi}{\hbar} \end{aligned} \qquad (8.42)$$

The current can be seen to vary with Φ and has a maxima when

$$\frac{e\Phi}{\hbar} = n\pi \qquad (8.43)$$

where n is an integer.

The control of the current through a superconducting loop (called a Josephson loop) by a magnetic field is the basis of many important superconducting devices.

It is important to note that the magnetic flux needed to alter the current through a loop is very small.

Superconducting loops can be used in the fabrication of magnetometers, digital logic devices, signal processing devices, detectors and for measurement standards. We will briefly discuss applications in magnetometers and digital logic devices later in this chapter.

8.6 DIAMAGNETIC AND PARAMAGNETIC EFFECTS

We will now start developing the basic theory behind magnetic response in solids. When an electronic system is subjected to a magnetic field, in general, a magnetic moment is induced in the system. The response of the electronic system to the magnetic field can be described via the magnetic susceptibility χ. In this section we will examine the susceptibility for several different systems. Depending upon the nature of the problem, the magnetic susceptibility can be negative (such systems are diamagnetic) or positive (these systems are paramagnetic).

In Section 8.4 we discussed the general hamiltonian of an electronic system in a magnetic field. If we examine the right-hand side of Eq. 8.29 the third term (second order in field B) leads to diamagnetic effects in atoms. The effect is calculated by using perturbation theory. The second term in the hamiltonian leads to paramagnetic effects. Let us start with a simple calculation for the diamagnetic effect.

8.6.1 Diamagnetic effect

Some materials have diamagnetic properties arising from how electron atomic orbits are influenced by an applied magnetic field. Let us consider an electron atomic state subject to a uniform magnetic field in the z-direction. We can write, for the components of the vector potential

$$A_x = -\frac{1}{2}yB, \quad A_y = \frac{1}{2}xB, \quad A_z = 0 \qquad (8.44)$$

The perturbation (i.e., the third term of Eq. 8.29) now becomes

$$H' = \frac{e^2 B^2}{8m}(x^2 + y^2) \qquad (8.45)$$

First-order perturbation theory then gives us the energy shift due to the perturbation as

$$E' = \frac{e^2 B^2}{8m}\langle x^2 + y^2 \rangle \qquad (8.46)$$

For a spherically symmetric system, where $\langle x^2 \rangle = \langle y^2 \rangle = \langle z^2 \rangle = \langle r^2 \rangle/3$, we have

$$E' = \frac{e^2 B^2}{12m}\langle r^2 \rangle \qquad (8.47)$$

The expectation value $\langle r^2 \rangle$ can be calculated if the wavefunction for the unperturbed electronic system is known.

8.6. Diamagnetic and paramagnetic effects

The magnetic moment is now given by

$$\mu = -\frac{\delta E'}{\delta B} = -\frac{e^2 \langle r^2 \rangle}{6m} B \tag{8.48}$$

In general, if Z electrons are present in the atom, the magnetic moment contribution has to be the sum of the contributions from all of these electrons. In a simplistic model, we can multiply the result above by Z, and $\langle r^2 \rangle$ then represents the average expectation value for all electrons.

From Eq. 8.50, we see that the magnetic susceptibility is diamagnetic in nature. The susceptibility is for N atoms per unit volume

$$\chi = \frac{\mu_0 M}{B} = -\frac{\mu_0 N Z e^2}{6m} \langle r^2 \rangle \tag{8.49}$$

Since in most data reviews the values of χ are given in cgs units, we note the value in cgs units

$$\chi = -\frac{N Z e^2}{6mc^2} \langle r^2 \rangle \tag{8.50}$$

Magnetic susceptibility is usually given for a mole of a material, in which case it is called the *molar susceptibility*, χ_M.

EXAMPLE 8.4 The molar susceptibility of He is (in cgs units) 1.9×10^{-6} cm^3/mole. Calculate the expectation value $\langle r^2 \rangle$.

Using Eq. 8.49, we have ($Z = 2$)

$$\langle r^2 \rangle = \frac{(1.9 \times 10^{-6} \text{ cm}^3/\text{mole}) \, 6 \, (9.1 \times 10^{-28} \text{ gm}) \, (3 \times 10^{10} \text{ cm/s})^2}{(6.022 \times 10^{23}) \, 2 \, (4.8 \times 10^{-10} \text{ esu})^2}$$

$$= 3.36 \times 10^{-17} \text{ cm}^2$$

If we were to take the square root of this value, it gives us a typical electron–nucleus distance of 0.58 Å.

8.6.2 Paramagnetic effect

As noted in the previous subsection, if electrons form close orbits, as in atomic orbitals, a magnetic field induces a diamagnetic material. However, in some materials additional effects lead to the paramagnetic effect. The second term in the full hamiltonian describing electrons in a magnetic field given by Eq. 8.29 leads to paramagnetic effects in materials. The perturbation is

$$H' = \frac{\mu_B}{\hbar} (\mathbf{L} + 2\mathbf{S}) \cdot \mathbf{B} \tag{8.51}$$

Electrons in atomic states are described by their orbital angular momentum $\mathbf{L}$ and spin $\mathbf{S}$. The total angular momentum is $\mathbf{J} (= \mathbf{L} + \mathbf{S})$. According to quantum mechanisms the orbital angular momentum $\mathbf{L}$ can take values (magnitude) $\ell = 0, \hbar, 2\hbar, 3\hbar$, etc., with a projection (in units of $\hbar$), $m = -\ell, -\ell+1 \cdots, 0, \cdots, \ell-1, \ell$. The total angular momentum can take values (magnitude) $j = 0, \hbar \cdots (\ell+s)\hbar$ with a projection going from $-j, -j+1 \cdots 0, \cdots j-1, j$. In Fig. 8.7 the choices in angular momentum are summarized. In the absence of a magnetic field an electronic level with a particular value of **J** has $(2j+1)$ degeneracy. This degeneracy is lifted by a magnetic field as discussed below and as shown in Fig. 8.7b. According to first-order perturbation theory, the effect of this additional term is evaluated by taking its expectation value in the original unperturbed states. The connection leads to the Zeeman effect and is given by

$$\Delta E = \mu_B g m B \tag{8.52}$$

where

$$g = 1 + \frac{j(j+1) + s(s+1) - l(l+1)}{2j(j+1)} \tag{8.53}$$

The quantum number m has values $j, j-1 \ldots, -j$, and as a result the level that may be originally $2j+1$ degenerate looses its degeneracy, as seen in Fig. 8.7b. *When electrons are now distributed in these levels, according to equilibrium statistics, there is a net magnetic moment in the material which is paramagnetic in nature.* Let us examine how this occurs for the case where $j = s = 1/2$.

In the case where $\ell = 0$, we have

$$g = 2; \quad m = \pm \frac{1}{2} \tag{8.54}$$

This leads to splitting, shown in Fig. 8.8a. The unperturbed level that is doubly degenerate (corresponding to the two spin states) is split. The lower energy state has its spin antiparallel to the field, while the magnetic moment is parallel to the field, as shown in Fig. 8.8b.

From thermodynamics, the relative equilibrium populations of the two levels is, at equilibrium

$$\frac{N_1}{N} = \frac{\exp(\mu_B B/k_B T)}{\exp(\mu_B B/k_B T) + \exp(-\mu_B B/k_B T)}$$
$$\frac{N_2}{N} = \frac{\exp(-\mu_B B/k_B T)}{\exp(\mu_B B/k_B T) + \exp(-\mu_B B/k_B T)} \tag{8.55}$$

where N_1 and N_2 are the populations of the lower energy and upper energy states and N is the total population. The populations are plotted in Fig. 8.8b as a function of the field. We see that, depending upon the temperature, there is a greater number of electrons with magnetic moment along the field. The net magnetization is given by (writing $x = \mu_B B/k_B T$)

$$M = (N_1 - N_2)\mu_B = N\mu_B \frac{e^x - e^{-x}}{e^x + e^{-x}} = N\mu_B \tanh x \tag{8.56}$$

If

$$\frac{\mu_B B}{k_B T} \ll 1 \tag{8.57}$$

8.6. Diamagnetic and paramagnetic effects

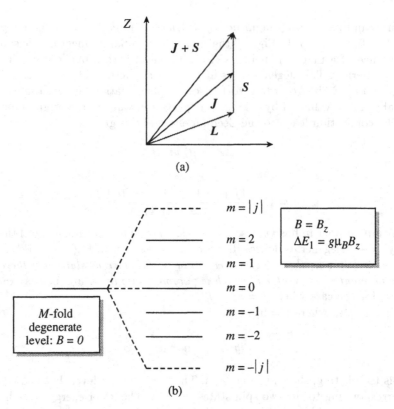

Figure 8.7: (a) A schematic of the vector addition of the total angular momentum J, which is the sum of the orbital angular momentum and spin. (b) Splitting of degenerate states in a magnetic field.

we have
$$M \cong \frac{N\mu_B^2 B}{k_B T} \tag{8.58}$$

This is the Curie law for paramagnetism applied to the case where $\ell = 0$. For a general case with $\ell \neq 0$, it can be shown by similar arguments that the magnetization is given by

$$M = Ngj\mu_B B_j(x); \quad x = \frac{gj\mu_B B}{k_B T} \tag{8.59}$$

where the function B_j, called the *Brillouin function*, is given by

$$B_j(x) = \frac{2j+1}{2j} \operatorname{ctnh}\left(\frac{(2j+1)x}{2j}\right) - \frac{1}{2j} \operatorname{ctnh}\left(\frac{x}{2j}\right) \tag{8.60}$$

For $x \ll 1$, we have
$$\operatorname{ctnh} x = \frac{1}{x} + \frac{x}{3} - \frac{x^3}{45} + \cdots \tag{8.61}$$

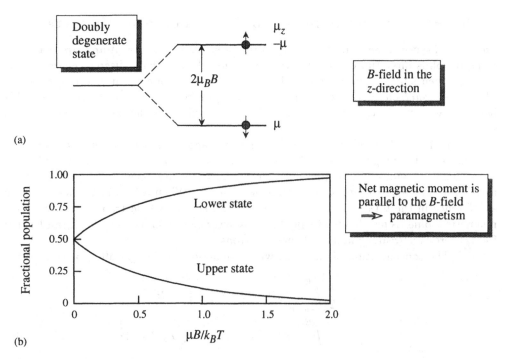

Figure 8.8: (a) The splitting of electronic levels with $L = 0$. The magnetic field is along the z-direction. Note that, for electrons, the magnetic moment is opposite in sign to the spin. In the lower energy state, the magnetic moment is parallel to the field. (b) Occupation of the two-level system as a function of the ratio of the field to the thermal energy.

8.6. Diamagnetic and paramagnetic effects

and the paramagnetic susceptibility is

$$\chi = \frac{\mu_0 M}{B} \cong \frac{\mu_0 N j(j+1) g^2 \mu_B^2}{3 k_B T} = \frac{C}{T} \tag{8.62}$$

where the *Curie constant* C is given by

$$C = \frac{\mu_0 N g^2 \mu_B^2 j(j+1)}{3 k_B} \tag{8.63}$$

8.6.3 Paramagnetism in the conduction electrons in metals

From the discussion on paramagnetism in the previous subsection, it appears that electrons in solids obey the Curie law. This happens when we have discrete energy levels that are then split by the magnetic field. However, it is found that *free electrons display paramagnetism which has essentially no temperature dependence*. A theory for this behavior was put forth first by Pauli, who explained it on the basis of Fermi–Dirac statistics.

In Fig. 8.9, we show the density of states and occupation of electron levels in the absence of a magnetic field. The Fermi level is at E_F, as shown. In the absence of the field, the spin-up and spin-down electronic states are degenerate and, therefore, there is no net magnetic moment in the electron gas. When a field is applied, there is a difference in the total energy of the spin-up and spin-down electrons. Electrons with magnetic moments parallel to the field have a lower energy and, since the Fermi level is fixed, there is a greater number of such electrons.

The concentration of electrons with magnetic moments parallel to the field is

$$N_+ = \frac{1}{2} \int_{-\mu_B B}^{E_F} dE\, f(E)\, N(E + \mu_B B)$$

$$\cong \frac{1}{2} \int_0^{E_F} dE\, f(E)\, N(E) + \frac{1}{2} \mu_B B N(E_F) \tag{8.64}$$

Here $f(E)$ is the Fermi–Dirac distribution function and $N(E)$ is the density of states.

The concentration of electrons with magnetic moment antiparallel to the field is

$$N_- = \frac{1}{2} \int_{-\mu_B B}^{E_F} dE\, f(E)\, N(E - \mu_B B)$$

$$\cong \frac{1}{2} \int_0^{B} dE\, f(E)\, N(E) - \frac{1}{2} \mu_B B N(E_F) \tag{8.65}$$

The net magnetization is now

$$\begin{aligned} M &= \mu_B (N_+ - N_-) \\ &= \mu_B^2 N(E_F) B \end{aligned} \tag{8.66}$$

For the conduction electrons in metals (the band is degenerate; i.e., the occupation factor is approximately unity up to Fermi energy and approximately zero above), we have

$$N(E_F) = \frac{3N}{2E_F} = \frac{3N}{2k_B T_F} \tag{8.67}$$

where N is the density of the conduction electrons and T_F defines the Fermi temperature. This gives us the temperature-independent Pauli paramagnetism

$$\chi = \frac{\mu_0 M}{B} = \frac{3\mu_0 N \mu_B^2}{2k_B T_F} \tag{8.68}$$

8.7 FERROMAGNETIC EFFECTS

In most solids, in the absence of an applied magnetic field, the net spin of electrons in any direction cancels, so that the net magnetization of the system is zero. As we have seen in the previous section, application of an external magnetic field orients the spin, leading to magnetization. In some materials the arrangement of electrons is such that it is possible to have a net magnetization in the absence of an external magnetic field. Materials containing Fe, Ni, and Mn have such magnetic properties. With advances in fabrication techniques, magnetic ceramics and magnetic semiconductors can be grown in thin film form for applications in power transmission, information storage, and microwave applications.

8.7.1 Exchange interaction and ferromagnetism

In a solid where there are a large number of atoms with non-zero spins, the spacing of the atoms can be small enough (~ 1 Å) that interaction between spins on neighboring sites is significant. The origins of this interaction lie in relativistic quantum mechanics, but it essentially arises because each spin has a magnetic moment associated with it and these moments interact with each other. For an arrangement of atoms on a lattice a general form of this interaction can be written as

$$H_{\text{spin}} = -J \sum_{j,m} S_j \cdot S_{j+m} \tag{8.69}$$

where the subscripts $j, j+m$ represent the position of the two spins. The constant J is known as the exchange integral. Usually the interaction is taken only between nearest neighbor spin so that $m = \pm 1$ in Eq. 8.69. If the exchange integral is positive, the solution to the problem defined by this hamiltonian leads to ferromagnetic phenomena where a system of atoms can have a spontaneous magnetization in the absence of an external magnetic field as long as the temperature is below a critical temperature (Curie temperature). If, however, J is negative, the system displays antiferromagnetism where the atoms have their spins arranged in an antiparallel manner.

Techniques have been developed to solve the problem defined by the hamiltonian given. However, the basic physics can be captured by mean field theory. In a

8.7. Ferromagnetic effects

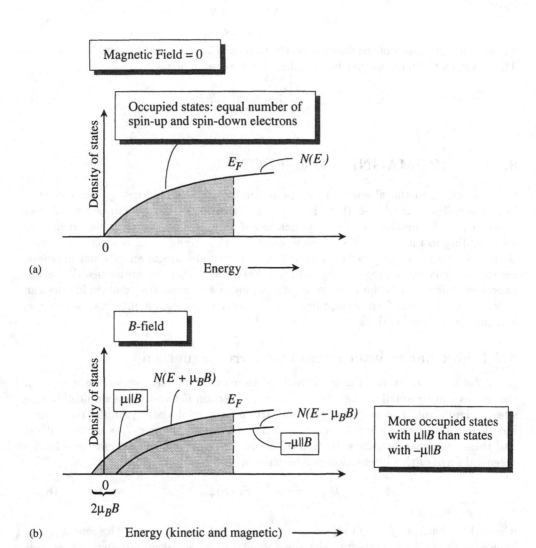

Figure 8.9: (a) Conduction electrons distribution in a metal in the absence of any field. (b) Density of states and occupation of electrons with magnetic moment parallel and antiparallel to the magnetic field.

simplistic mean field model, we postulate that the spin–spin interaction results in an effective exchange field B_E. This field acts as a real magnetic field, although it is only a mathematical representation of the spin–spin interaction. The exchange field causes a magnetization which is paramagnetic in nature. In the mean field approach, we assume that the exchange field is proportional to the macroscopic magnetization

$$B_E = \lambda \mu_0 M \tag{8.70}$$

where λ is a temperature-independent constant.

In addition, if there is an external applied field B_a, we have, for the magnetization

$$\mu_0 M = \chi_P (B_a + B_E) \tag{8.71}$$

where χ_P is the paramagnetic susceptibility of the atoms.

In the previous section, we derived the Curie law for paramagnetic materials, which gives us

$$\chi_P = \frac{C}{T} \tag{8.72}$$

Substituting this, we have, the *effective susceptibility*

$$\chi = \frac{\mu_0 M}{B_a} = \frac{C}{T - C\lambda} \tag{8.73}$$

Writing $T_c = C\lambda$, we have the Curie–Weiss law

$$\chi = \frac{C}{T - T_c} \tag{8.74}$$

This expression provides a reasonable description of susceptibility in ferromagnets. We see that at $T = T_c$ the susceptibility has a singularity. At $T = T_c$ and below, there is spontaneous magnetization in the system; i.e., the spin–spin interaction causes preferential alignment of the magnetic moments in the absence of an external field. At temperatures greater than T_c the thermal fluctuations destroy this alignment. In Fig. 8.10 we show the critical temperature T_c for several ferromagnets. It is important to note that this discussion is somewhat simplistic. More accurate calculations show that

$$\chi = \frac{C}{(T - T_c)^\gamma} \tag{8.75}$$

with $\gamma \sim 1.33$.

8.7.2 Antiferromagnetic ordering

There are some materials in which the nature of the spins is such that, instead of the exchange interaction leading to adjacent spins aligning the same way, they align opposite to each other. This phenomena is called *antiferromagnetism* and to understand it we assume a two-atom basis describing the unit cell of such a material. If A and B denote the two sublattices of the material, as shown in Fig. 8.11a, let us denote by C_A and C_B

8.7. Ferromagnetic effects

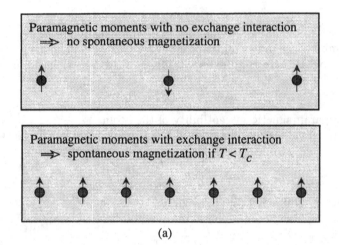

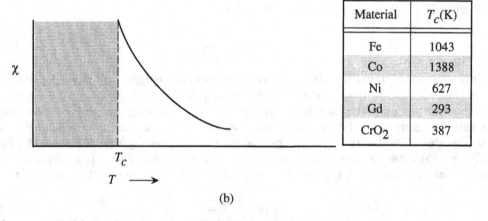

Figure 8.10: (a) A schematic of a collection of paramagnetic magnetic moments with no interactions between the moments (top figure) and a spin–spin interaction which tends to align the moments. (b) Susceptibility of a ferromagnetic material as a function of temperature. Values of T_c for some materials are given.

the Curie constants for the two kinds of atoms. Let us assume that all spin interactions are zero, except for an *antiparallel* interaction between A and B sites. In general, we write

$$B_A = -\alpha \mu_0 M_B$$
$$B_B = -\alpha \mu_0 M_A \qquad (8.76)$$

Once again, in the mean field approach, we have

$$\mu_0 M_A T = C_A(B_a - \alpha \mu_0 M_B)$$
$$\mu_0 M_B T = C_B(B_a - \alpha \mu_0 M_B) \qquad (8.77)$$

where B_a is the applied field. At zero applied fields, these equations have a non-zero value of M_A and M_B if

$$\begin{vmatrix} T & \alpha C_A \\ \alpha C_B & T \end{vmatrix} = 0 \qquad (8.78)$$

or

$$T = T_c = \alpha(C_A C_B)^{1/2} \qquad (8.79)$$

The susceptibility is

$$\chi = \frac{\mu_0(M_A + M_B)}{B_a} = \frac{(C_A + C_B)T - 2\alpha C_A C_B}{T^2 - T_c^2} \cdot \mu_0 \qquad (8.80)$$

The case where $C_A = C_B$ leads to antiferromagnets. Here, the spins are ordered in an antiparallel arrangement, as shown in Fig. 8.11b, with zero net moment below the ordering temperature T_c. This ordering temperature is called the *Neel temperature*.

It may be pointed out for the reader's reference that in the above discussion, if $C_A \neq C_B$, the resulting arrangement of spins, while antiparallel, does not have a net zero magnetism. Such materials are called *ferromagnetic*.

EXAMPLE 8.6 The critical temperature T_c for iron is 1000 K. Calculate the parameter λ and the exchange field B_E. Compare this field to the field $\mu_0 \mu_B/a^3$ produced by a magnetic moment μ_B per site in a material where a is the interatomic spacing. Assume that the magnetization for iron is 2×10^5 A/m.

The constant λ is given by (using $j = 1, g = 2, N \sim 10^{29}$ m^{-3}, $k_B T_c = 86.7$ meV)

$$\lambda = \frac{T_c}{C} = \frac{3 k_B T_c}{\mu_0 N g^2 \mu_B^2 j(j+1)}$$

$$\cong \frac{3 \left(8.67 \times 10^{-2} \times 1.6 \times 10^{-19} \text{ J}\right)}{(4\pi \times 10^{-7} \text{ H/m})(10^{29} \text{ m}^{-3})(4)(2)(9.274 \times 10^{-24} \text{ J/T})}$$

$$\cong 480$$

The exchange magnetic field is then

$$B_E = \lambda \mu_0 M$$
$$= 120 \text{ T}$$

8.7. Ferromagnetic effects

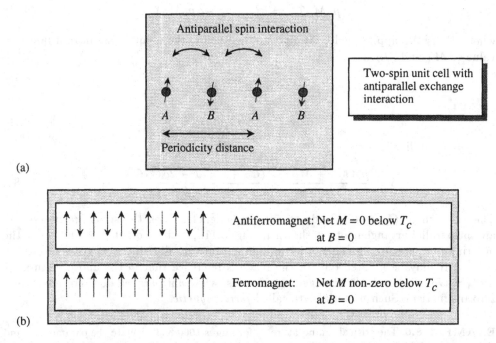

Figure 8.11: (a) A schematic showing a two-spin unit cell in which neighboring spins have interactions that tend to force them to be antiparallel. (b) A comparison of the arrangement of spins in an antiferromagnet and a ferromagnet.

This is an enormous field. The field due to a Bohr magneton is ($a \sim 2$ Å)

$$B = \frac{\left(4\pi \times 10^{-7}\right)\left(9.274 \times 10^{-24}\right)}{\left(2 \times 10^{-10}\right)^3} = 1.45 \text{ T}$$

8.8 APPLICATIONS IN MAGNETIC DEVICES

We will now review some of the important applications of magnetic materials as well as the use of magnetic fields in information technology. These applications exploit one or more of the phenomenon discussed in this chapter. We will discuss the following applications: (i) Magneto-optic devices, which exploit the changes in wave propagation of differently polarized light in magnetic materials. Such devices are very useful for microwave technology.(ii) Quantum interference devices that can be exploited for very sensitive magnetic field measurements (e.g., for sensing brain waves) as well as for logic applications. (iii) Use of magnetic materials to produce extremely low temperatures. (iv) Use of magnetic materials for recording and memory applications.

8.8.1 Quantum interference devices

In Section 8.5 we have seen how a magnetic field can change the phase difference between electron waves traveling along two different paths enclosing the field flux. Such an effect can be exploited to design superconducting devices (electron waves travel without scattering in superconductors), which can be used for sensing applications.

SQUID magnetometers

The superconducting quantum interference device (SQUID) is an extremely sensitive device for measuring changes in magnetic flux. Clever coupling of the device with other circuits also allows it to be used as a very sensitive voltmeter to measure tiny voltages in, say, Hall effect measurements, or as a gradiometer to measure field gradients.

The SQUID is used in either a dc or ac configuration as shown in Fig. 8.12. In the dc configuration shown in Fig. 8.12a, the Josephson loop encloses the flux Φ to be detected. The operation depends upon the fact that the maximum dc supercurrent as well as the I-V relations depend upon the flux Φ. This is shown schematically in Fig. 8.12a. The dc device uses a constant current source in which case the voltage across the device oscillates with changes in the flux through the loop.

In the rf SQUID, the device consists of a single Josephson junction, incorporated into a superconducting loop, and the circuit operates with an rf bias. The SQUID is coupled to the inductor of an LC circuit, excited at its resonant frequency. The rf voltage across the circuit versus the rf current is shown in Fig. 8.12b and oscillates with the applied flux.

Due to the extreme sensitivity of SQUID, the device (in various configurations) finds use in biomagnetism, geophysical exploration, gravitational experiments, Hall effect, magnetic monopole detection, relativity, and many other fields.

8.8. Applications in magnetic devices

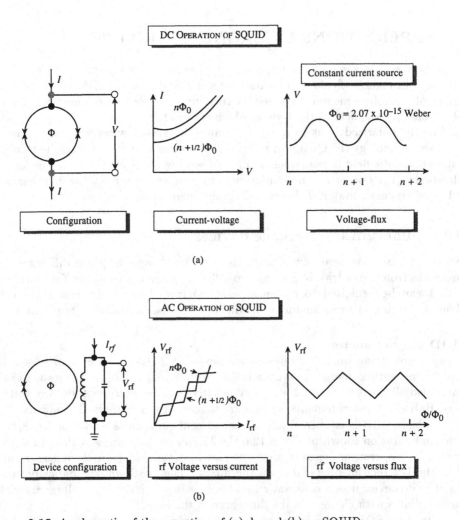

Figure 8.12: A schematic of the operation of (a) dc and (b) ac SQUIDs.

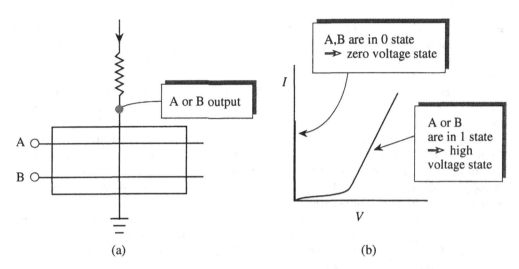

Figure 8.13: Use of Josephson devices in logic implementation. (a) A schematic of a circuit to perform A or B operation and (b) the states of the I–V characteristics of the circuit.

Digital devices

An important application of Josephson tunneling junctions is in the area of microelectronics for computers. Although it is not clear whether Josephson devices will ever compete with semiconductor devices due to manufacturing and cost concerns, in principle, the superconducting devices offer high-speed and low- power operation. The devices depend upon the use of magnetic flux (induced by a current flowing in an input) to alter the critical current that can flow in a superconducting loop.

The general operation of superconducting logic circuits can be appreciated by examining Fig. 8.13a, which shows an OR gate. Initially, the device is biased in the zero voltage state with a current level below the critical current. A current pulse in either of the inputs A or B couples magnetic flux into the loop, so that the critical current is reduced. The device thus switches into a non-zero voltage state as shown in Fig. 8.13b.

8.8.2 Application example: cooling by demagnetization

We have seen that in the presence of a magnetic field, originally unperturbed electronic levels split. As a result, there is a higher occupation of electronic states with magnetic moments parallel to the magnetic field. What happens if the field is switched off without changing the entropy of the system? As we will see in this subsection, this results in a cooling of the sample. This technique for cooling materials has resulted in temperatures as low as sub-millikelvins. This technique, when used in nuclear demagnetization, has resulted in temperatures of microkelvins.

To understand how cooling by isentropic demagnetization works, let us remind ourselves of the relationship between entropy and temperature in a system. Entropy is a measure of the disorder in a system. Thus, in an electronic system, if all the electrons have the same magnetic moment, the system has low entropy, while, if they are randomly arranged the entropy is high. Thus, when the system is in a magnetic field, there is a greater order in the magnetic moment and the entropy is lower. Note that if the temperature of a system is lowered, the entropy decreases, since more electrons occupy the ordered states.

The discussion above tells us that *the entropy of a system at temperature T_i in the presence of a field corresponds to the entropy of a system at a lower temperature T_f in the absence of the field.* Figure 8.14a gives a schematic view of the spin entropy of the system. In this figure, we show three states of the spin system discussed. When the magnetic field is removed under constant entropy conditions (adiabatically with the system insulated), the entropy of the spin system increases at the expense of the lattice entropy, as shown in Fig. 8.14b. The decrease in the lattice entropy corresponds to a reduction in the lattice temperature.

The arguments given for the electronic system also apply to nuclear paramagnets. However, nuclear magnetic moments are much smaller than electron magnetic moments. This results in the possibility of reaching much lower temperatures with nuclear demagnetization experiments.

8.8.3 Magneto-optic modulators

In Chapter 7 we have seen how directional couples and switches can be made by varying the dielectric constants of materials. The effect that is exploited for such devices is the difference in the propagation vectors of light with different polarization. In magnetic materials the Faraday effect can be exploited to create similar devices. In Section 8.2 we examined how light (or microwaves) with right- and left-circular polarization have different wave-vectors. This difference can be altered by a magnetic field–the Faraday effect–and thus be used for device applications.

In general the rotation of the polarization vector of an electromagnetic field is given by

$$\psi = VBL \tag{8.81}$$

where V is called the Verdet constant, B is the magnetic flux density, and L is the length of the device.

In Fig. 8.15 we show a schematic of a magneto-optic modulator. A typical material used is yttrium–iron–garnet (YIG). A constant magnetic field is applied in a direction normal to the device. The modulating magnetic field is applied along the device length. As a result the polarization vector of an incoming wave is altered. By placing a polarizer and analyzer in the path of the incoming and outgoing waves, a modulator can be built.

In microwave technology, devices based on ferrites and utilizing Faraday rotation are used as isolators (i.e., blocking a wave from reading certain regions), couplers (feeding a wave from one waveguide path to another path), and phase shifters. Phase shifters are devices that are very useful for phased array radars, where multiple antennas radiate a directed microwave beam. To emit a directed beam each antenna must radiate em waves at a different phase. By altering the phase difference between successive antennas the direction of the beam can be controlled.

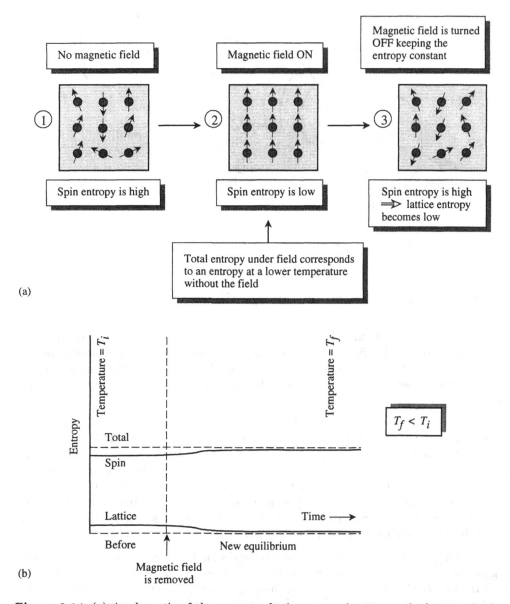

Figure 8.14: (a) A schematic of three states of spin entropy in a magnetic demagnetization experiment. (b) The variation of spin and lattice entropy with and without a magnetic field.

8.8. Applications in magnetic devices

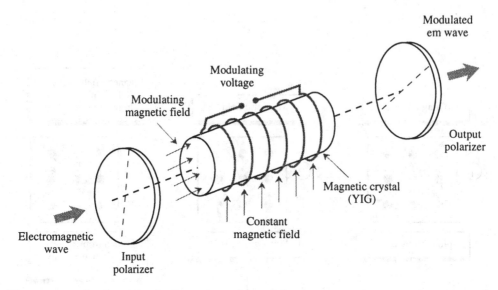

Figure 8.15: A schematic of a mangeto-optic modulator.

8.8.4 Application example: magnetic recording

Ferromagnetic materials find a wide range of applications as permanent magnets. In this section, we will briefly review their role as memory devices. Magnetic materials have found an extremely important application in the area of information storage and retrieval. From audio and video cassettes to computer disks, these materials provide low-cost, high-density memory.

A typical audio cassette is shown in Fig. 8.16a and consists of the magnetic tape on a spool, an erase head, a recording head, and a playback head. To see how the erase and recording process works, we briefly review the hysteresis curve exhibited by ferromagnets. Because of the possibility of spontaneous magnetization, the magnetization of a ferromagnet depends upon the past hystory of the material. In very strong magnetic fields, we get a saturated hysteresis, as shown in Fig. 8.16b, while, if the field is smaller, we get an unsaturated loop. The erase head creates a magnetic field profile that brings the tape to the demagnetized state.

The recording head converts the electrical signal to be recorded to a corresponding magnetic field which magnetizes the tape at a level that has a one-to-one correlation with the electrical signal. Finally, when the magnetized tape passes under the playback head, it generates an electrical signal that is then amplified and converted to sound.

The recording medium for the tape is a plastic base (~ 20 μm thick) on which a magnetic coating of gamma ferric oxide (γ-Fe_2O_3) is applied. In some tapes, chromium dioxide (CrO_2) is used. The ferromagnetic particles (~ 1 μm in size) are dispersed in a binder and coated on the plastic tape.

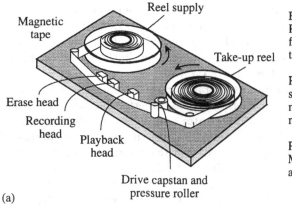

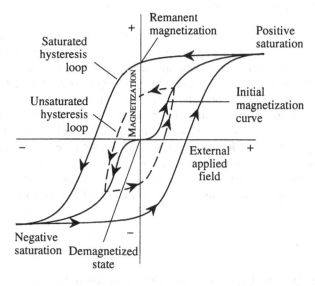

Figure 8.16: (a) A schematic of the audio cassette tape. (b) The hysteresis loop of a ferroelectric material.

8.8.5 Giant magnetic resistance (GMR) devices

Ferromagnetic materials have been used for information storage devices for decades. As noted above, the two stable states of a ferromagnetic element can be used to store digital data. Magnetic storage must compete with semiconductor-based storage. There are some areas where magnetic devices have advantages over semiconductor transistors. Magnetic devices store information in the absence of power and are radiation hard; i.e., are relatively unaffected by high energy particles in space. Additionally, recent advances in magnetic material technology have allowed fabrication of very dense magnetic memories with fast access times. An important development in the magnetic storage field is the observation of giant magnet resistance (GMR).

In Chapter 4 we have discussed transport of electrons in solids. As electrons move in a material they suffer scattering due to imperfections and lattice vibrations. In non-magnetic materials the scattering rates and, hence, conductivity or resistivity are independent of the spin of the electron. However, in magnetic materials, the scattering rates for electrons with spin aligned with the internal magnetization are lower than those for electrons with spin opposing it. In thick magnetic films, the differences in the resistivity of spin-up and spin-down electrons is not very large. However, in heterostructures containing magnetic layers, the difference can be very large (hence the term "giant magneto resistance"). GMR is now widely used in magnetic disk storage technology since sensing heads can be extremely sensitive, allowing for high speed and high memory density.

In Fig. 8.17 we show a schematic of a multilayered structure used for sensing magnetic field direction in a magnetic storage element. The heterostructure shown consists of thin (10 Å–50 Å) layers of ferromagnetic/noble metal layer. Multilayers of NiFe and NiFeCo are incorporated in these thin structures. As shown in Fig. 8.17 there are four layers in the structure. The role of the exchange layer is to ensure that the magnetization in the next layer (pinning layer) is always in the same direction. The three thin layers – sensing, conduction, and pinning – have electrons that carry the current. The magnetic state of the sensing layer can be switched by a small magnetic field. When the state of the sensing layer and the pinning layer is the same, the device offers very little resistance. However, when the sensing layer state is reversed, the resistance increases. Thus this device (called spin valve) can be used to read the state of the magnetic memory.

8.9 SUMMARY

The topics discussed in this chapter are summarized in Table 8.1.

8.10 PROBLEMS

Section 1.3
8.1 A Faraday rotation device has the following properties

$$M_0 = 7.8 \times 10^4 \text{ A/m}; \quad \omega = 2 \times 10^{10} \text{ rad/s}$$

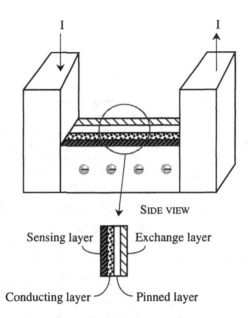

Figure 8.17: (a) A schematic of a device used to sense magnetic orientation in disk memories. A heterostructure containing a magnetic film uses GMR to sense field orientation through current flow.

$$\epsilon_r = 10; \qquad H_0 = 2 \times 10^5 \text{ A/m}$$

Calculate the Faraday rotation angle when waves travel over a distance of 1 cm.

8.2 An electronic system with orbital angular momentum $L = \hbar$ and spin $= 0$ is subjected to a magnetic field. At what magnetic field is the splitting of the various states 1 meV?

8.3 Consider electrons in hydrogen atoms in the ground state. Show that the molar susceptibility is -2.36×10^{-6} cm^{3}mole^{-1}.

8.4 Calculate the Curie constant for a system of free electrons with zero angular momentum. The electron density is 10^{23} cm^{-3}.

8.5 Discuss the role of scattering of electrons in the Aharonov–Bohm effect.

8.6 Consider a digital device based on the Aharonov–Bohm effect. The area enclosed by the two arms of the device is 5 μm×5 μm. Estimate the minimum switching energy needed to switch the device from ON to OFF if the volume over which the B-field is to be altered is 10^{-9} cm^3.

8.7 Discuss why magnetic flux quantization occurs in a superconducting ring, but not in a normal metallic ring.

8.8 Discuss how a magnetometer (like SQUID) can be used to monitor activities of various kinds in biological systems.

8.9 The exchange field B_E for a ferromagnet is estimated to be 10^3 T. Estimate the critical temperature T_c for this material. Use the following values:

$$N = 10^{29} \text{ m}^{-3}$$

8.10. Problems

Topics studied	Key findings
Electromagnetic wave propagation in magnetic materials	In materials with magnetization and a dc magnetic field, electromagnetic waves with different polarizations propagate with different wavevectors. Thus a phase difference develops between such waves.
Diamagnetic effects	A negative magnetic susceptibility is produced in electronic levels in atoms as a result of the precession of electrons.
Paramagnetic effects	In many materials a positive magnetic susceptibility occurs due to the spin-splitting of electronic levels and redistribution of electrons.
Ferromagnetic effects	In materials with net spins, interactions between neighboring spins can create long range alignment of spin. In ferromagnetic materials the spins are all aligned parallel to each other below the Curie temperature. Above Curie temperature the spins are disordered.
Applications of magnetic fields and materials	• Microwave devices such as phase shifters, modulators, couplers, etc., exploit wavevector control in magnetic materials. • Electron wave quantum interference devices can be built in superconducting materials for sensitive magnetometers and switching devices. • Adiabatic switching of magnetic fields can allow cooling devices to be designed. Extremely low temperatures can be reached. • Ferromagnetic materials can be used to store and need information. Very high density storage systems can be produced.

Table 8.1: Summary table.

$$g = 2$$
$$j = 1$$

8.10 Magnetic oxide are used as strips on cards to store information. Discuss the advantages and disadvantages of using materials with low coercive field (like iron oxide) and high coercive fields (like Barium iron oxide).

8.11 Estimate the threshold magnetic field that can be detected by a SQUID device with an area of 10^{-2} m^2. Note that the magnetic field associated with the heart is 10^{-10} T and that with the brain is 10^{-13} T. For reference, the Earth's magnetic field is 0.5×10^{-4} T.

8.12 A magneto-optic modulator uses certain flint glass with Verdet constant of 3.2×10^4 minutes of arc per meter per tesla. The magnetic flux density is 0.5 tesla and the device length is 2 cm. Calculate the light rotation at a wavelength of 5890 Å.

8.11 FURTHER READING

- **General**

    - B. Heinrich and J.A.C. Bland, editors, *Ultra-Thin Magnetic Structures II*, Springer Verlag, Berlin (1994).
    - J.D. Jackson, *Classical Electrodynamics*, J. Wiley (1975).
    - C. Kittel, *Introduction to Solid State Physics*, J. Wiley (1986).
    - A.J. Moulson and J.M. Herbert, *Electroceramics, Materials, Properties, Applications*, Chapman & Hall (1990).

APPENDIX A

IMPORTANT PROPERTIES OF SEMICONDUCTORS

The data and plots shown in this Appendix are extracted from a number of sources. A list of useful sources is given below.

- S. Adachi, *J. Appl. Phys.*, **58**, R1 (1985).

- H.C. Casey, Jr. and M.B. Panish, *Heterostructure Lasers*, Part A, "Fundamental Principles;" Part B, "Materials and Operating Characteristics," Academic Press, New York (1978).

- Landolt-Bornstein, *Numerical Data and Functional Relationship in Science and Technology*, Vol. 22, Eds. O. Madelung, M. Schulz, and H. Weiss, Springer Verlag, New York (1987).

- S.M. Sze, *Physics of Semiconductor Devices*, J. Wiley, New York (1981). This is an excellent source of a variety of useful information on semiconductors.

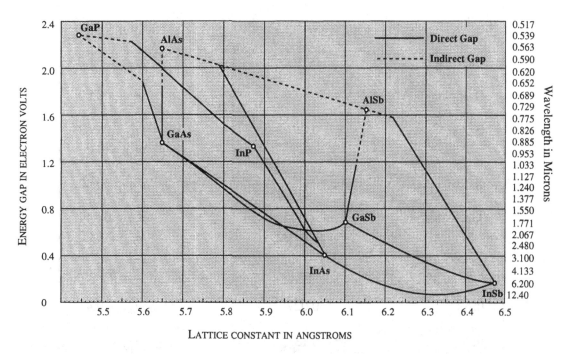

Figure A.1: Lattice constants and bandgaps of semiconductors at room temperature.

Important properties of semiconductors

Material	Electron Mass (m_0)	Hole Mass (m_0)
AlAs	0.1	
AlSb	0.12	$m_{dos} = 0.98$
GaN	0.19	$m_{dos} = 0.60$
GaP	0.82	$m_{dos} = 0.60$
GaAs	0.067	$m_{lh} = 0.082$ $m_{hh} = 0.45$
GaSb	0.042	$m_{dos} = 0.40$
Ge	$m_l = 1.64$ $m_t = 0.082$	$m_{lh} = 0.044$ $m_{hh} = 0.28$
InP	0.073	$m_{dos} = 0.64$
InAs	0.027	$m_{dos} = 0.4$
InSb	0.13	$m_{dos} = 0.4$
Si	$m_l = 0.98$ $m_t = 0.19$	$m_{lh} = 0.16$ $m_{hh} = 0.49$

Table A.1: Electron and hole masses for several semiconductors. Some uncertainty remains in the value of hole masses for many semiconductors.

Compound	Direct Energy Gap E_g (eV)
$Al_xIn_{1-x}P$	$1.351 + 2.23x$
$Al_xGa_{1-x}As$	$1.424 + 1.247x$
$Al_xIn_{1-x}As$	$0.360 + 2.012x + 0.698x^2$
$Al_xGa_{1-x}Sb$	$0.726 + 1.129x + 0.368x^2$
$Al_xIn_{1-x}Sb$	$0.172 + 1.621x + 0.43x^2$
$Ga_xIn_{1-x}P$	$1.351 + 0.643x + 0.786x^2$
$Ga_xIn_{1-x}As$	$0.36 + 1.064x$
$Ga_xIn_{1-x}Sb$	$0.172 + 0.139x + 0.415x^2$
GaP_xAs_{1-x}	$1.424 + 1.150x + 0.176x^2$
$GaAs_xSb_{1-x}$	$0.726 + 0.502x + 1.2x^2$
InP_xAs_{1-x}	$0.360 + 0.891x + 0.101x^2$
$InAs_xSb_{1-x}$	$0.18 + 0.41x + 0.58x^2$

Table A.2: Compositional dependence of the energy gaps of the binary III–V ternary alloys at 300 K. (After Casey and Panish (1978).)

Important properties of semiconductors

Semiconductor	Bandgap (eV) 300 K	Mobility at 300 K (cm^2/V·s) Electrons	Holes
C	5.47	800	1200
Ge	0.66	3900	1900
Si	1.12	1500	450
α-SiC	2.996	400	50
GaSb	0.72	5000	850
GaAs	1.42	8500	400
GaP	2.26	110	75
InSb	0.17	8000	1250
InAs	0.36	33000	460
InP	1.35	4600	150
CdTe	1.56	1050	100
PbTe	0.31	6000	4000

Table A.3: Bandgaps, electron, and hole mobilities of some semiconductors.

APPENDIX B

P–N DIODE: A SUMMARY

B.1 INTRODUCTION

In this appendix we will review some important aspects of the diode, which forms the basis of many of the optoelectronic devices discussed in Chapter 5. Most optical detectors and light emitters are based on p–n diodes.

B.2 P–N JUNCTION

As noted above the p–n diode is one of the most important optoelectronic devices. It forms the basis of most detectors and light emitting devices. Light detection occurs when photons create electrons and holes, while light emission occurs by e–h recombination.

Unbiased P–N junction
The p–n junction is one of the most important junctions in solid-state electronics. The fabrication techniques used to form p- and n-type regions involve (i) epitaxial procedures, where the dopant species are simply switched at a particular instant in time: (ii) ion-implantation in which the dopant ions are implanted at high energies into the semiconductor (the junction is obviously not as abrupt as in the case of epitaxial techniques); and (iii) diffusion of dopants into an oppositely doped semiconductor.

We will assume in our analysis that the p–n junction is abrupt, even though this is really only true for epitaxially grown junctions. Let us first discuss the properties of the junction in the absence of any external bias where there is no current flowing in the diode.

What happens when the p- and n-type materials are made to form a junction and there is no externally applied field? We know that, in absence of any applied bias, there is no current in the system and the Fermi level is uniform throughout the structure.

B.2. P–N junction

This gives the schematic view of the junction shown in Fig. B.1a. Three regions can be identified:

(i) The p-type region at the far left where the material is neutral and the bands are flat. The density of acceptors exactly balances the density of holes;

(ii) The n-type region in the far right where again the material is neutral and the density of immobile donors exactly balances the free electron density;

(iii) The depletion region where the bands are bent and a field exists which has swept out the mobile carriers leaving behind negatively charged acceptors in the p-region and positively charged donors in the n-region as shown in Fig. B.1a.

In the depletion region, which extends a distance W_p in the p-region and a distance W_n in the n-region, an electric field exists. Any electrons or holes in the depletion region are swept away by this field. Thus a drift current exists which counterbalances the diffusion current which arises because of the difference in electron and hole densities across the junction.

In the absence of any applied bias, there is a built-in potential between the n and the p side as shown in Fig. B.2. Denoting p_p and p_n as the hole densities in the p-type and n-type neutral regions the built-in potential is

$$V_{bi} = \frac{k_B T}{e} \ln \frac{p_p}{p_n} \qquad (B.1)$$

The built-in potential can also be written as

$$V_{bi} = \frac{k_B T}{e} \ln \frac{n_n}{n_p} \qquad (B.2)$$

where n_n and n_p are the electron densities in the n-type and p-type regions. Remember that the law of mass action tells us that

$$n_n p_n = n_p p_p = n_i^2 \qquad (B.3)$$

We can thus write the following equivalent expressions

$$\frac{p_p}{p_n} = e^{eV_{bi}/k_B T} = \frac{n_n}{n_p} \qquad (B.4)$$

We will now give the widths of the depletion regions on the n- and p-side in the absence of an external bias. The values in presence of a bias are simply given by replacing the built-in bias by the total bias across the p and n regions.

$$W_p(V_{bi}) = \left\{ \frac{2\epsilon V_{bi}}{e} \left[\frac{N_d}{N_a(N_a + N_d)} \right] \right\}^{1/2} \qquad (B.5)$$

$$W_n(V_{bi}) = \left\{ \frac{2\epsilon V_{bi}}{e} \left[\frac{N_a}{N_d(N_a + N_d)} \right] \right\}^{1/2} \qquad (B.6)$$

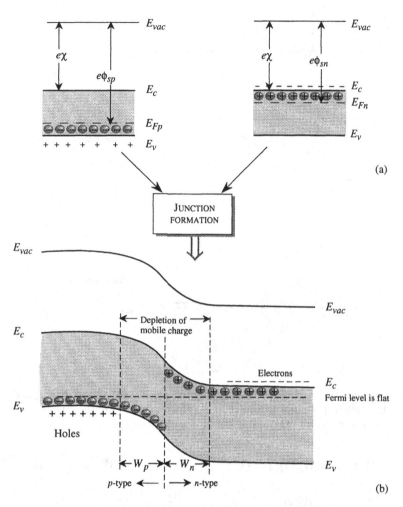

Figure B.1: (a) An idealized model of the p–n junction without bias, showing the neutral and the depletion areas. (b) A schematic showing various current and particle flow components in the p–n diode at equilibrium. For electrons, the current flow is in the direction opposite to that of the particle flow. Electrons that enter the depletion region from the p-side and holes that enter the depletion region from the n-side are swept away and are the source of the drift components.

B.2. P–N junction

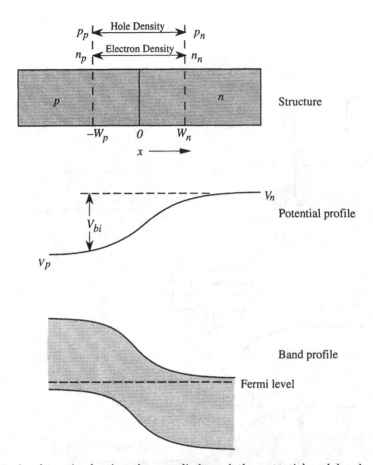

Figure B.2: A schematic showing the p–n diode and the potential and band profiles. The voltage V_{bi} is the built-in potential at equilibrium. The expressions derived in the text can be extended to the cases where an external potential is added to V_{bi}.

$$W(V_{bi}) = W_p(V_{bi}) + W_n(V_{bi}) = \left(W_n^2(V_{bi}) + W_p^2(V_{bi}) + 2W_n(V_{bi})W_p(V_{bi})\right)^{1/2}$$

$$W(V_{bi}) = \left[\frac{2\epsilon V_{bi}}{e}\left(\frac{N_a + N_d}{N_a N_d}\right)\right]^{1/2} \tag{B.7}$$

From these discussions we can draw the following important conclusions about the diode:

(i) The electric field in the depletion region peaks at the junction and decreases linearly towards the depletion region edges.

(ii) The potential drop in the depletion region has a quadratic form.

We remind ourselves that this procedure can be extended to find the electric fields, potential and depletion widths for arbitrary values of V_p and V_n under certain approximations to be discussed next. Thus we can directly use these equations when the diode is under external bias V, by simply replacing V_{bi} by $V_{bi} + V$. The applied bias can increase the total potential or decrease it as will be discussed later.

The results of the calculations carried out above are schematically shown in Fig. D.3. Shown are the charge density and the electric field profiles. Notice that the electric field is nonuniform in the depletion region, peaking at the junction with a peak value (the sign of the field simply reflects the fact that in our study the field is pointing towards the negative x-axis)

$$F_m = -\frac{eN_d W_n}{\epsilon} = -\frac{eN_a W_p}{\epsilon} \tag{B.8}$$

Notice that the depletion in the p- and n-sides can be quite different. If $N_a \gg N_d$, the depletion width W_p is much smaller than W_n. Thus a very strong field exists over a very narrow region in the heavily doped side of the junction. In such abrupt junction (p^+n or n^+p) the depletion region exists primarily on the lightly doped side.

B.2.1 P–N junction under bias

Let us now consider the situation where an external potential is applied across the p and n regions. In the presence of the applied bias, the balance between the drift and diffusion currents will no longer exist and a net current flow will occur. In general, we need a numerical treatment to understand the behavior of the p–n diode under bias. However, under quasi-equilibrium conditions, we can use the previous results for the biased diode as well.

In Fig. B.4 we show the schematic profiles of the depletion region, potential profile, and the band profiles in equilibrium, forward bias, and reverse bias. In forward bias V_f, the p-side is at a positive potential with respect to the n-side. In the reverse bias case, the p-side is at a negative potential $-V_r$ with respect to the n-side. Remember that the way we plot the energy bands includes the negative electron charge so the energy bands have the opposite sign of the potential profile.

In the forward bias case, the potential difference between the n- and p-side is (V_f is taken as having a positive value)

$$V_{Tot} = V_{bi} - V_f \tag{B.9}$$

while for the reverse biased case it is (V_r is taken as having a positive value)

$$V_{Tot} = V_{bi} + V_r \tag{B.10}$$

Under the quasi-equilibrium approximations, the equations for the electric field profile, the potential profile, and the depletion widths, we have calculated in the previous discussion, are directly applicable, except that V_{bi} *is replaced by* V_{Tot}. Thus the depletion width and the peak electric field at the junction decrease under forward bias, while they increase under reverse bias, as can be seen from Eqs. B.5 and B.6 if V_{bi} is replaced by V_{Tot}.

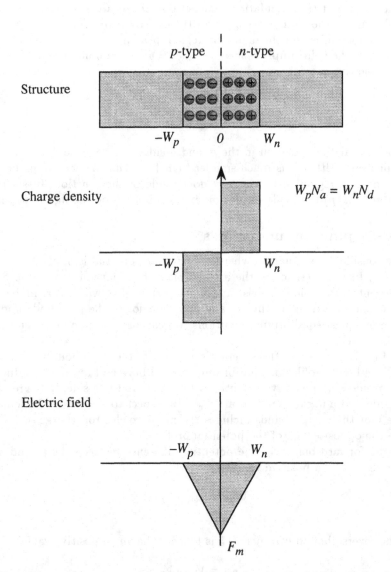

Figure B.3: The p–n structure, with the charge and the electric field profile in the depletion region. Note that in the depletion approximation there is no charge or electric field outside the depletion region. The electric field peaks at the junction as shown.

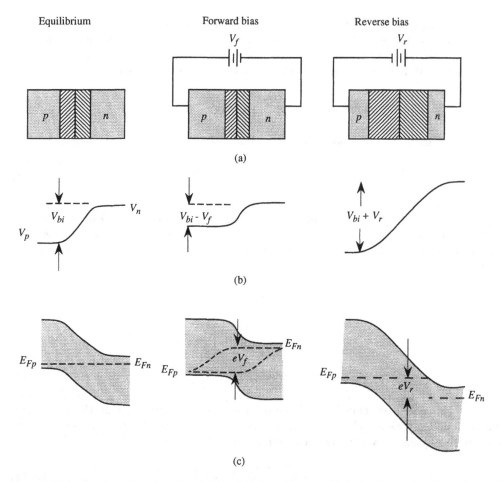

Figure B.4: A schematic showing: (a) the biasing of a p–n diode in the equilibrium, forward and reverse bias; (b) the voltage profile, and (c) the energy band profiles. In the forward bias, the potential across the junction decreases, while in reverse bias it increases. The quasi-Fermi levels are shown in the depletion region.

Charge injection and current flow

We will now discuss the current flow in presence of an applied bias. The presence of the bias increases or decreases the electric field in the depletion region. However, under moderate external bias, the electric field in the depletion region is always higher than the field for carrier velocity saturation ($E \gtrsim 10$ kV cm^{-1}). Thus the change in electric field does not alter the drift part of the electron or hole current in the depletion region. Regardless of the bias, electrons or holes that come into the depletion region are swept out and contribute to the same current independent of the field. The situation is quite different for the diffusion current. Remember that the diffusion current depends upon the gradient of the carrier density. As the potential profile is greatly altered by the applied

B.2. P–N junction

bias, the carrier profile changes accordingly, greatly affecting the diffusion current. In the presence of the applied bias, the change in current is due to the change in the hole current injected into the n-side and the electron current injected into the p-side, as shown in Fig. B.5. The hole current injected into the n-side is given by

$$I_p(W_n) = e \frac{AD_p}{L_p} p_n \left(e^{eV/k_BT} - 1\right) \qquad (B.11)$$

Similarly the total electron current injected into the p-side region is given by

$$I_n(-W_p) = \frac{eAD_n}{L_n} n_p \left(e^{eV/k_BT} - 1\right) \qquad (B.12)$$

We assume initially that in the ideal diode there is no recombination of the electron and hole injected currents in the depletion region. Thus the total current can be simply obtained by adding the hole current injected across W_n and electron current injected across $-W_p$. The diode current is then

$$\begin{aligned} I(V) &= I_p(W_n) + I_n(-W_p) \\ &= eA\left[\frac{D_p}{L_p}p_n + \frac{D_n}{L_n}n_p\right]\left(e^{eV/k_BT} - 1\right) \\ I(V) &= I_0\left(e^{eV/k_BT} - 1\right) \end{aligned} \qquad (B.13)$$

This equation, called the diode equation, gives us the current through a p–n junction under forward ($V > 0$) and reverse bias ($V < 0$). Under reverse bias, the current simply goes towards the value $-I_0$, where

$$I_0 = eA\left(\frac{D_p p_n}{L_p} + \frac{D_n n_p}{L_n}\right) \qquad (B.14)$$

Under forward bias the current increases exponentially with the applied forward bias. This strong asymmetry in the diode current is what makes the p–n diode attractive for many applications.

We see from the discussions of this section that the current flow through the simple p–n diode has some very interesting properties. We do not have the simple linear Ohm's law type behavior, but a strongly nonlinear and rectifying behavior. The current, as shown in Fig. B.6, saturates to a value I_0 given by Eq. B.14 when a reverse bias is applied. Since this value is quite small, the diode is essentially nonconducting. However, when a positive bias is applied, the diode current increases exponentially and the diode becomes strongly conducting. The forward bias voltage at which the diode current becomes significant ($\sim$ mA) is called the cut-in voltage. This voltage is ~ 0.8V for Si diodes and ~ 1.2 V for GaAs diodes.

Real diode: effects of defects
In the calculations above we have assumed that the semiconductor is perfect; i.e., there

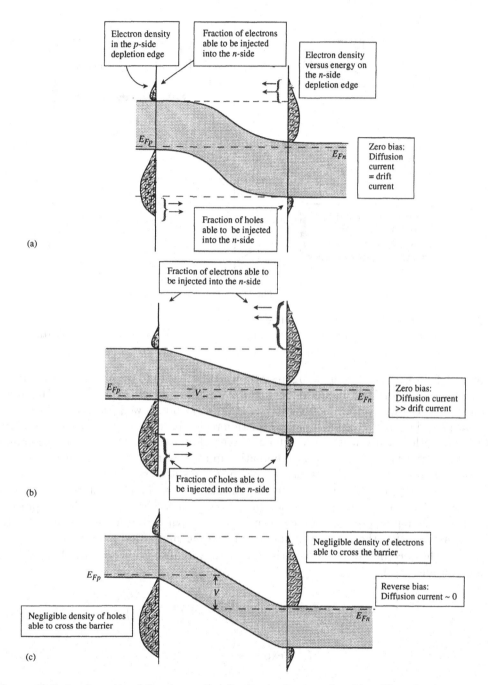

Figure B.5: A schematic of the charge distribution in the n- and p-sides. The minority carrier injection (electrons from n-side to p-side or holes from p-side to n-side) is controlled by the applied bias as shown.

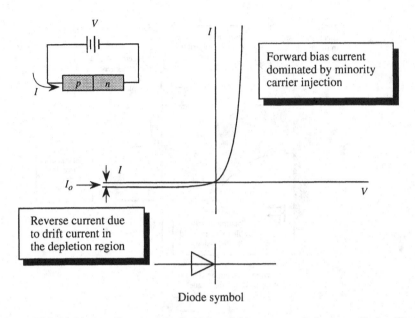

Figure B.6: The highly nonlinear and rectifying I–V current of the p–n diode. The strong nonlinear response makes the diode a very important device for a number of applications.

are no defects and associated bandgap states which may lead to trapping, recombination, or generation terms. In Chapter 3, Section 3.7 we have discussed the effects of bandgap states produced by defects. In our analysis of the diode ideal, we have assumed that the electrons and holes injected across the depletion region barrier, are not able to recombine with each other. Only when they enter the neutral regions are they able to recombine with the majority carriers. This recombination in the neutral region is described via the diffusion lengths L_n and L_p that appear in the expression for I_0.

In a real diode, a number of sources may lead to bandgap states. The states may arise if the material quality is not very pure so that there are chemical impurities present. The doping process itself can cause defects, such as vacancies, interstitials, etc. Let us assume that the deep level states lead to a recombination time τ.

The recombination current is now simply (current is equal to charge times volume times rate)

$$\begin{aligned} I_R &= eAWR_t = \frac{eAWn_i}{2\tau} \exp\left(\frac{eV}{2k_BT}\right) \\ &= I_{GR}^o \exp\left(\frac{eV}{2k_BT}\right) \end{aligned} \quad (B.15)$$

where W is the depletion width. At zero applied bias, a generation current of I_G balances out the recombination current.

The generation–recombination current has an exponential dependence on the voltage as well, but the exponent is different. The generation-recombination current is

$$I_{GR} = I_R - I_G = I_R - I_R(V=0)$$
$$= I_{GR}^o \left[\exp\left(\frac{eV}{2k_BT}\right) - 1 \right] \quad (B.16)$$

The total device current now becomes

$$I = I_0 \left[\exp\left(\frac{eV}{k_BT}\right) - 1\right] + I_{GR}^o \left[\exp\left(\frac{eV}{2k_BT}\right) - 1\right]$$

or

$$I \cong I'_o \left[\exp\left(\frac{eV}{mk_BT}\right) - 1\right] \quad (B.17)$$

The prefactor I_{GR}^o can be much larger than I_0 for real devices. Thus at low applied voltages the diode current is often dominated by the second term. However, as the applied bias increases, the injection current starts to dominate. We thus have two regions in the forward I–V characteristics of the diode, as shown in Fig. B.7.

At low applied bias the plot of $\frac{eV}{k_BT}$ and $\log(I)$ has a slope of 1/2, which turns over to 1.0 at higher voltages. The parameter m of Eq. B.17 is called the *diode ideality factor*. If the diode is of high quality, m is close to unity, otherwise it approaches a value of 2.

Diodes as optoelectronic devices

In Chapter 5 we have discussed how the p–n diode is used for a variety of applications in optoelectronics. The diode is the basis for detectors, light emitters, and modulators. When used as a detector the diode is usually reverse biased so that there is very little current flowing in the absence of an optical signal (the dark current is just the reverse bias current). As discussed in Section 5.5, when an optical signal with $\hbar\omega > E_g$ impinges, electron–hole pairs are created. If these photocarriers are in the depletion region where there is a large electric field, they are swept to create a photocurrent.

Light emitters are designed by using diodes that are forward biased. Electrons are injected into the p-region and holes into the n-region. These carriers recombine through radiative and non-radiative paths. The radiative path causes a photon to be emitted, while the non-radiative path leads to phonon emission (i.e., heat generation). The photons emitted have an energy equal to (approximately) the bandgap of the material.

In semiconductor lasers, usually the region where e–h recombination occurs (the active region), has a smaller bandgap. In modern devices the active region is made of quantum wells. Electrons and holes recombine from these regions and the emission energy (or wavelength) can be tailored by changing the bandgap of the active region.

B.2. P–N junction

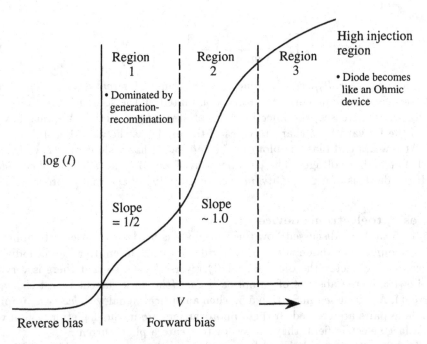

Figure B.7: The I–V characteristics of a real diode. At low biases, the recombination effects are quite pronounced leading to a curve with slope 1/2. At higher biases the slope becomes closer to unity. At still higher biases the behavior becomes more ohmic.

APPENDIX C

FERMI GOLDEN RULE

In the chapters on transport and optical properties we have seen that scattering of electrons from one state to another is critical in the understanding of almost all physical properties of materials. Optical and transport phenomena are linked to scattering processes. We have used the Fermi golden rule to evaluate several important scattering processes in materials. In this appendix we will give a derivation of this important equation in quantum physics. The general Hamiltonian of interest is of the form

$$H = H_0 + H' \tag{C.1}$$

where H_0 is a simple Hamiltonian with known solutions

$$H_0 u_k = E_k u_k \tag{C.2}$$

and E_k, u_k are known. For example the solutions to H would be the bandstructure of electrons in a crystal. In the absence of $H'(t)$, if a particle is placed in a state u_k, it remains there forever. The effect of H' is to cause time-dependent transitions between the states u_k. The time-dependent Schrödinger equation is

$$i\hbar \frac{\partial \psi}{\partial t} = H\psi \tag{C.3}$$

The approximation will involve expressing ψ as an expansion of the eigenfunctions $u_n \exp(-iE_n t/\hbar)$ of the unperturbed time-dependent functions

$$\psi = \sum_n a_n(t) u_n e^{-iE_n t/\hbar} \tag{C.4}$$

The time-dependent problem is solved when the coefficients $a_n(t)$ are known. In the spirit of the perturbation approach, these coefficients are determined to different orders. Hopefully, the first- or second-order terms would suffice and higher-order terms would be negligible.

Substituting for ψ (given by Eq. C.4) in Eq. C.3, using Eq. C.2, we get

$$\sum_n i\hbar \dot{a}_n(t) u_n e^{-iE_n t/\hbar} + \sum_n a_n E_n u_n e^{-iE_n t/\hbar}$$
$$= \sum_n a_n (H_0 + H') u_n e^{-iE_n t/\hbar} \quad (C.5)$$

Multiplying by u_k^* and integrating over space, we get (using orthogonality of u_ks)

$$i\hbar \dot{a}_k e^{-iE_k t/\hbar} = \sum_n a_n e^{-iE_n t/\hbar} \langle k|H'|n\rangle \quad (C.6)$$

Note that we are using the Dirac notation; i.e., u_k and u_n are written as $|k\rangle$ and $|n\rangle$ here. Writing

$$\omega_{kn} = \frac{E_k - E_n}{\hbar} \quad (C.7)$$

The time derivative of a_k is

$$\dot{a}_k = \frac{1}{i\hbar} \sum_n \langle k|H'|n\rangle a_n e^{i\omega_{kn} t} \quad (C.8)$$

To find the corrections to various orders in H', we can write the perturbation as $\lambda H'$, where λ is a parameter that goes from 0 (no perturbation) to 1

$$H' \to \lambda H'$$
$$a_n = a_n^{(0)} + \lambda a_n^{(1)} + \lambda^2 a_n^{(2)} + \cdots \quad (C.9)$$

Here $a_n^{(0)}, a_n^{(1)}, \ldots$ are the different orders of the expansion coefficients of the wavefunction. Substituting this expansion in Eq. C.8 and comparing the coefficients of the same powers of λ, we get for the time dependence of the coefficients

$$\dot{a}_k^{(0)} = 0$$
$$\dot{a}_k^{(s+1)} = \frac{1}{i\hbar} \sum_n \langle k|H'|n\rangle a_n^{(s)} e^{i\omega_{kn} t} \quad (C.10)$$

In principle, these equations can be integrated to any order to obtain the desired solution. To study the time evolution of the problem, we assume that the perturbation is absent at time $t < 0$ and is turned on at $t = 0$. With this assumption, the system is in a time-independent state up to $t = 0$. From Eq. C.10, we see that the zeroth-order coefficients $a_n^{(0)}$ are constant time and are simply given by the initial conditions of the problem, before the perturbation is applied. We assume that initially the system is in a single, well-defined state $|m\rangle$

$$a_m^{(0)} = 1$$
$$a_k^{(0)} = 0 \text{ if } k \neq m \quad (C.11)$$

Integration of the first-order term in Eq. C.10 gives

$$a_k^{(1)}(t) = \frac{1}{i\hbar} \int_{-\infty}^{t} \langle k|H'(t')|m\rangle e^{i\omega_{km}t'} dt' \qquad (C.12)$$

We choose the constant of integration to be zero since $a_k^{(1)}$ is zero at time $t \to -\infty$ when the perturbation is not present.

We see from Eq. C.12 that, if the perturbation is of finite duration, the amplitude of finding the system in a state $|k\rangle$ different from the initial state $|m\rangle$ is proportional to the Fourier component of the matrix element of the perturbation between the two states.

Time-dependent perturbation

A number of important problems in quantum mechanics involve a perturbation which has time dependence with a harmonic form. Examples include interaction of electrons with electromagnetic radiation (or photons), electrons in crystals interacting with lattice vibrations (or phonons), etc. For such perturbations, time-dependent perturbation theory gives some simple results that have been widely applied in understanding and designing experiments. Consider the case where the perturbation is harmonic, except that it is turned on at $t = 0$ and turned off at $t = t_0$. Let us assume that the time dependence is given by

$$\langle k|H(t')|m\rangle = 2\langle k|H'(0)|m\rangle \sin \omega t' \qquad (C.13)$$

Carrying out the integration until time $t \geq t_0$ in Eq. C.12, we get

$$a_k^{(1)}(t \geq t_0) = -\frac{\langle k|H'(0)|m\rangle}{i\hbar} \left(\frac{\exp[i(\omega_{km} + \omega)t_0] - 1}{\omega_{km} + \omega} - \frac{\exp[i(\omega_{km} - \omega)t_0] - 1}{\omega_{km} - \omega} \right) \qquad (C.14)$$

The structure of this equation tells us that the amplitude is appreciable only if the denominator of one term or the other is close to zero. The first term is important if $\omega_{km} \approx -\omega$ or $E_k \approx E_m - \hbar\omega$. The second term is important if $\omega_{km} \approx \omega$ or $E_k \approx E_m + \hbar\omega$. Thus in the first-order, the effect of a harmonic perturbation is to transfer, or to receive from the system, the quantum of energy $\hbar\omega$.

If we focus on a system where $|m\rangle$ is a discrete state, $|k\rangle$ is one of the continuous states, and $E_k > E_m$, so that only the second term of Eq. C.14 is important, the first-order probability of finding the system in the state k after the perturbation is removed is

$$\left|a_k^{(1)}(t \geq t_0)\right|^2 = 4|\langle k|H'(0)|m\rangle|^2 \frac{\sin^2\left[\frac{1}{2}(\omega_{km} - \omega)t_0\right]}{\hbar^2(\omega_{km} - \omega)^2} \qquad (C.15)$$

The probability function has an oscillating behavior, as shown in Fig. C.1. The probability is maximum when $\omega_{km} = \omega$ and the peak is proportional to t_0^2. However, the uncertainty in frequency $\Delta\omega = \omega_{km} - \omega$ is non-zero if the time t_0 over which the perturbation is applied is finite. This uncertainty is in accordance with the Heisenberg uncertainty principle

$$\Delta\omega \, \Delta t = \Delta\omega \, t_0 \sim 1 \qquad (C.16)$$

Fermi golden rule

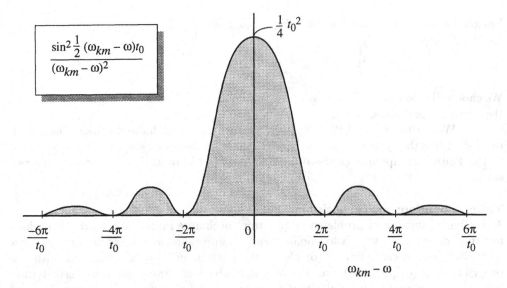

Figure C.1: The ordinate is proportional to the probability of finding the system in a state k after the perturbation has been applied to time t_0.

If the perturbation extends over a long time, the function plotted in Fig. C.2 approaches the Dirac δ-function and the probability is non-zero only for $E_k = E_m + \hbar\omega$.

Transition probability

An important class of problems falls in the category where either the perturbing potential or the unperturbed states are described by continuous spectra. For example, as shown in Fig. C.2a, the perturbation may have a spread of frequencies as in the case of electromagnetic radiation with a finite frequency spread. Or the states $|k\rangle$ or $|m\rangle$ may be in a continuum. In either case, this leads to a spread in the allowed values of $(\omega_{km} - \omega)$. In such cases, it is possible to define a scattering rate per unit time. We can see from Fig. C.2b that the probability of finding the system in a state $|k\rangle$ has a shape where the peak is proportional to t_0^2, while the width of the main peak decreases inversely as t_0. Thus the area under the curve is proportional to t_0. Thus, if we were to define the probability of finding the system anywhere in a spread of states covering the width of the main peak, the total probability will be proportional to t_0. This would allow us to define a transition rate; i.e., transition probability per unit time. The total rate per unit time for scattering into any final state, is given by

$$W_m = \frac{1}{t_0} \sum_{\text{final states}} \left|a_k^{(1)}(t \geq t_0)\right|^2$$

If t_0 is large, the sum over the final states includes only the final states where $\omega_{km} - \hbar\omega \cong 0$.

In summing over the final states, we can use the concept of density of states,

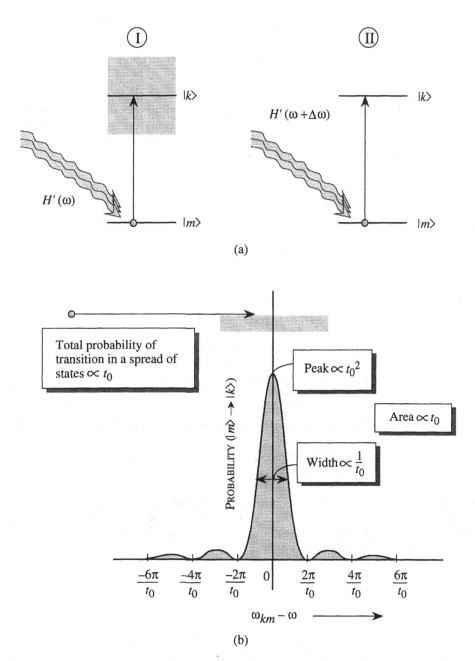

Figure C.2: (a) Cases where the states involved in a transition are part of a continuum. Case I shows the case where the unperturbed state is in a continuum. In case II, the perturbation has a continuum of frequencies. (b) A schematic of transitions in continuous spectra. The total transition probability in the continuum is proportional to the area under the curve.

Fermi golden rule

which gives us the number of states per unit volume per unit energy

$$W_m = \frac{1}{t_0} \int |a_k^{(1)}(t \geq t_0)|^2 \, \rho(k) \, dE_k \tag{C.17}$$

where $\rho(k)$ is the density of states near the final state.

From Fig. C.2b we see that the peak of the probability function becomes very narrow as t_0 becomes large. As a result, we can assume that the matrix element $\langle k|H'(0)|m\rangle$ does not vary over the width of the peak and can be taken outside the integral in Eq. C.17. We write

$$x = \frac{1}{2}(\omega_{km} - \omega) t_0$$

and use the integral (extending the limits of the integral in Eq. C.17 to $\pm\infty$)

$$\int_{-\infty}^{\infty} x^{-2} \sin^2 x = \pi \tag{C.18}$$

to get

$$W_m = \frac{2\pi}{\hbar} \rho(k) \, |\langle k|H'|m\rangle|^2$$

or equivalently (we will denote the time-independent amplitude $H'(0)$ by H')

$$W_m = \frac{2\pi}{\hbar} \sum_{\text{final states}} |\langle k|H'|m\rangle|^2 \, \delta(\hbar\omega_{km} - \hbar\omega) \tag{C.19}$$

This is the Fermi golden rule. A similar calculation for the case where $\omega_{km} = -\omega$ gives

$$W_m = \frac{2\pi}{\hbar} \sum_{\text{final states}} |\langle k|H'|m\rangle|^2 \, \delta(\hbar\omega_{km} + \hbar\omega) \tag{C.20}$$

According to the Fermi golden rule the scattering rate depends upon: (i) the matrix element coupling the initial and final state and (ii) density of electrons in the final state. It is possible to alter the density of states by changing the dimensionality of the electronic system as discussed in the text.

APPENDIX D

LATTICE VIBRATIONS AND PHONONS

In Chapter 4 we have discussed transport of electrons in crystalline and disordered materials. We have discussed how electrons suffer scattering during transport. An important source of scattering is due to the vibrations of atoms in the solid. In crystalline materials, these scattering processes hinder transport and reduce mobility. However, in disordered materials where electrons are in localized states (where they cannot move in the material), lattice vibrations help increase the mobility (conductivity).

In Chapter 1 we have discussed how atoms are arranged in a crystalline material. The reason a particular crystal structure is chosen by a material has to do with the minimum energy of the system. As atoms are brought together to form a crystal, there is an attractive potential that tends to bring the atoms closer and a repulsive potential which tends to keep them apart. As a result the overall energy-configuration profile for the system has a schematic form, shown in Fig. D.1. The total energy of the system is minimum when the atomic spacing becomes R_0 as shown in the figure.

In general we can expand the crystal binding energy around the point R_0 as follows

$$U(R) = U(R_0) + \left(\frac{dU}{dR}\right)_{R_0} \Delta R + \frac{1}{2}\left(\frac{d^2U}{dR^2}\right)_{R_0} \Delta R^2 + \ldots \tag{D.1}$$

The second term is zero since R_0 is the equilibrium interatomic separation. Retaining terms to the second order in ΔR (this is called the harmonic approximation), we get

$$U(R) = U(R_0) + \frac{1}{2}C(\Delta R)^2 \tag{D.2}$$

where

$$C = \frac{\partial^2 U}{\partial R^2} \tag{D.3}$$

Lattice vibrations and phonons

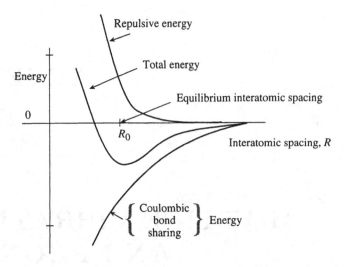

Figure D.1: General form of the binding energy versus atomic distance of a crystal. In the case of most semiconductors, the long-range attraction is due to either electrostatic interactions of the ions or the bond-sharing energy of the covalent bond.

is the force constant of the material. The restoring force is then

$$\text{Force} = -C\Delta R \tag{D.4}$$

Due to this restoring force the atoms in the crystal vibrate as a particle attached to a spring would do. We will now discuss such vibrations for semiconductors. Let us consider a diatomic lattice (two atoms per basis) as shown in Fig. D.2. The atoms are at an equilibrium position around which they vibrate. There is a restoring force (let us assume this force is between the nearest neighbors only). We assume that the atoms have masses M_1 and M_2.

If u_s and v_s represent the displacements of the two kinds of atoms of the unit cell s (see Fig. D.2), we get the following equations of motion for the atoms in the unit cell s

$$M_1 \frac{d^2 u_s}{dt^2} = C(v_s + v_{s-1} - 2u_s) \tag{D.5}$$

$$M_2 \frac{d^2 v_s}{dt^2} = C(u_{s+1} + u_s - 2v_s) \tag{D.6}$$

We look for solutions of the traveling wave form, but with different amplitudes u and v on alternating planes

$$\begin{aligned} u_s &= u \exp(iska) \exp(-i\omega t) \\ v_s &= v \exp(iska) \exp(-i\omega t) \end{aligned} \tag{D.7}$$

We note that a is the distance between nearest identical planes and not nearest planes, i.e. it is the minimum distance of periodicity in the crystal as shown in Fig. D.3. Eq.

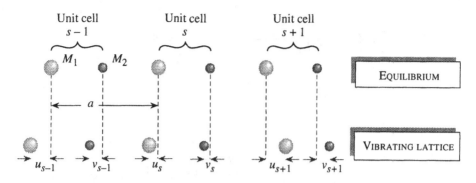

Figure D.2: Vibrations in a crystal with two atoms per unit cell with masses M_1, M_2 connected by force constant C between adjacent planes.

D.7, when substituted in Eqs. D.5 and D.6 gives

$$-\omega^2 M_1 u = Cv[1 + \exp(-ika)] - 2Cu \quad (D.8)$$
$$-\omega^2 M_2 v = Cu[\exp(-ika) + 1] - 2Cv \quad (D.9)$$

These are coupled eigenvalue equations which can be solved by the matrix method. The equations can be written as the matrix vector product

$$\begin{vmatrix} -\omega^2 M_1 + 2C & -C[1 + \exp(-ika)] \\ -C[\exp(-ika) + 1] & -\omega^2 M_2 + 2C \end{vmatrix} \begin{vmatrix} u \\ v \end{vmatrix} = 0$$

Equating the determinant to zero, we get

$$\left| 2C - M_1 \omega^2 \quad -C[1 + \exp(-ika)] - C[1 + \exp(ika)] \quad 2C - M_2 \omega^2 \right| = 0$$

or

$$M_1 M_2 \omega^4 - 2C(M_1 + M_2)\omega^2 + 2C^2(1 - \cos ka) = 0 \quad (D.10)$$

This gives the solution

$$\omega^2 = \frac{2C(M_1 + M_2) \pm [4C^2(M_1 + M_2)^2 - 8C^2(1 - \cos ka)M_1 M_2]^{1/2}}{2M_1 M_2} \quad (D.11)$$

It is useful to examine the results at two limiting cases. For small k, we get the two solutions

$$\omega^2 \approx 2C \left(\frac{1}{M_1} + \frac{1}{M_2} \right) \quad (D.12)$$

and

$$\omega^2 \approx \frac{C/2}{M_1 + M_2} k^2 a^2 \quad (D.13)$$

Near $k = \pi/a$ we get (beyond this value the solutions repeat)

$$\omega^2 = 2C/M_2$$
$$\omega^2 = 2C/M_1 \quad (D.14)$$

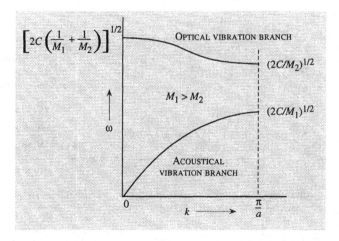

Figure D.3: Optical and acoustical branches of the dispersion relation for a diatomic linear lattice

The general dependence of ω on k is shown in Fig. D.3. Two branches of lattice vibrations can be observed in the results. The lower branch, which is called the acoustic branch, has the property as for the monatomic lattice, ω goes to zero as k goes to zero. The upper branch, called the optical branch, has a finite ω even at $k = 0$.

The acoustical branch represents the propagation of sound waves in the crystal. The sound velocity is

$$v_s = \frac{d\omega}{dk} = \sqrt{\frac{C}{2(M_1 + M_2)}}\, a \tag{D.15}$$

It is important to examine the eigenfunctions (i.e., u_s), for the optical branch and the acoustic branch of the dispersion relation. For $k = 0$, for the optical branch, we have, after substituting

$$\omega^2 = 2C\left(\frac{1}{M_1} + \frac{1}{M_2}\right) \tag{D.16}$$

in the equation of motion (say, Eq. D.8)

$$u = \frac{-M_2}{M_1} v \tag{D.17}$$

The two atoms vibrate against each other, but their center of mass is fixed. If we examine the acoustic branch, we get $u = v$ in the long wavelength limit. In Fig. D.4a we show the different nature of vibration of the acoustic and optical mode.

Note that for each wavevector, **k**, there will be a longitudinal mode and two transverse modes. The frequencies of these modes will, in general, be different, since the restoring force will be different. When optical vibration takes place in ionic materials like GaAs, polarization fields are set up that vibrate as well. These fields are important for longitudinal vibration, but not for translational vibration. As a result, in longitudinal vibrations there is an additional restoring force due to the long-range polarization. In

Fig. D.4b we show the lattice vibration frequency wavevector relation for GaAs. Notice that the longitudinal optical mode frequency is higher than that of the transverse mode frequency.

Phonons and scattering of electrons

The vibrations of atoms in a lattice are an important source of scattering of electrons. According to quantum mechanics the classical waves of lattice vibrations are represented by particles called phonons. This is similar to calling electromagnetic waves by the particle description, photons. The lattice vibrations cause local strains in the crystal which, in turn, translate into local variations in the electronic spectra. In particular the conduction and valence band energies see small perturbations. These perturbations are the source of electron scattering.

In the electron–phonon scattering an electron can absorb a phonon of energy $\hbar\omega$ and change its energy from E_i to E_f where

$$E_f = E_i + \hbar\omega \tag{D.18}$$

An electron can also emit a phonon of energy $\hbar\omega$ and in this case

$$E_f = E_i - \hbar\omega \tag{D.19}$$

In addition to energy conservation in the scattering process, momentum has to be conserved as well

$$k_f = k_i + q_{\text{phonon}} + G \tag{D.20}$$

where k_i, k_f and q_{phonon} represent the wave vectors for the initial electron, final electron, and the phonon. The vector G is called the reciprocal lattice vector and arises from the periodicity of the system.

An important question in relation to phonon scattering is the phonon number or occupation of phonon states. This is given by Bose Einstein statistics.

$$\langle n(\hbar\omega) \rangle = \frac{1}{\exp\left(\hbar\omega/k_B T\right) - 1} \tag{D.21}$$

It can be seen that as temperature increases the phonon number increases. Here are some important outcomes of electron–phonon scattering:

- At low temperatures (T<77 K) and low electric fields ($E \lesssim 1\,\text{kV/cm}$) the dominant scattering is due to the lower energy acoustic phonons. The higher energy optical phonons have occupation numbers that are too low to contribute to scatterings.

- At low temperatures and high electric fields ($E \gtrsim 10\,\text{kV/cm}$) electrons have high energies and can emit phonons. Phonon emission is the dominant source of scattering.

- At high temperatures (T∼ 300 K) optical phonon emission and absorption are dominant.

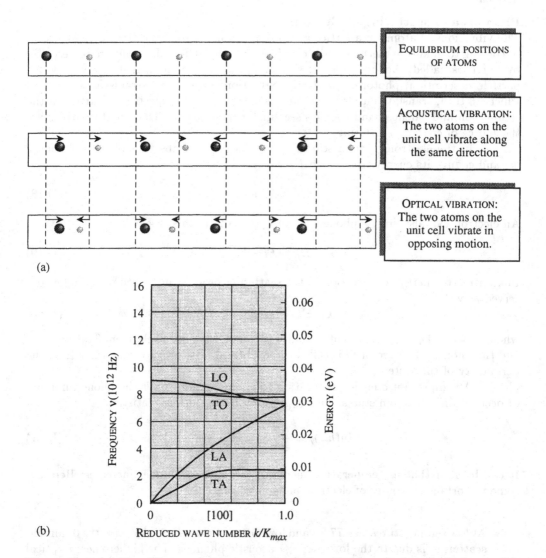

Figure D.4: (a) The difference between as acoustical mode an optical mode is shown. (b) Phonon dispersion relation in GaAs. The longitudinal (LO, LA) and transverse (TO, TA) optical and acoustical modes are shown.

- In indirect bandgap semiconductors (like Si) bandgap energy photons (or higher) are absorbed through phonon participation. This is due to momentum conservation issues discussed in Chapter 5. In these materials optical absorption increases as temperature increases, since more phonons are present.

APPENDIX E

DEFECT SCATTERING AND MOBILITY

In this appendix we will discuss scattering of electrons from defect in materials. The models used here to represent alloy scattering and ionized impurity scattering can be extended to cover a variety of other defects as well.

E.1 ALLOY SCATTERING

An interesting scattering problem is the scattering of a particle from a spherically symmetric square well potential. An important application of such scattering is in the understanding of carrier transport in alloys. Alloys are used in a number of technologies and are based on a material synthesis process that can randomly mix two or more different materials. By using a mixture of materials it is possible to obtain new materials that have properties intermediate between those of the components. Ideal alloys represent a random arrangement of atoms on a lattice and, therefore, even if the lattice is periodic the potential seen by electrons is non-periodic. Thus when an electron is in an alloy, it sees a randomly varying potential. To simplify the problem the random potential is separated into a periodic potential and one with random fluctuations. We write

$$H = H_0 + H' \qquad (E.1)$$

where H_0 results from the average potential of the alloy and H' arises from the difference of the real and average potentials.

Let us consider the problem of a perfectly random alloy where the smallest physical size over which the crystal potential fluctuates randomly is the unit cell. An electron moving in the alloy $A_x B_{1-x}$ will see a random potential, schematically shown in Fig. E.1.

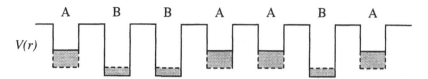

Figure E.1: A schematic of the actual atomic potential (solid line) and the average virtual crystal potential (dashed line) of an A–B alloy. The shaded area shows the difference between the real potential and the virtual crystal approximation.

The average potential and the average bandstructure of the alloy are described to the lowest order by the *virtual crystal approximation*. In this approximation, the averaging of the atomic potentials

$$\{M\}_{\text{all}} = x\{M\}_A + (1-x)\{M\}_B \tag{E.2}$$

gives an average *periodic* potential represented by the dashed line in Fig. E.1. An important approximation is now made. The difference between the real potential and the assumed virtual crystal potential is represented within each unit cell by a highly localized potential. For example, for A-type atoms, the difference is

$$\begin{aligned} E_{\text{all}} - E_A &= xE_A + (1-x)E_B - E_A \\ &= (1-x)[E_B - E_A] \\ &= (1-x)U_{\text{all}} \end{aligned} \tag{E.3}$$

Similarly, for the B atom, the difference is

$$\begin{aligned} E_B - E_{\text{all}} &= x[E_B - E_A] \\ &= x\, U_{\text{all}} \end{aligned} \tag{E.4}$$

The scattering potential is chosen to be of the form

$$U(\mathbf{r}) = \begin{array}{ll} U_0 & \text{for } |\mathbf{r}| \leq r_0 \\ = 0 & \text{for } |\mathbf{r}| > r_0 \end{array} \tag{E.5}$$

where r_0 is the interatomic distance. If we use the Born approximation to calculate the scattering rate, we have

$$W(\mathbf{k}) = \frac{2\pi}{\hbar} \sum_{\mathbf{k}'} |M_{\mathbf{kk}'}|^2 \delta(E_\mathbf{k} - E_{\mathbf{k}'})$$

and

$$M_{\mathbf{kk}'} = \int e^{i(\mathbf{k}-\mathbf{k}')\cdot \mathbf{r}} \Delta U(\mathbf{r})\, d^3 r$$

We will now use the fact that the scattering potential only extends over a unit cell, and over this small distance

$$e^{i(\mathbf{k}-\mathbf{k}')\cdot\mathbf{r}} \approx 1 \tag{E.6}$$

E.1. Alloy scattering

Thus
$$M_{\mathbf{kk'}} = \frac{4\pi}{3} r_0^3 U_0 \tag{E.7}$$

and
$$W(\mathbf{k}) = \frac{2\pi}{\hbar} \left(\frac{4\pi}{3} r_0^3 U_0\right)^2 \frac{1}{(2\pi)^3} \int \delta(E_\mathbf{k} - E_{\mathbf{k'}}) \, d^3k'$$
$$= \frac{2\pi}{\hbar} \left(\frac{4\pi}{3} r_0^3 U_0\right)^2 N(E_\mathbf{k}) \tag{E.8}$$

Let us consider the face-centered cubic (fcc) lattice (the lattice for most semiconductors). We may write
$$r_0 = \frac{\sqrt{3}}{4} a$$
where a is the cube edge for the fcc lattice. This gives
$$\left(\frac{4\pi}{3} r_0^3\right)^2 = \frac{3\pi^2}{16} V_0^2 \tag{E.9}$$

where $V_0 = a^3/4$ is the volume of the unit cell. We finally obtain for the scattering rate:
$$W(\mathbf{k}) = \frac{2\pi}{\hbar} \left(\frac{3\pi^2}{16} V_0^2\right) U_0^2 N(E_\mathbf{k}) \tag{E.10}$$

We will now assume that all scattering centers scatter independently, so that we can simply sum the scattering rates. For A-type atoms, the scattering rate is (using $U_0 = (1-x)U_{all}$ from Eq. E.2)
$$W_A(\mathbf{k}) = \frac{2\pi}{\hbar} \left(\frac{3\pi^2}{16} V_0^2\right) (1-x)^2 U_{all}^2 \, N(E_\mathbf{k}) \tag{E.11}$$

For B-type atoms, the rate is (using $U_0 = xU_{all}$ from Eq. E.3)
$$W_A(\mathbf{k}) = \frac{2\pi}{\hbar} \left(\frac{3\pi^2}{16} V_0^2\right) x^2 U_{all}^2 \, N(E_\mathbf{k}) \tag{E.12}$$

There are x/V_0 A-type atoms and $(1-x)/V_0$ B-type atoms in the unit volume, so that the total scattering rate is
$$W_{tot} = \frac{2\pi}{\hbar} \left(\frac{3\pi^2}{16} V_0\right) U_{all}^2 \, N(E_\mathbf{k}) \left[x(1-x)^2 + (1-x)x^2\right]$$
$$= \frac{3\pi^3}{8\hbar} V_0 U_{all}^2 \, N(E_\mathbf{k}) \, x(1-x) \tag{E.13}$$

E.2 SCREENED COULOMBIC SCATTERING

As another example of the use of the Fermi golden rule or Born approximation, we will examine the scattering of an electron from a charged particle. The scattering potential is Coulombic in nature. This scattering plays a very important role in many important applications. Problems that require an understanding of this scattering process include:

- *Scattering of α particles in matter*: When a thin film of metal is bombarded with α particles (He-nuclei), the properties of the outgoing particles are understood on the basis of Coulombic scattering.

- *Mobility in devices*: Semiconductor devices have regions that are doped with donors or acceptors. These dopants provide excess carriers in the conduction or the valence band. Without these carriers most devices will not function. When a dopant provides a free carrier, *the remaining ion provides a scattering center for the free carriers*. This causes scattering which is understood on the basis of electron-ion scattering. Additionally, at high densities one can have electron-electron scattering as well as electron–hole scattering, which is also understood for the general problem discussed in this subsection.

Before starting our study of scattering from a Coulombic interaction, it is important to note that in most materials there is a finite *mobile carrier density*. These carriers can adjust their spatial position in response to a potential and thus *screen* the potential. The screening is due to the dielectric response of the material and includes the effect that the background ions as well as the other free electrons have on the potential. A number of formalisms have been developed to describe the dielectric response function. We will use a form given by the Thomas-Fermi formalism.

Let us consider an electron scattering from a charged particle in a crystalline material. We will assume that the electron is described by the effective mass theory. We also asume that the density of free carriers is n_0. In the Thomas-Fermi formalism, the background free carriers modify their carrier concentration near the impurity so that when the scattering electron is far from the impurity it sees a potential much weaker than the Coulombic potential. Very close to the impurity the potential is not affected much by the screening.

The real-space behavior of the screened potential is given by

$$\phi_{tot}(\mathbf{r}) = \frac{q}{4\pi\epsilon r}e^{-\lambda r} \tag{E.14}$$

where q is the charge of the impurity and ϵ is the dielectric constant. The quantity λ, which represents the effect of the background free carriers is given for a non-degenerate carrier gas (i.e., a carrier distribution where the Fermi statistics is reasonably approximated by the Boltzmann statistics) as

$$\lambda^2 = \frac{n_0 e^2}{\epsilon k_B T} \tag{E.15}$$

E.2. Screened Coulombic scattering

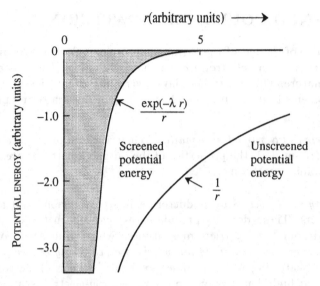

Figure E.2: Comparison of screened and unscreened Coulomb potentials of a unit positive charge as seen by an electron. The screening length is λ^{-1}.

When the free carrier density is high so that the carriers are degenerate,

$$\lambda^2 = \frac{3n_0 e^2}{2\epsilon E_F} \quad (E.16)$$

where E_F is the Fermi energy.

As noted, the effect of screening is to reduce the range of the potential from a $1/r$ variation to a $\exp(-\lambda r)/r$ variation. This is an extremely important effect and is shown schematically in Fig. E.2.

We now calculate the matrix element for the screened Coulombic potential

$$U(\mathbf{r}) = \frac{Ze^2}{4\pi\epsilon} \frac{e^{-\lambda r}}{r} \quad (E.17)$$

where Ze is the charge of the impurity. We choose the initial normalized state to be $|k\rangle = \exp(i\mathbf{k} \cdot \mathbf{r})/\sqrt{V}$ and the final state to be $|k'\rangle = \exp(i\mathbf{k'} \cdot \mathbf{r})/\sqrt{V}$, where V is the volume of the crystal. The matrix element is then

$$M_{\mathbf{kk'}} = \frac{Ze^2}{4\pi\epsilon V} \int e^{-i(\mathbf{k'}-\mathbf{k})\cdot\mathbf{r}} \frac{e^{-\lambda r}}{r} r^2 dr \sin\theta' d\theta' d\phi'$$

Carrying out the ϕ' integration which gives a factor of 2π, we have

$$M_{\mathbf{kk'}} = \frac{Ze^2}{4\pi\epsilon V} 2\pi \int_0^\infty r\, dr \int_{-1}^1 d(\cos\theta') e^{-\lambda r} e^{-i|\mathbf{k'}-\mathbf{k}|r\cos\theta'}$$

$$= \frac{Ze^2}{4\pi\epsilon V} 2\pi \frac{2}{|\mathbf{k'}-\mathbf{k}|^2 + \lambda^2}$$

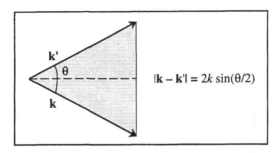

Figure E.3: As a consequence of the elastic scattering, there is a simple relation between the magnitude of the scattered wavevector and the scattering angle θ.

We note that $|\mathbf{k}'| = |\mathbf{k}|$ since the scattering is elastic. Then, as can be seen from Fig. E.3,

$$\left|\mathbf{k} - \mathbf{k}'\right| = 2k\sin(\theta/2) \tag{E.18}$$

where θ is the polar scattering angle.

$$M_{\mathbf{k}\mathbf{k}'} = \frac{Ze^2}{V\epsilon} \frac{1}{4k^2 \sin^2(\theta/2) + \lambda^2}$$

The scattering rate is now given by the Born approximation:

$$W(\mathbf{k}, \mathbf{k}') = \frac{2\pi}{\hbar} \left(\frac{Ze^2}{V\epsilon}\right)^2 \frac{\delta(E_\mathbf{k} - E_{\mathbf{k}'})}{\left(4k^2 \sin^2(\theta/2) + \lambda^2\right)^2} \tag{E.19}$$

One can see that in the two extremes of no screening ($\lambda \to 0$) and strong screening ($\lambda \to \infty$), the rate becomes, respectively,

$$W(\mathbf{k}, \mathbf{k}') = \frac{2\pi}{\hbar} \left(\frac{Ze^2}{V\epsilon}\right)^2 \frac{\delta(E_\mathbf{k} - E_{\mathbf{k}'})}{16k^4 \sin^4(\theta/2)} \tag{E.20}$$

and

$$W(\mathbf{k}, \mathbf{k}') = \frac{2\pi}{\hbar} \left(\frac{Ze^2}{V\epsilon}\right)^2 \frac{\delta(E_\mathbf{k} - E_{\mathbf{k}'})}{\lambda^4} \tag{E.21}$$

The angular dependence of the scattering process is very important. One can intuitively see that scattering that produces a large angle scattering is much more effective in altering the motion of electrons than small-angle scattering. In fact, since the scattering is elastic, a forward-angle scattering (scattering angle is zero) has no effect on the motion of the initial electron. Thus it is important to examine the angular dependence of $W(\mathbf{k}, \mathbf{k}')$. This is plotted as a function of the scattering angle θ in Fig. E.4. The ionized impurity scattering has a strong forward angle bias, as can be seen.

Relation between τ and $W(\mathbf{k}, \mathbf{k}')$

In general, the relations between τ to be used for mobility, and $W(\mathbf{k}, \mathbf{k}')$ calculated by

E.2. Screened Coulombic scattering

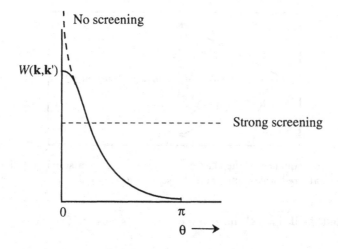

Figure E.4: Angular dependence of the scattering by ionized impurities. The scattering has a strong forward angle preference.

the Fermi golden rule (or Born approximation) is quite complicated. However, if the scattering is elastic, i.e., no energy is lost in the scattering process, the relation becomes quite simple. We find that if the angle between **k** and **k'** is θ, we have

$$\frac{1}{\tau} = \int W(\mathbf{k}, \mathbf{k}')(1 - \cos\theta)\frac{d^3k'}{(2\pi)^3} \tag{E.22}$$

The factor $(2\pi)^3$ in the denominator comes from the definition of the density of states in a k-space volume d^3k'. From this expression we see that elastic scattering processes which do not change the angle of a particle are ineffective in transport. Scattering where $\theta = 0$ is called forward-angle scattering, while if $\theta = \pi$ it is called back-scattering.

In the next section we will apply the expression given above to ionized impurity and alloy scattering. For inelastic scattering processes, there is no simple relation between τ and $W(\mathbf{k}, \mathbf{k}')$ and the problem has to be solved numerically. However, an approximate value of τ can be obtained by taking the inverse of $W(\mathbf{k})$, i.e., of the integral of $W(\mathbf{k}, \mathbf{k}')$ over all final states.

Averaging over electron energies

The expression given above for the scattering time τ will, in general, depend on the energy of the electron. Since the electrons are distributed with different energies, how does one find the averaged τ to be used in transport? Time averaging is given by

$$\langle\langle\tau\rangle\rangle = \frac{\int \tau E f^\circ(E) dE}{\int E f^\circ(E) dE} \tag{E.23}$$

where $f^\circ(E)$ is the Fermi–Dirac distribution function. For the case where the value of $f^\circ(E)$ is small, i.e. we can use the Boltzmann distribution function, the averaging essentially means that $\tau(E)$ is approximately replaced by $\tau(k_B T)$; i.e., the average energy of the electrons is taken as $\sim k_B T$.

Effect of various scattering processes
In general, as electrons move in a semiconductor they will suffer from a number of distinct scattering mechanisms. These may involve ionized impurities, alloy disorder, lattice vibrations, etc. These scattering processes can be assumed to be independent and, to a reasonable approximation, the following approximate rule (called Mathieson's rule) can be applied

$$\frac{1}{\mu_{tot}} = \sum_i \frac{1}{\mu_i} \qquad (E.24)$$

where μ_i is the mobility limited by the scattering process i.

In modern semiconductor devices, scattering is an integral part of device operation. However, as device dimensions shrink, it will be possible for electrons to move without scattering.

E.3 IONIZED IMPURITY LIMITED MOBILITY

Based on the brief discussion of transport theory in the previous subsection, let us evaluate the relaxation time weighted with the $(1 - \cos\theta)$ factor (this time is known as the *momentum relaxation time*), which is used to obtain the low field carrier mobility.

$$\begin{aligned}
\frac{1}{\tau} &= \frac{V}{(2\pi)^3} \int (1 - \cos\theta)\, W(\mathbf{k}, \mathbf{k}')\, d^3k' \\
&= \frac{2\pi}{\hbar} \left(\frac{Ze^2}{\epsilon}\right)^2 \frac{1}{V^2} \frac{V}{(2\pi)^3} \\
&\quad \times \int (1 - \cos\theta) \frac{\delta(E_\mathbf{k} - E_{\mathbf{k}'})}{(4k^2 \sin^2(\theta/2) + \lambda^2)^2} k'^2\, dk'\, \sin\theta\, d\theta\, d\phi \\
&= \frac{1}{2\hbar} \left(\frac{Ze^2}{\epsilon}\right)^2 \frac{1}{V} \\
&\quad \times \int (1 - \cos\theta)\, (4k^2 \sin^2(\theta/2) + \lambda^2)^2\, N(E_\mathbf{k})\, dE_\mathbf{k}\, d(\cos\theta)\, d\phi \\
&= \frac{1}{\hbar} \left(\frac{Ze^2}{\epsilon}\right)^2 \frac{1}{V} \frac{N(E_\mathbf{k})}{32k^4} \\
&\quad \times \int (1 - \cos\theta) \frac{1}{\left[\sin^2(\theta/2) + \left(\frac{\lambda}{2k}\right)^2\right]^2}\, d(\cos\theta)\, d\phi \\
&= F \int (1 - \cos\theta) \frac{1}{\left[\sin^2(\theta/2) + \left(\frac{\lambda}{2k}\right)^2\right]^2}\, d(\cos\theta)\, d\phi \qquad (E.25)
\end{aligned}$$

E.3. Ionized impurity limited mobility

Finally

$$\frac{1}{\tau} = \frac{\pi}{4\hbar}\left(\frac{Ze^2}{\epsilon}\right)^2 \frac{N(E_{\mathbf{k}})}{Vk^4}\left[\ln\left(1+\left(\frac{2k}{\lambda}\right)^2\right) - \frac{1}{1+(\lambda/2k)^2}\right]$$

$$N(E_{\mathbf{k}}) = \frac{m^{*3/2}E^{1/2}}{\sqrt{2}\pi^2\hbar^3} \quad (E.26)$$

Note that the spin degeneracy is ignored, since the ionized impurity scattering cannot alter the spin of the electron. In terms of the electron energy, $E_{\mathbf{k}}$, we have

$$\frac{1}{\tau} = \frac{1}{V16\sqrt{2\pi}}\left(\frac{Ze^2}{\epsilon}\right)^2 \frac{1}{m^{*1/2}E_{\mathbf{k}}^{3/2}}$$

$$\times \left[\ln\left(1+\left(\frac{8m^*E_{\mathbf{k}}}{\hbar^2\lambda^2}\right)\right) - \frac{1}{1+(\hbar^2\lambda^2/8m^*E_{\mathbf{k}})}\right] \quad (E.27)$$

To calculate the mobility limited by ionized impurity scattering, we have to find the ensemble averaged τ. To a good approximation, the effect of this averaging is essentially to replace $E_{\mathbf{k}}$ by $k_B T$ in the expression for $1/\tau$. A careful evaluation of the average $\langle\langle\tau\rangle\rangle$ gives (see Eq. E.23)

$$\frac{1}{\langle\langle\tau\rangle\rangle} = \frac{1}{V128\sqrt{2\pi}}\left(\frac{Ze^2}{\epsilon}\right)^2 \frac{1}{m^{*1/2}(k_BT)^{3/2}}$$

$$\times \left[\ln\left(1+\left(\frac{24m^*k_BT}{\hbar^2\lambda^2}\right)\right) - \frac{1}{1+\left(\frac{\hbar^2\lambda^2}{24m^*k_BT}\right)}\right] \quad (E.28)$$

If there are N_i impurities per unit volume, and if we assume that they scatter electrons independently, the total relaxation time is simply obtained by multiplying the above results by $N_i V$,

$$\frac{1}{\langle\langle\tau\rangle\rangle} = \frac{N_i}{128\sqrt{2\pi}}\left(\frac{Ze^2}{\epsilon}\right)^2 \frac{1}{m^{*1/2}(k_BT)^{3/2}}$$

$$\times \left[\ln\left(1+\left(\frac{24m^*k_BT}{\hbar^2\lambda^2}\right)\right) - \frac{1}{1+\left(\frac{\hbar^2\lambda^2}{24m^*k_BT}\right)}\right] \quad (E.29)$$

The mobility is then

$$\mu = \frac{e\langle\langle\tau\rangle\rangle}{m^*}$$

Mobility limited by ionized impurity scattering has the special $\mu \sim T^{3/2}$ behavior that is represented in Eq. E.29. This temperature dependence (the actual temperature dependence is more complex due to the other T-dependent terms present) is a special

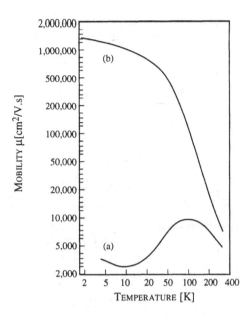

Figure E.5: (a) A typical plot of electron mobility as a function of temperature in a uniformly doped GaAs with $N_D = 10^{17}$ cm^{-3}. The mobility drops at low temperature due to ionized impurity scattering, becoming very strong. In contrast, the curve (b) shows a typical plot of mobility in a modulation-doped structure where ionized impurity is essentially eliminated.

signature of the ionized impurity scattering. One can understand this behavior physically by realizing that at higher temperatures the electrons are traveling faster and are less affected by the ionized impurities.

Ionized impurity scattering plays a very central role in controlling the mobility of carriers in semiconductor devices. This is especially true at low temperatures where the other scattering processes (due to lattice vibrations) are weak. To avoid impurity scattering, the concept of *modulation doping* has been developed. In this approach, the device is made from two semiconductors—a large bandgap *barrier* layer and a smaller bandgap *well* layer. The barrier layer is doped so that the free carriers spill over into the well region where they are physically separated from the dopants. This essentially eliminates ionized impurity scattering. Fig. E.5 compares the mobilities of conventionally doped and modulation doped GaAs channels. As can be seen, there is a marked improvement in the mobility, especially at low temperatures. Modulation doping forms the basis of the highest performance semiconductor devices in terms of speed and noise.

In Fig. E.6 we show how the mobility in Ge, Si and GaAs varies as a function of doping density. The mobility shown includes the effects of lattice scattering, as well as ionized impurity scattering.

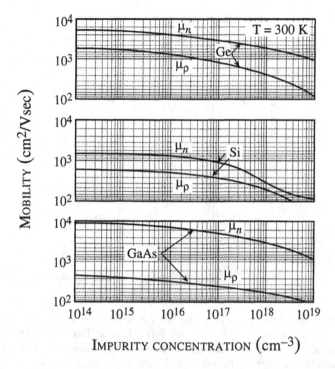

Figure E.6: Drift mobility of Ge, Si, and GaAs as 300 K versus impurity concentration. (After S. M. Sze, *Physics of Semiconductor Devices*, 2nd ed., John Wiley and Sons, New York, 1981.)

E.4 ALLOY SCATTERING LIMITED MOBILITY

The ensemble averaged relaxation time for the alloy scattering is quite simple (see Eqs. E.12 and E.23):

$$\frac{1}{\langle\langle\tau\rangle\rangle} = \frac{3\pi^3}{8\hbar} V_0 U_{all}^2 x(1-x) \frac{m^{*3/2}(k_B T)^{1/2}}{\sqrt{2}\pi^2 \hbar^3} \frac{1}{0.75} \qquad (E.30)$$

according to which the mobility due to alloy scattering is

$$\mu_{all} \propto T^{-1/2} \qquad (E.31)$$

Thus in 3D systems the mobility decreases with temperature. This temperature dependence should be contrasted to the situation for the ionized impurity scattering.

Index

Absorption coefficient, 214
Absorption coefficient, in indirect semiconductors, 216
Accelerometer, 288
Acceptor level, 128
Acoustic phonon, mobility, 159
Acoustic vibrations, 157, 388
Acoustic waves, 388
Acousooptic device, 322
Actuator, piezoelectric, 288
Aharonov–Bohm effect, 335
Alloy, 12, 131
Alloy, GaAs/AlAs, 146, 366
Alloy, HgTe/CdTe, 146
Alloy, InAs/GaAs, 146, 366
Alloy scattering, 161, 393
Aluminum nitride, 116
Amorphous materials, 26
Angular momentum, 45, 341
Anisotropy, 279, 299
Antenna, phased array, 355
Antiferromagnetic, 348
Artificial structures, 16
Atomic spectra, 43
Atomic, theory, 45
Averaging procedures for scattering time, 164, 399

Ballistic transport, 154
Band, tailing, 138
Band lineups in heterostructures, 133
Band theory of solids, 84, 85
Bandedge states, 136
Bandgap, strain effects, 159
Bandgap, temperature dependence of, 116
Bandstructure, 85, 110
Bandstructure, effects on devices, 184

Bandstructure, in quantum wells, 133
Bandstructure, of AlAs, 115
of GaAs, 113
of Ge, 115
of nitrides, 116
Bandstructure,
of alloys, 131
of Si, 112
Bandstructure, valence band, in quantum wells, 135
Bandtail states, 138
Barium titanate, 281
Basis, 2
BCS theory, 109
Binding energy of crystals, 387
Birefringence, 299
Blackbody radiation, 39
Bloch function, 85
Bloch theorem, 84
Body centered cubic (bcc) lattice, 3
Bohr, magneton, 332
radius, 59
theory of atoms, 45
Boltzmann, statistics, 102
Bound state problem, 132
Bragg's law, 322
Breakdown, electric, 171
Breakdown, in devices, 184

Cantilever, sensor, 288
Capacitance, dielectric response, 270
Carrier, extrinsic, 126
Carriers, intrinsic, 122
Carrier, freezeout, 130
Cavity gain, 247
Circular polarization, 301
Coercive field,
ferroelectric, 274

magnetic, 328
Compliance, 279
Conduction band, 106
Conduction bandedge states, 106
Conduction, hopping, 190
Conductivity, 163
Conservation of momentum, 211
Cooper pairs, 109, 338
Critical thickness, 23
Crystal, binding, 378
Crystal growth, bulk, 23
    epitaxial, 23
Crystal, restoring force, 387
Crystal structure, 5
Curie temperature, 274, 349
Cutoff wavelength, 204
Coupled wells, 67

Defect, interstitial, 21
Defect, substitutional, 21
Defects,
    in crystals, 20
Deformation potential theory, 157
Demagnetization for cooling, 354
Density of states, 80
    effective, 122
    in 2D systems, 81
    in one dimension, 81
    in semiconductors, 122
    in three dimensions, 80
Depolarization, 268
Detection of light, 204, 231
Diamagnetism, 328, 340
Diamond lattice, 9
Dielectric,
    in AC fields, 270
Dielectric response, 269, 271
Diffusion, 173
Diffusion, coefficient, 173
Dipole,
    electric, 265
Dislocation generation, 23
Disordered semiconductors, 138
    extended states, 139
    localized states, 139

Disordered system, transport, 188
Direct gap, 111
Distribution function, 102
Domains, 278
Donor, energy levels, 127
Doping, 126,128

Eddy currents, 326
Effective mass, 88
Effective mass, equation, 88
Einstein model, 40
Einstein relation, 176
Ehrenfest theorem, 54
Elastic collisions, 156
Elastic constants, 284
Elastic strain, 279
Electric fields, built-in, from
    strained epitaxy, 285
Electromagnetic fields,
    in magnetic materials, 329
Electron–hole recombination, 222
Electro-absorption, 309
Electro-optic,
    coefficients, 305
    effect, 303, 316
    image storage, 317
Electrons, in a magnetic field,
    quantum theory, 332
Elliptical polarization, 300
Epitaxial crystal growth, 23
Epitaxy, coherent, 23, 281
Epitaxy, incoherent, 23
Epitaxy, lattice matched and
    dislocations, 23
Equation of motion, for k, 90
Exchange interaction, 346
Exciton, 254, 310
Exciton, absorption spectra, 256
Exciton, binding energy, 254
Extraordinary ray, 299
Extrinsic carriers, 129

Face centered cubic (fcc) lattice, 5
Faraday rotation, 331
Fermi energy, 102,118

Fermi vector, 118
Fermi velocity, 118
Fermi, Golden Rule, 380
Fermi–Dirac distribution, 102
Ferrite,
    microwave devices, 357
Ferroelectric materials,
    coercive field, 275
    domains, 278
Field effect transistor FET, 179
    operation, 179
    high performance issues, 184
Field, local, 268
Free carriers, 122
Freezeout, carrier, 130

Gain in a semiconductor, 225
Garnet, 327
Gas sensor, 195
Giant magnetoresistance, 359
Grain boundaries, 25
Group velocity, for lattice vibrations, 389
Gyromagnetic ratio, 330

Harmonic oscillator, 66
Heavy hole states, 112
Heisenberg uncertainty relation, 54
Heterointerface polar charge, 283
Heterostructures, bandlineup, 133
Heterostructures, structural, 23
Hexagonal close packed (hcp)
    structure, 9
Hexagonal structure, simple, 5
Holes, 104
Hole, effective mass, 111
Hole, energy, 106
Hole, momentum, 106
HOMO bonds, 108, 252
Hopping conductivity, 188
Humidity sensor, 195
Hydrogen atom, 56
Hydrogen molecule, 72
Hysteresis,
    ferroelectric, 273
    ferromagnetic, 328

Ideal surfaces, 17
Identical particle, 101
Impact ionization, 170
Impact ionization, coefficient, 171
Impact ionization, threshold, 171
Impurity, scattering, 161, 394
Incoherent light, 243
Indirect gap, 111
Inelastic collisions, 156
Information storage, 317, 357
Infrared,
    detection, 205
    pyroelectric devices, 289
    semiconductor devices, 178, 231
Insulators, simple description, 104
Interband transitions, bulk
    semiconductors, 212
Interband transitions, quantum wells, 224
Interface, 19
Interface roughness, 19
Interference, quantum, 336
Intrinsic carriers, 122
Ionic conduction, 191
Ionized impurity scattering, 161, 396

Joyce–Dixon approximation, 130
Junctions, 359

k-vector, significance of, 90
Kerr effect, 304
Kronig–Penney model, 85

Laser diode, 244
Laser, optical confinement, 246
Laser, threshold, 251
Lattice, 2
Lattice constant, for selected
    semiconductors, 13
Lattice types, 2
Lattice vibrations, 157, 386
Law of mass action, 122
Lead,
    PLZT, 281, 286
Light emitting diode, 238
Light hole states, 112
Line defects, 21

Liquid crystals, 27
Liquid crystal devices, 313, 321
Lithium niobate, 305
Lithium tantalate, 305
Localized states, 136
Loss factor in dielectrics, 270
LUMO bands, 108, 255

Magnetic effects, 327
    applications, 552
Magnetic semiconductors, 316
Magnetoresistance, 359
    giant, 359
Mass action, law of, 122
Maxwell equations, 38
Memory,
    magnetic, 357, 359
    semiconductor, 179
Metal–organic chemical vapor deposition (MOCVD), 23
Metals, electrons in, 117
Metals, simple description, 104
Microwave devices, 179, 355
Miller indices, 12
Mobility, 163
Mobility, edge, 140
Mobility, in GaAs, 164
Mobility, in Si, 164
Mobility, in modulation doped structures, 182, 402
Mobility, in selected semiconductors, 164
Modulation of light, 301, 312
Molecular beam epitaxy (MBE), 23
Molecular semiconductors, 107,
Mott conductivity model, 191

Newton's equation of motion, 38, 154
Nitrides, spontaneous polarization, 283
    bandstructure, 116
    piezoelectric effect, 282
Non-parabolic band, 114
Non-radiative processes, 239

OLEDs, 255
Optical axis, 299
Optical confinement, 245

Optical interband transitions, 213
Optical lattice vibrations, 389
Optical phonon, scattering, 157
Ordinary ray, 299

Paramagnetism, 327, 341
Pauli principle, 101
Permeability, 327
Perovskite structure, 12
Perturbation theory, 154, 381
Phonons, 157, 389
Phonon, acoustic scattering, 157
Phonon, conservation laws for scattering, 390
Phonon, dispersion, 158, 389
Phonon, optical mode, 157, 389
Phonon, optical scattering, 157
Phonon, statistics, 156
Photoelectric effect, 40
Piezoelectric effect, 279
    acoustic power, 289
    coefficient, 282, 284
    direct effect, 280
    sensor, 288
Planck constant, 40
Plasma, frequency, 329
PLZT, 282, 286
p–n diode,
    theory of, 368
    devices, 233, 240, 245
Pockel's effect, 306
Polar charge at interfaces, 284
Polar materials, 274
Polarization electric, 268
    control of, 273, 275, 285
Polarization of light 299
Poling, 278
Polycrystalline materials, 25
Polymers, 107, 255
Potential, screened Coulomb, 161
Power dissipation, 270
Poynting vector, 298
Pyroelectric,
    coefficients, 285
    devices, 288

Quantum confined Stark effect, 319
Quantum interference, 335, 352
Quantum wells, 62, 132, 310
Quasi-Fermi level, 219

Radiative lifetime, 215, 222
Radiative transitions,
       recombination time, 215, 222
Recording tape, 358
Reduced mass, 214
Refractive index, 301, 305
Remnant,
    magnetization, 328
    polarization, 274
Restoring force, crystal, 387

SAW, 323
Scattering, acoustic phonon, 159
Scattering, alloy, 161, 393
Scattering, ionized impurity, 161, 395
Scattering, phonon, 159
Scattering, polar optical phonon, 159
Scattering, time, 163
Schrödinger equation, 47
Screened Coulomb potential, 161, 397
Screening, length, 161, 397
Second quantization, phonons,
Semiconductor
      material properties, 116, 164, 363
Semiconductors, simple description, 104
Sensors,
    gas, 195
    humidity, 195
    strain, 288
Simple cubic lattice, 3
Solar cells, 235
Spinal ferrites, 327
Spintronics, 326
Spontaneous emission rate, 214
Squid Magnetometer, 352
Statistics, electrons, 102
Statistics, phonon, 159, 390
Stefan's law, 39
Stimulated emission, 215

Strain tensor,
    in epitaxy, 282
Strained heterostructures, 281
Superconducting state, 109
Superlattice structure, 18
Surface acoustic wave, 323
Surface reconstruction, 18
Surfaces, ideal and real, 18

$\tan \delta$,
    dielectric response, 270
Temperature, coefficient,
    bandgap change, 116
    pyroelectric materials, 286
Tensor, 279
    contracted notation, 279
Thermal sensors, 289
Transitions, radiative 210
Transport, in GaAs and Si, 164
Transport, high field, in GaAs, 164
Transport, high field in Si, 164
Transport, overview, 151
Tunneling, in semiconductors, 173
Twisted nematic, 320

Ultrasonic energy, 289
Uncertainty relation, 54

Variable range hopping, 191
Vegard's law, 207
Verdet constant, 35
Vertical transitions, 212
Vibration, crystal with diatomic basis, 38
Virtual crystal approximation, 131

Wave amplitude and probability, 51
Wien's law, 40
Work function, 42
Wurtzite structure, 11

Zinc-blende structure, 9

MATERIAL	CRYSTAL STRUCTURE	BANDGAP (eV)	STATIC DIELECTRIC CONSTANT	LATTICE CONSTANT (Å)	DENSITY (gm-cm^{-3})
C	DI	5.50, I	5.570	3.5668	3.5153
Si	DI	1.1242, I	11.9	5.431	2.3290
SiC	ZB	2.416, I	9.72	4.3596	3.166
Ge	DI	0.664, I	16.2	5.658	5.323
AlN	W	6.2, D	$\bar{\varepsilon} = 9.14$	$a = 3.111$ $c = 4.981$	3.255
AlP	ZB	2.45, I	9.8	5.4635	2.401
AlAs	ZB	2.153, I	10.06	5.660	3.760
GaN	W	3.44, D	$\varepsilon_\| = 10.4$ $\varepsilon_\perp = 9.5$	$a = 3.175$ $c = 5.158$	6.095
GaP	ZB	2.272, I	11.11	5.4505	4.138
GaAs	ZB	1.424, D	13.18	5.653	5.318
GaSb	ZB	0.75, D	15.69	6.0959	5.6137
InN	W	1.89, D	—	$a = 3.5446$ $c = 8.7034$	6.81
InP	ZB	1.344, D	12.56	5.8687	4.81
InAs	ZB	0.354, D	15.15	6.058	5.667
InSb	ZB	0.230, D	16.8	6.479	5.775
ZnS	ZB	3.68, D	8.9	5.4102	4.079
ZnS	W	3.9107, D	$\bar{\varepsilon} = 9.6$	$a = 3.8226$ $c = 6.6205$	4.084
ZnSe	ZB	2.822, D	9.1	5.668	5.266
ZnTe	ZB	2.394, D	8.7	6.104	5.636
CdS	W	2.501, D	$\bar{\varepsilon} = 9.38$	$a = 4.1362$ $c = 6.714$	4.82
CdS	ZB	2.50, D	—	5.818	—
CdSe	W	1.751, D	$\varepsilon_\| = 10.16$ $\varepsilon_\perp = 9.29$	$a = 4.2999$ $c = 7.0109$	5.81
CdTe	ZB	1.475, D	10.2	6.482	5.87
PbS	R	0.41, D*	169.	5.936	7.597
PbSe	R	0.278, D*	210.	6.117	8.26
PbTe	R	0.310, D*	414.	6.462	8.219

Data given are room temperature values (300 K).
KEY: DI: diamond; R: rocksalt; W: wurtzite; ZB: zinc-blende
*: gap at L point; D: direct; I: indirect; $\varepsilon_\|$: parallel to c-axis; $\varepsilon_\perp$: perpendicular to c-axis

FREQUENTLY USED QUANTITIES

QUANTITY	SYMBOL	VALUE
Planck's constant	h	6.626×10^{-34} J-s
	$\hbar = h/2\pi$	1.055×10^{-34} J-s
Velocity of light	c	2.998×10^8 m/s
Electron charge	e	1.602×10^{-19} C
Electron volt	eV	1.602×10^{-19} J
Mass of an electron	m_0	9.109×10^{-31} kg
Permittivity of vacuum	$\varepsilon_0 = \dfrac{10^7}{4\pi c^2}$	8.85×10^{-14} F cm^{-1} = 8.85×10^{-12} F m^{-1}
Boltzmann constant	k_B	8.617×10^{-5} eVK^{-1}
Thermal voltage at 300 K	$k_B T/e$	0.026 V

Wavelength – energy relation: $\lambda(\mu m) = \dfrac{1.24}{E(eV)}$

Linewidth $\delta E = 1$ meV $\Rightarrow$ $\delta \nu = 0.243$ THz

Linewidth $\delta E = 1$ meV $\Rightarrow$ $\begin{cases} \delta\lambda = 19.4 \text{ Å} @ \lambda = 1.55\ \mu m \\ \delta\lambda = 6.2 \text{ Å} @ \lambda = 0.88\ \mu m \end{cases}$

MATERIAL	CONDUCTION BAND EFFECTIVE DENSITY (N_c)	VALENCE BAND EFFECTIVE DENSITY (N_v)	INTRINSIC CARRIER CONCENTRATION ($n_i = p_i$)
Si (300 K)	2.78×10^{19} cm^{-3}	9.84×10^{18} cm^{-3}	1.5×10^{10} cm^{-3}
Ge (300 K)	1.04×10^{19} cm^{-3}	6.0×10^{18} cm^{-3}	2.33×10^{13} cm^{-3}
GaAs (300 K)	4.45×10^{17} cm^{-3}	7.72×10^{18} cm^{-3}	1.84×10^6 cm^{-3}

Printed in the United States
By Bookmasters